GENERALIZED ORDINARY DIFFERENTIAL EQUATIONS

Series in Real Analysis Volume 5

GENERALIZED ORDINARY DIFFERENTIAL EQUATIONS

Štefan Schwabik
Mathematical Institute of the Czechoslovak Academy of Sciences

World Scientific
Singapore • New Jersey • London • Hong Kong

Published by

World Scientific Publishing Co. Pte. Ltd.
P O Box 128, Farrer Road, Singapore 9128
USA office: Suite 1B, 1060 Main Street, River Edge, NJ 07661
UK office: 73 Lynton Mead, Totteridge, London N20 8DH

Library of Congress Cataloging-in-Publication data is available.

GENERALIZED ORDINARY DIFFERENTIAL EQUATIONS

ISBN 981-02-1225-9

Printed in Singapore by Utopia Press.

CONTENTS

Typeset by $\mathcal{A}_{\mathcal{M}}\mathcal{S}$-TEX

PREFACE

The intention of this book is to survey some topics related to the study of ordinary differential equations in connection with the theory of nonabsolutely convergent integral.

Although the nonabsolutely convergent Perron integral is present in mathematics from the second decade of our century, the main step to establish this connection was done by Jaroslav Kurzweil (1957) and Ralph Henstock (1961) when they independently arrived at a construction of the Perron integral which is a simple extension of the Riemann theory of integration based on lucid and transparent Riemann integral sums. The approach of J. Kurzweil presented in [68] and in the subsequent papers [69], [70], [71], [75] shows how the new approach to the general integration theory was growing up from the needs of ordinary differential equations. In particular, the presence of rapidly oscillating external forces was the main impulse for introducing a nonabsolutely convergent integral into the theory of ordinary differential equations instead of the Lebesgue integral or the classical Riemann integral.

In Chapter I of the book the generalized Perron integral is described in a relatively complete form for use in the theory of ordinary differential equations. In Chapter II some aspects of using the Perron integral in the integral form of an ordinary differential equation are discussed from a recent point of view.

A special class of generalized ordinary differential equations is introduced and fundamental results concerning existence and uniqueness of their solutions are derived in Chapters III and IV. The solutions of this class of generalized ordinary differential equations turn out to be functions which are locally of bounded variation on their domain of definition.

Chapter V contains the comparison of the generalized ordinary differential equations with other types of differential systems. Besides the classical Carathéodory concept we pay special attention to measure differential equations and to the concept of differential systems with impulses. The theory of differential systems with impulses has been developed to a great extent by the Kiev school represented by A.M. Samojlenko, N.A. Perestjuk [120] and many others.

Linear problems are dealt with in Chapters VI and VII. There is a full parallelism with the classical theory of linear ordinary differential equations and also with linear boundary value problems. Only a short introduction to the field of boundary value problems is given here, the main results being presented in [152] and [155]. It should be pointed out that a more general theory of integral equations of Volterra and Fredholm type in the space of functions of bounded variation can be found in [155]. This theory is also based on the general Perron – Stieltjes integral.

Chapter VII is devoted to the fundamental properties of multiplicative integrals. The origins of product integration belong to V. Volterra [169], [170]. Integral products replace the Riemann integral sums and the lines of the theory of generalized Perron integral from Chapter I are followed for the definition of the Perron product integral of a matrix valued point – interval function. This idea was used for the first time by Jiří Jarník and Jaroslav Kurzweil in [53]. Evolutions described by linear systems from Chapter VI are characterized via Perron product integrals in this part of the book.

The final three chapters are devoted to problems of dependence of solutions on the right hand sides of the generalized ordinary differential equations. Continuous dependence results are given and in Chapter X the variational stability of a solution of a generalized ordinary differential equation is described. Variational stability is a general concept which is suitable for the class of generalized ordinary differential equations described in Chapter III because of the local finiteness of the variation of a solution.

The references given at the end of the book contain in addition to the necessary bibliographical sources also a relatively complete list of relevant items originated in the working group of J. Kurzweil in Prague. These entries concern not only generalized ordinary differential equations but also the theory of the Perron integral and its applications. This presentation represents an addition to the exhaustive list of references given in the recent book [41] of Ralph Henstock on integration.

Now, when the text of the book is almost finished I must express my thanks to Prof. Lee Peng Yee who convinced me that it is worth to prepare this text for the Real Analysis series of World Scientific even if some of its parts are covered by my two issues [150] and [151] on the same topic. I would like to thank my colleagues and friends in the seminar on differential equations in the Mathematical Institute of the Czech. Acad. of Sciences for their patience and useful comments.

Especially I am indebted to my colleague Prof. Jiří Jarník for reading the proofs and helping me to improve substantially the text as well as its typeset form resulting from my long and tough discussions with $\mathcal{AMS}$-TEX.

Prague, July 1992 *Štefan Schwabik*

CHAPTER I

THE GENERALIZED PERRON INTEGRAL

Definition of the integral

Let an interval $[a,b] \subset \mathbb{R}$, $\quad -\infty < a < b < +\infty$ be given. A pair (τ, J) of a point $\tau \in \mathbb{R}$ and a compact interval $J \subset \mathbb{R}$ is called a *tagged interval,* τ is the *tag* of J.

A finite collection $\Delta = \{(\tau_j, J_j),\ j = 1, \dots, k\}$ of tagged intervals is called a *system* in $[a,b]$ if $\tau_j \in J_j \subset [a,b]$ for every $j = 1, \dots, k$ and the intervals J_j are nonoverlapping, i.e. $Int(J_i) \cap Int(J_j) = \emptyset$ for $i \neq j$ where $Int(J)$ denotes the interior of an interval J.

A system $\Delta = \{(\tau_j, J_j),\ j = 1, \dots, k\}$ is called a *partition* of $[a,b]$ if

$$\bigcup_{j=1}^{k} J_j = [a,b].$$

Given a positive function $\delta : [a,b] \to (0, +\infty)$ called a *gauge* on $[a,b]$, a tagged interval (τ, J) with $\tau \in [a,b]$ is said to be *δ–fine* if

$$J \subset [\tau - \delta(\tau), \tau + \delta(\tau)].$$

A *system* (in particular, a *partition*) $\Delta = \{(\tau_j, J_j),\ j = 1, \dots, k\}$ is δ–fine if the point–interval pair (τ_j, J_j) is δ–fine for every $j = 1, \dots, k$.

1.1 Definition. Let $h(\tau, J)$ be a finite real-valued function of point-interval pairs with $\tau \in [a,b]$, $J \subset [a,b]$. The function h is called *integrable over* $[a,b]$ if there is an $I \in \mathbb{R}$ such that given $\varepsilon > 0$, there is a gauge δ on $[a,b]$ such that

$$|\sum_{j=1}^{k} h(\tau_j, J_j) - I| < \varepsilon$$

for every δ-fine partition $\Delta = \{(\tau_j, J_j),\ j = 1,\dots,k\}$ of $[a,b]$. The number I is called the *integral* of the function h over $[a,b]$.

This is a general definition of the *Henstock–Kurzweil integral* that can be found in a still growing series of books on integration (see e.g. [40], [41], [84], [92], [93], [96], [98]) where this idea of integration is explained from various aspects and in different generality.

For our purposes we will consider a specific situation that has its origin in the theory of ordinary differential equations.

Assume that $U : [a,b] \times [a,b] \to \mathbb{R}$ is a function of two variables $\tau, t \in [a,b]$. We define the integral of this function as the Henstock–Kurzweil integral for the point-interval function h given by

$$h(\tau, t) = U(\tau, \beta) - U(\tau, \alpha) \tag{1.1}$$

for $\tau \in [a,b]$ and $J = [\alpha, \beta] \subset [a,b]$.

1.2 Definition. A function $U : [a,b] \times [a,b] \to \mathbb{R}$ is called *integrable over* $[a,b]$ if there is an $I \in \mathbb{R}$ such that given $\varepsilon > 0$, there is a gauge δ on $[a,b]$ such that

$$|S(U,D) - I| = |\sum_{j=1}^{k}[U(\tau_j, \alpha_j) - U(\tau_j, \alpha_{j-1})] - I| < \varepsilon \tag{1.2}$$

for every δ-fine partition

$$D = \{(\tau_j, [\alpha_{j-1}, \alpha_j]),\ j = 1,\dots,k\} =$$

$$= \{\alpha_0, \tau_1, \alpha_1, \ldots, \alpha_{k-1}, \tau_k, \alpha_k\} \tag{1.3}$$

of $[a,b]$.

The real number $I \in \mathbb{R}$ is called the *generalized Perron integral of* U over the interval $[a,b]$ and will be denoted by $\int_a^b DU(\tau,t)$.

If $\int_a^b DU(\tau,t)$ exists then we define $\int_b^a DU(\tau,t) = -\int_a^b DU(\tau,t)$ and set $\int_b^a DU(\tau,t) = 0$ when $a = b$.

We denote by $\mathcal{K}([a,b])$ the set of all functions U which are integrable over $[a,b]$.

We use the notation $S(U,D) = \sum_{j=1}^k [U(\tau_j,\alpha_j) - U(\tau_j,\alpha_{j-1})]$ for the Riemann–type sum corresponding to the function U and the partition D.

1.3 *Remark.* Let us mention that for a given gauge δ on $[a,b]$ the partition $D = \{\alpha_0, \tau_1, \alpha_1, \ldots, \alpha_{k-1}, \tau_k, \alpha_k\}$ is δ–fine if

$$a = \alpha_0 < \alpha_1 < \cdots < \alpha_k = b, \tag{1.4}$$

$$\alpha_{j-1} \le \tau_j \le \alpha_j,\ j = 1, 2, \ldots, k \tag{1.5}$$

and

$$[\alpha_{j-1}, \alpha_j] \subset [\tau_j - \delta(\tau_j), \tau_j + \delta(\tau_j)],\ j = 1, 2, \ldots, k. \tag{1.6}$$

Looking at Definition 1.2 and the properties of a δ–fine partition D of $[a,b]$ it is easy to see that for the function U defined on the square $[a,b] \times [a,b]$ only its values at points close to the diagonal $\tau = t$ play a role when defining the integral and that it is sufficient to have U defined on a set

$$S = \{(\tau,t) \in \mathbb{R}^2; \tau \in [a,b], t \in [\tau - \overline{\delta}(\tau), \tau + \overline{\delta}(\tau)] \cap [a,b]\}$$

with some gauge $\overline{\delta} : [a,b] \to (0,+\infty)$ on $[a,b]$. The gauges δ in Definition 1.2 then have to be chosen such that $\delta(\tau) \le \overline{\delta}(\tau)$, $\tau \in [a,b]$ and the result is the same.

For the integral we use the notation $\int_a^b DU(\tau,t)$ introduced by J. Kurzweil in [68]. This notation has a symbolic meaning only.

Clearly the above Definitions 1.1 and 1.2 are viable only if for a given gauge δ on $[a,b]$ there exists at least one δ–fine partition D of $[a,b]$. This fundamental question has an affirmative answer given by the following statement.

1.4 Lemma (Cousin). *Given a gauge δ on $[a,b]$, there is a δ–fine partition $D=\{\alpha_0,\tau_1,\alpha_1,\dots,\alpha_{k-1},\tau_k,\alpha_k\}$ of $[a,b]$.*

For the proof of this lemma see e.g. [40,Theorem 4.1] or any of the textbooks [41], [84], [92], [93], [96], [97]. The lemma was discovered often in various contexts. J.Mawhin [93] pointed out that the first result of this type belongs to Cousin (1895).

1.5 *Remark.* Assume that $f:[a,b]\to\mathbb{R}$ is a real function. Let us set $U(\tau,t)=f(\tau).t$ for $\tau,t\in[a,b]$. Then clearly by (1.1)

$$h(\tau,t)=U(\tau,\beta)-U(\tau,\alpha)=f(\tau)(\beta-\alpha)$$

and

$$\sum_{j=1}^{k}h(\tau_j,J_j)=\sum_{j=1}^{k}[U(\tau_j,\alpha_j)-U(\tau_j,\alpha_{j-1})]=\sum_{j=1}^{k}f(\tau_j)(\alpha_j-\alpha_{j-1})$$

represents the classical Riemann sum for the function f and a given partition D of $[a,b]$.

Moreover, if $g:[a,b]\to\mathbb{R}$ is given then for $U(\tau,t)=f(\tau).g(t)$, $\tau,t\in[a,b]$ we obtain the Riemann–Stieltjes sum

$$\sum_{j=1}^{k}h(\tau_j,J_j)=\sum_{j=1}^{k}f(\tau_j)[g(\alpha_j)-g(\alpha_{j-1})]$$

for the functions f,g and a given partition D of $[a,b]$.

Using Definition 1.2 (or equivalently Definition 1.1) we obtain in these cases certain concepts of integration based on Riemann–type sums.

If $U(\tau,t) = f(\tau).t$ then we write $\int_a^b f(s)\,ds$ instead of $\int_a^b D[f(\tau).t]$ and similarly also for the Stieltjes case $U(\tau,t) = f(\tau)g(t)$ we use the notation $\int_a^b f(s)\,dg(s)$ instead of $\int_a^b D[f(\tau).g(t)]$ provided the generalized Perron integrals $\int_a^b DU(\tau,t)$ exist in these cases.

It turns out that in the case $U(\tau,t) = f(\tau)t$ we have

$$\int_a^b DU(\tau,t) = \int_a^b D[f(\tau).t] = \int_a^b f(s)\,ds = (P)\int_a^b f(s)\,ds$$

where (P) stands for the classical Perron integral. This fact was shown for the first time by J. Kurzweil in 1957 in his fundamental paper [68] concerning differential equations. It is also worth mentioning that the primary definition in [68] for the integral $\int_a^b DU(\tau,t)$ is given using major and minor functions to U. These are of course given in a more general form because of the general form of the function U but they coincide with Perron's major and minor functions in the known form when $U(\tau,t) = f(\tau).t$. Definition 1.2 is given in [68] as a secondary one and it is shown that the underlying idea of the integral $\int_a^b DU(\tau,t)$ is the same as for the concept based on minor and major functions.

It is clear that Definition 1.2 can be given also for functions U with values in a general linear topological space. For the purposes of this text we give such a definition for $\mathbb{R}^n$–valued functions only.

1.2n Definition. A function $U : [a,b] \times [a,b] \to \mathbb{R}^n$ is called *integrable over* $[a,b]$ if there is an $I \in \mathbb{R}^n$ such that given $\varepsilon > 0$, there is a gauge δ on $[a,b]$ such that

$$\|S(U,D) - I\| = \|\sum_{j=1}^{k}[U(\tau_j,\alpha_j) - U(\tau_j,\alpha_{j-1})] - I\| < \varepsilon \quad (1.2)$$

for every δ–fine partition

$$D = \{(\tau_j, [\alpha_{j-1}, \alpha_j]),\ j = 1, \ldots, k\} =$$

$$= \{\alpha_0, \tau_1, \alpha_1, \ldots, \alpha_{k-1}, \tau_k, \alpha_k\} \tag{1.3}$$

of $[a, b]$.

The element $I \in \mathbb{R}^n$ is called the *generalized Perron integral of* U over the interval $[a, b]$ and will be denoted by $\int_a^b DU(\tau, t)$.

If $\int_a^b DU(\tau, t)$ exists then we define $\int_b^a DU(\tau, t) = -\int_a^b DU(\tau, t)$ and set $\int_b^a DU(\tau, t) = 0$ when $a = b$.

We denote by $\mathcal{K}([a, b])$ the set of all functions U which are integrable over $[a, b]$.

Here $\|.\|$ stands for the norm in $\mathbb{R}^n$ (e.g. the Euclidean one).

1.6 Theorem. *An $\mathbb{R}^n$–valued function $U : [a, b] \times [a, b] \to \mathbb{R}^n$, $U = (U_1, U_2, \ldots, U_n)$ is integrable if and only if every component U_m, $m = 1, 2, \ldots, n$ is integrable in the sense of Definition 1.2.*

Proof. Suppose the integral $\int_a^b DU(\tau, t) = I = (I_1, \ldots, I_n) \in \mathbb{R}^n$ exists. For every $m = 1, \ldots, n$ and any partition D of $[a, b]$ we evidently have $S(U_m, D) = (S(U, D))_m$ and the inequality

$$|S(U_m, D) - I_m| = |(S(U, D))_m - I_m| = |(S(U, D) - I)_m| \leq$$

$$\leq \|S(U, D) - I\|$$

holds. Therefore we get the existence of $\int_a^b DU_m(\tau, t)$ and the equality $\int_a^b DU_m(\tau, t) = I_m = (\int_a^b DU(\tau, t))_m$.

If we suppose that the integrals $\int_a^b DU_m(\tau, t) = I_m$ exist for every $m = 1, \ldots, n$ then for every $\varepsilon > 0$ there exists a gauge δ_m

on $[a, b]$ such that for every δ_m–fine partition D_m of $[a, b]$ we have $|S(U_m, D_m) - I_m| < \varepsilon$. Let us choose a gauge δ on $[a, b]$ such that

$$\delta(\tau) \leq \min(\delta_1(\tau), \ldots, \delta_n(\tau))$$

for $\tau \in [a, b]$. Then evidently every δ–fine partition D of $[a, b]$ is also δ_m–fine, $m = 1, \ldots, n$ and for $I = (I_1, \ldots, I_n) \in \mathbb{R}^n$ we get the inequality

$$\|S(U, D) - I\| = (\sum_{m=1}^{n} (S(U_m, D) - I_m)^2)^{\frac{1}{2}} \leq (n\varepsilon^2)^{\frac{1}{2}} = n^{\frac{1}{2}}\varepsilon$$

for every δ–fine partition D of $[a, b]$.

Hence the integral $\int_a^b DU(\tau, t)$ exists and its components equal to $\int_a^b DU_m(\tau, t) = I_m$ for $m = 1, \ldots, n$. □

1.7 Theorem. *The function $U : [a, b] \times [a, b] \to \mathbb{R}^n$ is integrable over $[a, b]$ if and only if for every $\varepsilon > 0$ there is a gauge δ on $[a, b]$ such that*

$$\|S(U, D_1) - S(U, D_2)\| < \varepsilon$$

for any δ–fine partitions D_1, D_2 of $[a, b]$, where

$$S(U, D) = \sum_{j=1}^{k} [U(\tau_j, \alpha_j) - U(\tau_j, \alpha_{j-1})]$$

is the Riemann–type sum corresponding to U and the partition $D = \{\alpha_0, \tau_1, \alpha_1, \ldots, \alpha_{k-1}, \tau_k, \alpha_k\}$ of $[a, b]$.

Proof. By the previous Theorem 1.6 it is sufficient to prove the statement for a real valued function $U : [a, b] \times [a, b] \to \mathbb{R}$.

Assume that the *Bolzano - Cauchy condition* of the theorem holds and that $\varepsilon > 0$ is given. Then there is a gauge δ on $[a,b]$ such that for any δ–fine partitions D_1, D_2 of $[a,b]$ we have

$$\|S(U,D_1) - S(U,D_2)\| < \tfrac{\varepsilon}{2}. \tag{1.7}$$

Denote by $\mathcal{M}$ the set of all $s \in \mathbb{R}$ such that there exists a gauge ω on $[a,b]$ such that for every ω–fine partition D we have $s \leq S(U,D)$. Assume that D_0 is an arbitrary δ–fine partition of $[a,b]$. By (1.7), for every δ–fine partition of $[a,b]$ the inequalities

$$S(U,D_0) - \tfrac{\varepsilon}{2} < S(U,D) < S(U,D_0) + \tfrac{\varepsilon}{2}$$

hold. Therefore $(-\infty, S(U,D_0) - \frac{\varepsilon}{2}) \subset \mathcal{M}$ and $\mathcal{M} \subset (-\infty, S(U,D_0) + \frac{\varepsilon}{2})$ i.e. the set $\mathcal{M}$ is nonempty and bounded from above. Consequently the supremum $\sup \mathcal{M}$ of the set $\mathcal{M}$ exists and by the above inclusions we have

$$S(U,D_0) - \tfrac{\varepsilon}{2} < \sup \mathcal{M} < S(U,D_0) + \tfrac{\varepsilon}{2}.$$

Hence $|S(U,D_0) - \sup \mathcal{M}| < \frac{\varepsilon}{2}$ and by (1.7) we have

$$|S(U,D) - \sup \mathcal{M}| \leq |S(U,D) - S(U,D_0)| +$$

$$+ |S(U,D_0) - \sup \mathcal{M}| < \varepsilon$$

for every δ–fine partition D of the interval $[a,b]$. Therefore by definition the function U is integrable over $[a,b]$ and $\int_a^b DU(\tau,t) = \sup \mathcal{M}$.

Conversely, if the integral $\int_a^b DU(\tau,t)$ exists then the Bolzano–Cauchy condition of the theorem can be deduced easily. □

1.8 *Remark.* The Bolzano–Cauchy condition of Theorem 1.7 can be used also for defining integrability of a function $U : [a,b] \times [a,b] \to \mathbb{R}^n$. In the more general case of U with values in a linear topological space the completeness of this space is of course necessary.

Fundamental properties of the integral

Now we turn our attention to some fundamental properties of the integral $\int_a^b DU(\tau,t)$.

From the evident identity

$$S(c_1U + c_2V, D) = c_1S(U, D) + c_2S(V, D)$$

for the Riemann sums of the functions $U, V : [a,b] \times [a,b] \to \mathbb{R}^n$, $c_1, c_2 \in \mathbb{R}$ and an arbitrary partition

$$D = \{\alpha_0, \tau_1, \alpha_1, \ldots, \alpha_{k-1}, \tau_k, \alpha_k\}$$

of $[a,b]$ we immediately have the following result.

1.9 Theorem. *If $U, V \in \mathcal{K}([a,b])$ and $c_1, c_2 \in \mathbb{R}$ then $c_1U + c_2V \in \mathcal{K}([a,b])$ and*

$$\int_a^b D[c_1U(\tau,t) + c_2V(\tau,t)] = c_1 \int_a^b DU(\tau,t) + c_2 \int_a^b DV(\tau,t).$$

1.10 Theorem. *If $U \in \mathcal{K}([a,b])$ then for every $[c,d] \subset [a,b]$ we have $U \in \mathcal{K}([c,d])$.*

Proof. Assume that $\varepsilon > 0$ is given. By the Bolzano–Cauchy condition for the existence of the integral $\int_a^b DU(\tau,t)$ (see Theorem 1.7) there exists a gauge δ on $[a,b]$ such that

$$\|S(U, D_1) - S(U, D_2)\| < \varepsilon \tag{1.8}$$

for every δ–fine partitions D_1, D_2 of $[a,b]$.

Let now $\widetilde{D_1}, \widetilde{D_2}$ be arbitrary δ–fine partitions of $[c,d]$. Assume that $a < c < d < b$. Let D_L be a δ–fine partition of $[a,c]$ and D_R a δ–fine partition of $[d,b]$. They exist by the Cousin Lemma 1.4.

If

$$\widetilde{D_1} = \{\alpha_0, \tau_1, \alpha_1, \dots, \alpha_{k-1}, \tau_k, \alpha_k\},$$

$$D_L = \{\alpha_0^L, \tau_1^L, \alpha_1^L, \dots, \alpha_{l-1}^L, \tau_l^L, \alpha_l^L\},$$

$$D_R = \{\alpha_0^R, \tau_1^R, \alpha_1^R, \dots, \alpha_{r-1}^R, \tau_r^R, \alpha_r^R\}$$

then $\alpha_0^L = a$, $\alpha_l^L = c = \alpha_0$, $\alpha_r = d = \alpha_0^R$, $\alpha_r^R = b$. Let us put together the partitions $D_L, \widetilde{D_1}, D_R$ to create a partition D_1 as follows

$$D_1 = \{\alpha_0^L, \tau_1^L, \alpha_1^L, \dots, \alpha_{l-1}^L, \tau_l^L, \alpha_0, \tau_1, \alpha_1, \dots$$

$$\dots, \alpha_{k-1}, \tau_k, \alpha_k, \tau_1^R, \alpha_1^R, \dots, \alpha_{r-1}^R, \tau_r^R, \alpha_r^R\}.$$

Then evidently D_1 is a partition of $[a,b]$ and it is δ–fine because the partitions $D_L, \widetilde{D_1}, D_R$ are δ–fine. Similarly we put together the partitions $D_L, \widetilde{D_2}, D_R$ forming a δ–fine partition D_2 of $[a,b]$. By (1.8) we have

$$\|S(U, \widetilde{D_1}) - S(U, \widetilde{D_2})\| = \|S(U, D_1) - S(U, D_2)\| < \varepsilon$$

because the terms in the sums $S(U, D_1) - S(U, D_2)$ that correspond to the common parts D_L, D_R appear in each of the partitions D_1, D_2 and hence cancel each other. Therefore by Theorem 1.7, the integral $\int_c^d DU(\tau,t)$ exists because $\widetilde{D_1}, \widetilde{D_2}$ have been arbitrary δ–fine partitions of $[c,d]$. □

1.11 Theorem. *If $c \in (a,b)$ and $U : [a,b] \times [a,b] \to \mathbb{R}^n$ is such that $U \in \mathcal{K}([a,c])$ and $U \in \mathcal{K}([c,b])$ then $U \in \mathcal{K}([a,b])$ and*

$$\int_a^b DU(\tau,t) = \int_a^c DU(\tau,t) + \int_c^b DU(\tau,t). \qquad (1.9)$$

Proof. Let $\varepsilon > 0$ be given. Denote $I_L = \int_a^c DU(\tau, t)$ and $I_R = \int_c^b DU(\tau, t)$. By the assumptions there exists a gauge δ_L on $[a, c]$ such that for every δ_L–fine partition D^L of $[a, c]$ we have

$$\|S(U, D^L) - I_L\| < \varepsilon.$$

Similarly there exists a gauge δ_R on $[c, b]$ such that for every δ_R–fine partition D^R of $[c, b]$ we have

$$\|S(U, D^R) - I_R\| < \varepsilon.$$

Let us define

$$\widetilde{\delta}(\tau) = \begin{cases} \delta_L(\tau) & \text{for } \tau \in [a, c), \\ \min(\delta_L(c), \delta_R(c) & \text{for } \tau = c, \\ \delta_R(\tau) & \text{for } \tau \in (c, d] \end{cases}$$

and choose $\delta : [a, b] \to (0, +\infty)$ such that

$$\delta(\tau) < \min(\widetilde{\delta}(\tau), |\tau - c|) \quad \text{if } \tau \neq c$$

$$\delta(c) = \widetilde{\delta}(c).$$

Evidently $\delta : [a, b] \to (0, +\infty)$ is a gauge on $[a, b]$. Assume that $D = \{\alpha_0, \tau_1, \alpha_1, \ldots, \alpha_{k-1}, \tau_k, \alpha_k\}$ is a δ–fine partition of $[a, b]$. Then there exists an index m such that $c \subset [\alpha_{m-1}, \alpha_m]$. Assume that $\tau_m \neq c$. Then we get the contradictory inequality

$$|\tau_m - c| \leq \delta(\tau_m) < |\tau_m - c|$$

which holds because by the definition we have $[\alpha_{m-1}, \alpha_m] \subset [\tau_m - \delta(\tau_m), \tau_m + \delta(\tau_m)]$ and $\delta(\tau) < |\tau - c|$ for every $\tau \neq c$. Therefore necessarily $\tau_m = c$. Moreover, we have

$$S(U, D) = \sum_{j=1}^{m-1} [U(\tau_j, \alpha_j) - U(\tau_j, \alpha_{j-1})] + U(c, \alpha_m) - U(c, \alpha_{m-1}) +$$

$$+\sum_{j=m+1}^{k}[U(\tau_j,\alpha_j)-U(\tau_j,\alpha_{j-1})]=$$

$$=\sum_{j=1}^{m-1}[U(\tau_j,\alpha_j)-U(\tau_j,\alpha_{j-1})]+$$

$$+U(c,c)-U(c,\alpha_{m-1})+U(c,\alpha_m)-U(c,c)+$$

$$+\sum_{j=m+1}^{k}[U(\tau_j,\alpha_j)-U(\tau_j,\alpha_{j-1})]=S(U,D_L)+S(U,D_R),$$

where $D_L = \{\alpha_0,\tau_1,\alpha_1,\dots,\alpha_{m-1},\tau_m^L = c,\alpha_m^L = c\}$, $D_R = \{\alpha_{m-1}^R = c,\tau_m^R = c,\alpha_m,\dots,\alpha_{k-1},\tau_k,\alpha_k\}$ are partitions of $[a,c]$, $[c,b]$, respectively, D_L is δ_L–fine and D_R is δ_R–fine in virtue of the choice of the gauge δ. Hence for a δ–fine partition D of $[a,b]$ we have

$$\|S(U,D)-I_L-I_R\| = \|S(U,D_L)+S(U,D_R)-I_L-I_R\| \le$$

$$\le \|S(U,D_L)-I_L\|+\|S(U,D_R)-I_R\| < 2\varepsilon.$$

By definition this yields the existence of the integral $\int_a^b DU(\tau,t)$ and also the equality (1.9). □

1.12 *Remark.* Besides the technique of joining two partitions of two intervals having a common endpoint as used in the proof of Theorem 1.10 we should point out the construction used in the proof of the previous Theorem 1.11. Namely, if δ is a gauge satisfying

$$\delta(\tau) < |\tau - c| \text{ for } \tau \neq c$$

then it forces every δ–fine partition of an interval $[a,b]$ with $c \in [a,b]$ to have the tag of one of its point–interval pairs at the point c. This fact can be helpful in various constructions of special integral sums.

The following statement provides an operative tool in the theory of generalized Perron integral. Its original version belongs to S. Saks and it was formulated for generalized integrals using Riemann–like sums by R. Henstock.

1.13 Lemma (Saks–Henstock). *Let $U : [a,b] \times [a,b] \to \mathbb{R}^n$ be integrable over $[a,b]$. Given $\varepsilon > 0$ assume that the gauge δ on $[a,b]$ is such that*

$$\|\sum_{j=1}^{k}[U(\tau_j,\alpha_j) - U(\tau_j,\alpha_{j-1})] - \int_a^b DU(\tau,t)\| < \varepsilon$$

for every δ–fine partition $D = \{\alpha_0,\tau_1,\alpha_1,\ldots,\alpha_{k-1},\tau_k,\alpha_k\}$ of $[a,b]$. If

$$a \le \beta_1 \le \xi_1 \le \gamma_1 \le \beta_2 \le \xi_2 \le \gamma_2 \le \ldots$$

$$\cdots \le \beta_m \le \xi_m \le \gamma_m \le b$$

represents a δ–fine system $\{(\xi_j,[\beta_j,\gamma_j]),\ j = 1,\ldots,m\}$, i.e.

$$\xi_j \in [\beta_j,\gamma_j] \subset [\xi_j - \delta(\xi_j), \xi_j + \delta(\xi_j)], \quad j = 1,\ldots,m$$

then

$$\|\sum_{j=1}^{m}[U(\xi_j,\gamma_j) - U(\xi_j,\beta_j) - \int_{\beta_j}^{\gamma_j} DU(\tau,t)]\| < \varepsilon. \qquad (1.10)$$

Proof. Without any loss of generality it can be assumed that $\beta_j < \gamma_j$ for every $j = 1,\ldots,m$. Denote $\gamma_0 = a$ and $\beta_{m+1} = b$. If $\gamma_j < \beta_{j+1}$ for some $j = 0,1,\ldots,m$ then Theorem 1.10 yields the existence of the integral $\int_{\gamma_j}^{\beta_{j+1}} DU(\tau,t)$ and therefore for every $\eta > 0$ there exists a gauge δ_j on $[\gamma_j,\beta_{j+1}]$ such that $\delta_j(\tau) < \delta(\tau)$

for $\tau \in [\gamma_j, \beta_{j+1}]$ and for every δ_j–fine partition D^j of $[\gamma_j, \beta_{j+1}]$ we have

$$\|S(U, D^j) - \int_{\gamma_j}^{\beta_{j+1}} DU(\tau, t)\| < \tfrac{\eta}{m+1}.$$

If $\gamma_j = \beta_{j+1}$ then we take $S(U, D^j) = 0$.

The expression

$$\sum_{j=1}^{m}[U(\xi_j, \gamma_j) - U(\xi_j, \beta_j)] + \sum_{j=1}^{m} S(U, D^j)$$

represents an integral sum which corresponds to a certain δ–fine partition and consequently

$$\|\sum_{j=1}^{m}[U(\xi_j, \gamma_j) - U(\xi_j, \beta_j) + \sum_{j=1}^{m} S(U, D^j) - \int_a^b DU(\tau, t)\| < \varepsilon.$$

Hence

$$\|\sum_{j=1}^{m}[U(\xi_j, \gamma_j) - U(\xi_j, \beta_j) - \int_{\beta_j}^{\gamma_j} DU(\tau, t)]\| \leq$$

$$\leq \|\sum_{j=1}^{m}[U(\xi_j, \gamma_j) - U(\xi_j, \beta_j)] + \sum_{j=1}^{m} S(U, D^j) - \int_a^b DU(\tau, t)\| +$$

$$+ \sum_{j=1}^{m} \|S(U, D^j) - \int_{\gamma_j}^{\beta_{j+1}} DU(\tau, t)\| < \varepsilon + (m+1)\tfrac{\eta}{m+1} = \varepsilon + \eta.$$

Since this inequality holds for every $\eta > 0$ we immediately obtain (1.10). □

1.14 Theorem. *Let a function $U : [a,b] \times [a,b] \to \mathbb{R}^n$ be given such that $U \in \mathcal{K}([a,c])$ for every $c \in [a,b)$ and let there exist a finite limit*

$$\lim_{c \to b-} \left[\int_a^c DU(\tau,t) - U(b,c) + U(b,b) \right] = I. \tag{1.11}$$

Then the function U is integrable over $[a,b]$ ($U \in \mathcal{K}([a,b])$) and

$$\int_a^b DU(\tau,t) = I.$$

Proof. Assume that $\varepsilon > 0$ is given. By (1.11) for every $\varepsilon > 0$ we can find a $B \in [a,b)$ such that for every $c \in [B,b)$ the inequality

$$\| \int_a^c DU(\tau,t) - U(b,c) + U(b,b) - I \| < \varepsilon \tag{1.12}$$

is satisfied. Assume that $a = c_0 < c_1 < \dots$ is an increasing sequence $(c_p)_{p=1}^{\infty}$ of points $c_p \in [a,b)$ with $\lim_{p\to\infty} c_p = b$. By the assumption we have $U \in \mathcal{K}([a,c_p])$ for every $p = 1,2,\dots$ and therefore for every $p = 1,2,\dots$ there exists a gauge $\delta_p : [a,c_p] \to (0,+\infty)$ such that for any δ_p-fine partition D of $[a,c_p]$ we have

$$\|S(U,D) - \int_a^{c_p} DU(\tau,t)\| < \frac{\varepsilon}{2^{p+1}}, \quad p = 1,2,\dots. \tag{1.13}$$

For any $\tau \in [a,b)$ there is exactly one $p(\tau) = 1,2,\dots$ for which $\tau \in [c_{p(\tau)-1}, c_{p(\tau)})$. Given $\tau \in [a,b)$ let us choose $\widehat{\delta}(\tau) > 0$ such that $\widehat{\delta}(\tau) \le \delta_{p(\tau)}(\tau)$ and $[\tau - \widehat{\delta}(\tau), \tau + \widehat{\delta}(\tau)] \cap [a,b) \subset [a, c_{p(\tau)})$.

Assume that $c \in [a,b)$ is given and that

$$\widehat{D} = \{\alpha_0, \tau_1, \alpha_1, \dots, \alpha_{k-2}, \tau_{k-1}, \alpha_{k-1}\}$$

is a $\widehat{\delta}$–fine partition of $[a,c]$. If $p(\tau_j)=p$ then $[\alpha_{j-1},\alpha_j]\subset[\tau_j-\widehat{\delta}(\tau_j),\tau_j+\widehat{\delta}(\tau_j)]\subset[a,c_p]$ and also $[\alpha_{j-1},\alpha_j]\subset[\tau_j-\delta_p(\tau_j),\tau_j+\delta_p(\tau_j)]$. Let

$$\sum_{j=1,\,p(\tau_j)=p}^{k-1}[U(\tau_j,\alpha_j)-U(\tau_j,\alpha_{j-1})-\int_{\alpha_{j-1}}^{\alpha_j}DU(\tau,t)]$$

be the sum of those terms in the corresponding "total" sum

$$\sum_{j=1}^{k-1}[U(\tau_j,\alpha_j)-U(\tau_j,\alpha_{j-1})-\int_{\alpha_{j-1}}^{\alpha_j}DU(\tau,t)]$$

for which the tags τ_j satisfy the relation $\tau_j\in[c_{p-1},c_p)$. Since (1.13) holds we obtain by the Saks–Henstock lemma 1.13

$$\|\sum_{j=1,\,p(\tau_j)=p}^{k-1}[U(\tau_j,\alpha_j)-U(\tau_j,\alpha_{j-1})-\int_{\alpha_{j-1}}^{\alpha_j}DU(\tau,t)]\|<\frac{\varepsilon}{2^{p+1}}$$

and finally

$$\|\sum_{j=1}^{k-1}[U(\tau_j,\alpha_j)-U(\tau_j,\alpha_{j-1})]-\int_a^c DU(\tau,t)]\|=$$

$$=\|\sum_{j=1}^{k-1}[U(\tau_j,\alpha_j)-U(\tau_j,\alpha_{j-1})-\int_{\alpha_{j-1}}^{\alpha_j}DU(\tau,t)]\|\le$$

$$\le\sum_{p=1}^{\infty}\|\sum_{j=1,\,p(\tau_j)=p}^{k-1}[U(\tau_j,\alpha_j)-U(\tau_j,\alpha_{j-1})-\int_{\alpha_{j-1}}^{\alpha_j}DU(\tau,t)]\|\le$$

$$\le\sum_{p=1}^{\infty}\frac{\varepsilon}{2^{p+1}}=\varepsilon.$$

Define now a gauge δ on the interval $[a, b]$ as follows. For $\tau \in [a, b)$ set

$$0 < \delta(\tau) < \min\{b - \tau, \widehat{\delta}(\tau)\}$$

while

$$0 < \delta(b) < b - B.$$

If $D = \{\alpha_0, \tau_1, \alpha_1, \ldots, \alpha_{k-1}, \tau_k, \alpha_k\}$ is an arbitrary δ–fine partition of $[a, b]$ then by the choice of the gauge δ we have $\tau_k = \alpha_k = b$ and $\alpha_{k-1} \in (B, b)$. Using (1.12) we get

$$\|S(U, D) - I\| =$$

$$= \|\sum_{j=1}^{k-1}[U(\tau_j, \alpha_j) - U(\tau_j, \alpha_{j-1})] + U(\tau_k, \alpha_k) - U(\tau_j, \alpha_{k-1}) - I\| \leq$$

$$\leq \|\sum_{j=1}^{k-1}[U(\tau_j, \alpha_j) - U(\tau_j, \alpha_{j-1})] - \int_a^{\alpha_{k-1}} DU(\tau, t)]\| +$$

$$+\| \int_a^{\alpha_{k-1}} DU(\tau, t)] - U(b, \alpha_{k-1}) + U(b, b) - I\| <$$

$$< \varepsilon + \|\sum_{j=1}^{k-1}[U(\tau_j, \alpha_j) - U(\tau_j, \alpha_{j-1})] - \int_a^{\alpha_{k-1}} DU(\tau, t)]\|.$$

Since $\alpha_{k-1} < b$ and $\widehat{D} = \{\alpha_0, \tau_1, \alpha_1, \ldots, \alpha_{k-2}, \tau_{k-1}, \alpha_{k-1}\}$ is a $\widehat{\delta}$–fine partition of $[a, \alpha_{k-1}]$, the second term on the right hand side of the last inequality can be esimated by ε as shown above. In this way we obtain

$$\|S(U, D) - I\| < 2\varepsilon$$

and this inequality yields the existence of the integral $\int_a^b DU(\tau, t)$ as well as the equality $\int_a^b DU(\tau, t) = I$. □

1.15 *Remark.* The "left endpoint" analog of Theorem 1.14 can be proved in a completely similar manner:

Let a function $U : [a,b] \times [a,b] \to \mathbb{R}^n$ *be given such that* $U \in \mathcal{K}([c,b])$ *for every* $c \in (a,b]$ *and let there exist a finite limit*

$$\lim_{c \to a+} \left[\int_c^b DU(\tau,t) + U(a,c) - U(a,a) \right] = I.$$

Then $U \in \mathcal{K}([a,b])$ *and*

$$\int_a^b DU(\tau,t) = I.$$

Theorem 1.14 and its "left endpoint" version given here represent the Cauchy extension of the generalized Perron integral. For more details see [40, p. 67] or [41,2.10, p.115].

1.16 Theorem. *Let* $U : [a,b] \times [a,b] \to \mathbb{R}^n$ *be such that* $U \in \mathcal{K}([a,b])$ *and* $c \in [a,b]$. *Then*

$$\lim_{s \to c} \left[\int_a^s DU(\tau,t) - U(c,s) + U(c,c) \right] = \int_a^c DU(\tau,t). \quad (1.14)$$

Proof. Let $\varepsilon > 0$ be given and let δ be a gauge on $[a,b]$ which corresponds to ε by Definition 1.2n, i.e. the inequality

$$\|S(U,D) - \int_a^b DU(\tau,t)\| < \varepsilon$$

holds for every δ–fine partition D of $[a,b]$. If $s \in [c - \delta(c), c + \delta(c)] \cap [a,b]$ then the Saks–Henstock lemma 1.13 gives

$$\|U(c,s) - U(c,c) - \int_c^s DU(\tau,t)\| < \varepsilon$$

that is

$$\|\int_a^s DU(\tau,t) - U(c,s) + U(c,c) - \int_a^c DU(\tau,t)\| =$$

$$= \|\int_c^s DU(\tau,t) - U(c,s) + U(c,c)\| < \varepsilon,$$

and this yields the relation (1.14). □

1.17 *Remark.* Theorem 1.16 shows that the function given by

$$s \in [a,b] \mapsto \int_a^s DU(\tau,t) \in \mathbb{R}^n,$$

i.e. the *indefinite generalized Perron integral to U* is not continuous in general. The indefinite integral is continuous at a point $c \in [a,b]$ if and only if the function $U(c,\cdot) : [a,b] \to \mathbb{R}^n$ is continuous at the point c.

We have to note that if $U : [a,b] \times [a,b] \to \mathbb{R}^n$ is such that $U \in \mathcal{K}([a,b])$ then by Theorem 1.10 the indefinite generalized Perron integral to the function U is well defined on the whole interval $[a,b]$.

Substitution theorem

1.18 Theorem (Substitution). *Assume that $-\infty < c < d < +\infty$ and that $\varphi : [c,d] \to \mathbb{R}$ is a continuous strictly monotone function on $[c,d]$.*

Let $U : [\varphi(c),\varphi(d)] \times [\varphi(c),\varphi(d)] \to \mathbb{R}^n$ be given.

If one of the integrals

$$\int_{\varphi(c)}^{\varphi(d)} DU(\tau,t), \quad \int_c^d DU(\varphi(\sigma),\varphi(s))$$

exists then also the other one exists and

$$\int_{\varphi(c)}^{\varphi(d)} DU(\tau,t) = \int_c^d DU(\varphi(\sigma),\varphi(s)) \tag{1.15}$$

holds.

Proof. Assume that the function φ is increasing.

Assume that $\int_c^d DU(\varphi(\sigma),\varphi(s))$ exists and that $\varepsilon > 0$ is given. Then there is a gauge ω on $[c,d]$ such that for every ω–fine partition

$$\{\beta_0,\sigma_1,\beta_1,\dots,\beta_{k-1},\sigma_k,\beta_k\}$$

of $[c,d]$ we have

$$\|\sum_{j=1}^{k}[U(\varphi(\sigma_j),\varphi(\beta_j)) - U(\varphi(\sigma_j),\varphi(\beta_{j-1}))] -$$

$$- \int_c^d DU(\varphi(\sigma),\varphi(s))\| < \varepsilon.$$

Since φ is increasing, $\varphi([c,d]) = [\varphi(c),\varphi(d)]$, the inverse φ^{-1} : $[\varphi(c),\varphi(d)] \to [c,d]$ exists and is continuous and increasing on $[\varphi(c),\varphi(d)]$. Hence for every $\tau \in [\varphi(c),\varphi(d)]$ there is exactly one $\sigma = \varphi^{-1}(\tau) \in [c,d]$. For $\tau \in [\varphi(c),\varphi(d)]$ let us define $\delta(\tau) > 0$ such that

$$[\tau-\delta(\tau),\tau+\delta(\tau)] \cap [\varphi(c),\varphi(d)] \subset \varphi([\sigma-\omega(\sigma),\sigma+\omega(\sigma)] \cap [c,d]) \tag{1.16}$$

where $\sigma = \varphi^{-1}(\tau) \in [c,d]$. This is possible since φ,φ^{-1} are continuous.

Let $\{\alpha_0,\tau_1,\alpha_1,\dots,\alpha_{k-1},\tau_k,\alpha_k\}$ be an arbitrary δ–fine partition of $[\varphi(c),\varphi(d)]$. Let us set $\beta_j = \varphi^{-1}(\alpha_j)$, $j = 0,\dots,k$,

$\sigma_j = \varphi^{-1}(\tau_j)$, $j = 1, \dots, k$. Then by the monotonicity of φ^{-1} we have

$$\beta_0 = \varphi^{-1}(\alpha_0) = c < \beta_1 < \beta_2 < \cdots < \beta_{k-1} < \beta_k = \varphi^{-1}(\alpha_k) = d$$

and

$$\beta_{j-1} \le \sigma_j \le \beta_j, \quad j = 1, 2, \dots, k.$$

This yields that $\{\beta_0, \sigma_1, \beta_1, \dots, \beta_{k-1}, \sigma_k, \beta_k\}$ is a partition of $[c, d]$. Since $\{\alpha_0, \tau_1, \alpha_1, \dots, \alpha_{k-2}, \tau_{k-1}, \alpha_{k-1}\}$ is a δ–fine partition of $[\varphi(c), \varphi(d)]$ we have

$$\tau_j - \delta(\tau_j) \le \alpha_{j-1} \le \tau_j \le \alpha_j \le \tau_j + \delta(\tau_j)$$

and (1.16) implies

$$\varphi(\sigma_j - \omega(\sigma_j)) \le \alpha_{j-1} \le \alpha_j \le \varphi(\sigma_j + \omega(\sigma_j)).$$

Therefore

$$\sigma_j - \omega(\sigma_j) = \varphi^{-1}(\varphi(\sigma_j - \omega(\sigma_j)) \le \varphi^{-1}(\alpha_{j-1}) =$$

$$= \beta_{j-1} < \beta_j \le \varphi^{-1}(\alpha_j) \le \sigma_j + \omega(\sigma_j)$$

and this shows that the partition $\{\beta_0, \sigma_1, \beta_1, \dots, \beta_{k-1}, \sigma_k, \beta_k\}$ is ω–fine.

Further, clearly

$$\|\sum_{j=1}^{k}[U(\tau_j), \alpha_j)) - U(\tau_j), \alpha_{j-1}))] - \int_c^d DU(\varphi(\sigma), \varphi(s))\| =$$

$$= \|\sum_{j=1}^{k}[U(\varphi(\sigma_j), \varphi(\beta_j)) - U(\varphi(\sigma_j), \varphi(\beta_{j-1}))] -$$

$$- \int_c^d DU(\varphi(\sigma), \varphi(s))\| < \varepsilon$$

because the partition $\{\beta_0, \sigma_1, \beta_1, \ldots, \beta_{k-1}, \sigma_k, \beta_k\}$ is ω–fine. By definition this yields the existence of the integral $\int_{\varphi(c)}^{\varphi(d)} DU(\tau, t)$ and shows that (1.15) is satisfied.

Now assume that the integral $\int_{\varphi(c)}^{\varphi(d)} DU(\tau, t)$ exists and that $\varepsilon > 0$ is given. Then there exists a gauge δ on $[\varphi(c), \varphi(d)]$ such that for every δ–fine partition $\{\alpha_0, \tau_1, \alpha_1, \ldots, \alpha_{k-1}, \tau_k, \alpha_k\}$ of $[\varphi(c), \varphi(d)]$ we have

$$\| \sum_{j=1}^{k} [U(\tau_j), \alpha_j)) - U(\tau_j), \alpha_{j-1})] - \int_{\varphi(c)}^{\varphi(d)} DU(\tau, t)\| < \varepsilon.$$

For $\sigma \in [c, d]$ there exists a value $\omega(\sigma) > 0$ such that

$$|\varphi(s) - \varphi(\sigma)| < \delta(\varphi(\sigma)) \tag{1.17}$$

if $s \in [\sigma - \omega(\sigma), \sigma + \omega(\sigma)] \cap [c, d]$.

If $\{\beta_0, \sigma_1, \beta_1, \ldots, \beta_{k-1}, \sigma_k, \beta_k\}$ is an ω–fine partition of $[c, d]$, then we set $\tau_j = \varphi(\sigma_j)$, $j = 1, 2, \ldots, k$, $\alpha_j = \varphi(\beta_j)$, $j = 0, 1, \ldots, k$. From the monotonicity of φ and from (1.17) we easily deduce that $\{\alpha_0, \tau_1, \alpha_1, \ldots, \alpha_{k-1}, \tau_k, \alpha_k\}$ is a δ–fine partition of $[\varphi(c), \varphi(d)]$. Hence

$$\| \sum_{j=1}^{k} [U(\varphi(\sigma_j), \varphi(\beta_j)) - U(\varphi(\sigma_j), \varphi(\beta_{j-1}))] - \int_{\varphi(c)}^{\varphi(d)} DU(\tau, t)\| < \varepsilon$$

for every ω–fine partition $\{\beta_0, \sigma_1, \beta_1, \ldots, \beta_{k-1}, \sigma_k, \beta_k\}$ of $[c, d]$ because

$$\sum_{j=1}^{k} [U(\varphi(\sigma_j), \varphi(\beta_j)) - U(\varphi(\sigma_j), \varphi(\beta_{j-1}))] =$$

$$= \sum_{j1}^{k} [U(\tau_j), \alpha_j)) - U(\tau_j), \alpha_{j-1})]$$

where $\{\alpha_0, \tau_1, \alpha_1, \ldots, \alpha_{k-1}, \tau_k, \alpha_k\}$ is a δ–fine partition of $[\varphi(c), \varphi(d)]$. This yields the existence of the integral

$$\int_c^d DU(\varphi(\sigma), \varphi(s))$$

as well as the equality (1.15). □

1.19 *Remark.* Theorem 1.18 holds also in a more general setting when φ is continuous and monotone only, i.e. nondecreasing or nonincreasing. In this case the proof of the substitution theorem is more technical. It can be given by generalizing the concept of the inverse to φ. If e.g. $\varphi : [c, d] \to \mathbb{R}$ is nondecrasing and $\tau \in [\varphi(c), \varphi(d)]$ then we define

$$\varphi^{-1}(\tau) = \{\sigma \in [c, d]; \varphi(\sigma) = \tau\}.$$

Clearly $\varphi^{-1}(\tau)$ is either a single point or a closed subinterval in $[c, d]$. Using this definition of an "inverse" to φ, partitions of $[c, d]$ and $[\varphi(c), \varphi(d)]$ can be mapped one to the other like in the proof of Theorem 1.18.

Integration by parts

Now we give a general form of the *integration by parts formula* for the generalized Perron integral. First of all we prove a simple lemma.

1.20 Lemma. *Assume that* $V : [a, b] \times [a, b] \to \mathbb{R}$ *is such that*

$$V(\tau, t) = V(t, \tau) \text{ for } t, \tau \in [a, b]. \tag{1.18}$$

Let us set $W(\tau,t)=V(t,\tau)$. Then if one of the integrals

$$\int_a^b DV(\tau,t),\ \int_a^b DW(\tau,t)$$

exists, then the other one exists as well and

$$\int_a^b DV(t,\tau)=\int_a^b DW(\tau,t)=\int_a^b DV(\tau,t).$$

Proof. It is easy to see that the integral sums $S(V,D)$, $S(W,D)$ coincide by (1.18) for any partition of $[a,b]$ and this yields the result. □

1.21 Theorem (Integration by parts). *Suppose that $U: [a,b]\times[a,b]\to\mathbb{R}$ is given. Let us set $U^\star(\tau,t)=U(t,\tau)$ and $V(t,\tau)=U(\tau,\tau)-U(\tau,t)-U(t,\tau)+U(t,t)$.*

If two of the integrals $\int_a^b DU(\tau,t)$, $\int_a^b DU^\star(\tau,t)$, $\int_a^b DV(\tau,t)$ exist then the third exists as well and the equality

$$\int_a^b DU(\tau,t)+\int_a^b DU^\star(\tau,t)=U(b,b)-U(a,a)-\int_a^b DV(\tau,t) \tag{1.19}$$

holds.

Proof. It can be checked immediately that V satisfies (1.18). Suppose that e.g. the integrals $\int_a^b DU(\tau,t)$, $\int_a^b DU^\star(\tau,t)$ exist. Then there is a gauge δ on $[a,b]$ such that for any δ–fine partition $D=\{\alpha_0,\tau_1,\alpha_1,\dots,\alpha_{k-1},\tau_k,\alpha_k\}$ of $[a,b]$ we have

$$\|\int_a^b DU(\tau,t)-\sum_{j=1}^k[U(\tau_j,\alpha_j)-U(\tau_j,\alpha_{j-1})]\|<\varepsilon$$

and

$$\| \int_a^b DU^\star(\tau,t) - \sum_{j=1}^{k}[U^\star(\tau_j,\alpha_j) - U^\star(\tau_j,\alpha_{j-1})]\| < \varepsilon.$$

Then for every δ–fine partition D of $[a,b]$ we have

$$\|S(V,D) - U(b,b) + U(a,a) + \int_a^b DU(\tau,t) + \int_a^b DU^\star(\tau,t)\| =$$

$$= \| \sum_{j=1}^{k}[(U(\tau_j,\tau_j) - U(\tau_j,\alpha_j) - U(\alpha_j,\tau_j) + U(\alpha_j,\alpha_j)) -$$

$$-(U(\tau_j,\tau_j) - U(\tau_j,\alpha_{j-1}) - U(\alpha_{j-1},\tau_j) + U(\alpha_{j-1},\alpha_{j-1}))] -$$

$$-U(b,b) + U(a,a) + \int_a^b DU(\tau,t) + \int_a^b DU^\star(\tau,t)\| \leq$$

$$\leq \| \sum_{j=1}^{k}[-U(\tau_j,\alpha_j) - U(\alpha_j,\tau_j) + U(\tau_j,\alpha_{j-1}) + U(\alpha_{j-1},\tau_j)] +$$

$$+ \int_a^b DU(\tau,t) + \int_a^b DU^\star(\tau,t) +$$

$$+ \sum_{j=1}^{k}[U(\alpha_j,\alpha_j) - U(\alpha_{j-1},\alpha_{j-1})] - U(b,b) + U(a,a)\| \leq$$

$$\leq \| \int_a^b DU(\tau,t) - \sum_{j=1}^{k}[U(\tau_j,\alpha_j) - U(\tau_j,\alpha_{j-1})]\| +$$

$$+\| \int_a^b DU^\star(\tau,t) - \sum_{j=1}^{k}[U^\star(\tau_j,\alpha_j) - U^\star(\tau_j,\alpha_{j-1})]\| < 2\varepsilon.$$

Hence the integral $\int_a^b DV(\tau,t)$ exists and the equality (1.19) holds by the definition of the integral.

The proof for the other cases is similar. □

1.22 *Remark.* The proof of Theorem 1.21 is based on purely algebraic manipulations of integral sums for the generalized integral. This approach to integration by parts goes back to the paper [74] of J. Kurzweil.

Now we give a corollary to Theorem 1.21 which contains the integration by parts formula in a more conventional form.

1.23 Corollary. *If $f,g:[a,b]\to\mathbb{R}$ are functions of bounded variation on $[a,b]$ then the integrals $\int_a^b f(s)\,dg(s)$ and $\int_a^b g(s)\,df(s)$ exist*) and*

$$\int_a^b f(s)\,dg(s)+\int_a^b g(s)\,df(s)=f(b)g(b)-f(a)g(a)-$$

$$-\sum_{a\le\tau<b}\Delta^+f(\tau)\Delta^+g(\tau)+\sum_{a<\tau\le b}\Delta^-f(\tau)\Delta^-g(\tau), \qquad (1.20)$$

where $\Delta^+f(\tau)=f(\tau+)-f(\tau)=lim_{\sigma\to\tau+}f(\sigma)-f(\tau)$, $\Delta^-f(\tau)=f(\tau)-f(\tau-)=f(\tau)-lim_{\sigma\to\tau-}f(\sigma)$ and similarly for $\Delta^+g(\tau)$, $\Delta^-g(\tau)$.

Proof. We postpone at this moment the proof of the fact that both integrals $\int_a^b f(s)\,dg(s)$ and $\int_a^b g(s)\,df(s)$ exist. This will be shown later (see Corollary 1.34).

Let us set $U(\tau,t)=f(\tau)g(t)$ and $U^\star(\tau,t)=g(\tau)f(t)$ for $\tau,t\in[a,b]$. Then

$$\int_a^b DU(\tau,t)=\int_a^b f(s)\,dg(s),\quad \int_a^b DU^\star(\tau,t)=\int_a^b g(s)\,df(s).$$

*)See Remark 1.5 for these integrals.

and clearly also $U^{\star}(\tau,t) = U(t,\tau)$. By Theorem 1.21 the integral $\int_a^b DV(\tau,t)$ exists where

$$V(\tau,t) = [f(\tau) - f(t)][g(\tau) - g(t)].$$

Now it is clear that for proving the formula (1.20) we have to show that

$$\int_a^b DV(\tau,t) = \sum_{a \leq \tau < b} \Delta^+ f(\tau)\Delta^+ g(\tau) - \sum_{a < \tau \leq b} \Delta^- f(\tau)\Delta^- g(\tau). \tag{1.21}$$

Let $\varepsilon > 0$ be given. Set

$$N = \{\tau \in (a,b); |f(\tau+) - f(\tau)| \geq \varepsilon, \; |f(\tau) - f(\tau-)| \geq \varepsilon\}.$$

Since the function f is of bounded variation, the set N is finite. Further, for every $\tau \in [a,b]$ there is $\delta_1(\tau) > 0$ such that

$$|f(s) - f(\tau+)| < \varepsilon, |g(s) - g(\tau+)| < \varepsilon \text{ for } s \in (\tau, \tau + \delta_1(\tau))$$

and

$$|f(s) - f(\tau-)| < \varepsilon, |g(s) - g(\tau-)| < \varepsilon \text{ for } s \in (\tau - \delta_1(\tau), \tau).$$

This is clear because the onesided limits for the functions f, g exist at every point in $[a,b]$; at the endpoints only the corresponding ones.

Let us set $\delta_2(\tau) = dist(\tau, N)$ for $\tau \notin N$ ($dist(\tau, N)$ stands here for the distance of the point τ from the set N) and $\delta_2(\tau) = \delta_1(\tau)$ for $\tau \in N$. Finally, define $\delta(\tau) = \min(\delta_1(\tau), \delta_2(\tau)) > 0$ for $\tau \in [a,b]$. The function δ represents a gauge on $[a,b]$. Let

$$D = \{\alpha_0, \tau_1, \alpha_1, \ldots, \alpha_{k-1}, \tau_k, \alpha_k\}$$

be an arbitrary δ–fine partition of $[a, b]$. If $c \in N$ then there is an index $j = 1, \ldots, k$ such that $c \in [\alpha_{j-1}, \alpha_j]$. Moreover, the form of the gauge δ forces in this case also $\tau_j = c$ (see Remark 1.12) and consequently we have $N \subset \{\tau_1, \tau_2, \ldots, \tau_k\}$.

For the δ–fine partition D we now get ($S(V, D)$ is the integral sum corresponding to the function V and the partition D)

$$|S(V,D) - \sum_{a\le\tau<b} \Delta^+ f(\tau)\Delta^+ g(\tau) + \sum_{a<\tau\le b} \Delta^- f(\tau)\Delta^- g(\tau)| =$$

$$= |\sum_{j=1}^{k}[(f(\tau_j) - f(\alpha_j))(g(\tau_j) - g(\alpha_j)) -$$

$$-(f(\tau_j) - f(\alpha_{j-1}))(g(\tau_j) - g(\alpha_{j-1}))] -$$

$$- \sum_{a\le\tau<b} \Delta^+ f(\tau)\Delta^+ g(\tau) + \sum_{a<\tau\le b} \Delta^- f(\tau)\Delta^- g(\tau)| \le$$

$$\le |\sum_{j=1}^{k}[(f(\tau_j) - f(\alpha_j))(g(\tau_j) - g(\alpha_j))] - \sum_{a\le\tau<b} \Delta^+ f(\tau)\Delta^+ g(\tau)| +$$

$$+ |\sum_{j=1}^{k}(f(\tau_j) - f(\alpha_{j-1}))(g(\tau_j) - g(\alpha_{j-1})) - \sum_{a<\tau\le b} \Delta^- f(\tau)\Delta^- g(\tau)|. \tag{1.22}$$

For the first term on the right hand side of this inequality we have

$$|\sum_{j=1}^{k}[(f(\tau_j) - f(\alpha_j))(g(\tau_j) - g(\alpha_j))] - \sum_{a\le\tau<b} \Delta^+ f(\tau)\Delta^+ g(\tau)| =$$

$$= |\sum_{j=1}^{k}[\Delta^+ f(\tau_j)\Delta^+ g(\tau_j) - \Delta^+ f(\tau_j)(g(\tau_j+) - g(\alpha_j)) -$$

$$-\Delta^+ g(\tau_j)(f(\tau_j+) - f(\alpha_j)) + (f(\tau_j+) - f(\alpha_j))(g(\tau_j+) - g(\alpha_j))]-$$

$$-\sum_{j=1}^{k} \Delta^+ f(\tau_j)\Delta^+ g(\tau_j) - \sum_{\substack{a \le \tau < b \\ \tau \neq \tau_j}} \Delta^+ f(\tau)\Delta^+ g(\tau)| \le$$

$$\le \sum_{j=1}^{k} |\Delta^+ f(\tau_j)||g(\tau_j+) - g(\alpha_j)| + \sum_{j=1}^{k} |\Delta^+ g(\tau_j)||f(\tau_j+) - f(\alpha_j)| +$$

$$+ \sum_{j=1}^{k} |f(\tau_j+) - f(\alpha_j)||g(\tau_j+) - g(\alpha_j)| + \sum_{\substack{a \le \tau < b \\ \tau \notin N}} |\Delta^+ f(\tau)||\Delta^+ g(\tau)| <$$

$$< \varepsilon \operatorname{var}_a^b f + \varepsilon \operatorname{var}_a^b g + \varepsilon \operatorname{var}_a^b g + \varepsilon \operatorname{var}_a^b g = \varepsilon(\operatorname{var}_a^b f + 3 \operatorname{var}_a^b g).$$

In these cumbersome but straightforward calculations we assumed that if $\tau_j \in N$ then $\alpha_{j-1} < \tau_j < \alpha_j$. The possibility of doing this is apparent from the fact that if for example $\alpha_{j-1} < \tau_j = \alpha_j = \tau_{j+1} < \alpha_{j+1}$ then

$$V(\tau_j, \alpha_j) - V(\tau_j, \alpha_{j-1}) + V(\tau_{j+1}, \alpha_{j+1}) - V(\tau_{j+1}, \alpha_j) =$$

$$= V(\tau_j, \alpha_{j+1}) - V(\tau_j, \alpha_{j-1})$$

and the two intervals $[\alpha_{j-1}, \alpha_j]$, $[\alpha_j, \alpha_{j+1}]$ of the partition D can be replaced by a single interval $[\alpha_{j-1}, \alpha_{j+1}]$ with $\alpha_{j-1} < \tau_j < \alpha_j$ without changing the integral sum $S(V, D)$.

In a completely analogous way the second term on the right hand side of (1.22) can be estimated which leads to the inequality

$$|S(V, D) - \sum_{a \le \tau < b} \Delta^+ f(\tau)\Delta^+ g(\tau) + \sum_{a < \tau \le b} \Delta^- f(\tau)\Delta^- g(\tau)| < c\varepsilon$$

where c is a constant, and this implies by the definition of the integral that (1.21) holds. □

1.24 *Remark.* It should be mentioned that the unusual additional terms

$$-\sum_{a\le\tau<b}\Delta^+f(\tau)\Delta^+g(\tau)+\sum_{a<\tau\le b}\Delta^-f(\tau)\Delta^-g(\tau)$$

on the right hand side of (1.20) vanish if f and g are functions with no common discontinuities.

The integration by parts result from Corollary 1.23 can be strengthened for the case when one of the functions f, g is of bounded variation and the other is regulated (i.e. has onesided limits at every point in $[a,b]$). In this case the integrals in question exist as was shown by M. Tvrdý in [164] and the proof is the same as ours because regulated functions have only a finite number of onesided discontinuity points at which their jumps exceed a given positive number.

For more information on the extensive literature concerning integration by parts see the expository papers [39] or [11].

Convergence theorems

1.25 Theorem. *Let functions $U, U_m : [a,b]\times[a,b]\to\mathbb{R}^n$, $m = 1,2,\ldots$ be given where $U_m\in\mathcal{K}([a,b])$ for $m=1,2,\ldots$. Assume that there is a gauge ω on $[a,b]$ such that*

$$\lim_{m\to\infty}[U_m(\tau,t_2)-U_m(\tau,t_1)]=U(\tau,t_2)-U(\tau,t_1) \tag{1.23}$$

for $\tau\in[a,b]$ and all $t_1,t_2\in\mathbb{R}$ such that $t_1\le\tau\le t_2$, $[t_1,t_2]\subset[\tau-\omega(\tau),\tau+\omega(\tau)]$. Assume further that

for every $\eta>0$ there is a gauge δ on $[a,b]$ such that

$$\|S(U_m,D)-\int_a^b DU_m(\tau,t)\|<\eta \tag{1.24}$$

for every δ–fine partition D of $[a,b]$ and every $m = 1,2,\ldots$.

Then $U \in \mathcal{K}([a,b])$ *and*

$$\lim_{m\to\infty}\int_a^b DU_m(\tau,t) = \int_a^b DU(\tau,t). \tag{1.25}$$

Proof. Let $\varepsilon > 0$ be given. By (1.24) there is a gauge δ on $[a,b]$, $\delta(\tau) \le \omega(\tau)$, $\tau \in [a,b]$ such that for every δ–fine partition

$$D = \{\alpha_0, \tau_1, \alpha_1, \dots, \alpha_{k-1}, \tau_k, \alpha_k\}$$

of $[a,b]$ we have

$$\|S(U_m,D) - \int_a^b DU_m(\tau,t)\| < \frac{\varepsilon}{2}$$

for $m = 1,2,\dots$. By (1.23) for every fixed partition D of $[a,b]$ there exists a positive integer m_0 such that for $m > m_0$ the inequality

$$\|S(U_m,D) - S(U,D)\| =$$

$$\|\sum_{j=1}^{k}[U_m(\tau_j,\alpha_j) - U(\tau_j,\alpha_j) - U_m(\tau_j,\alpha_{j-1}) + U(\tau_j,\alpha_{j-1})]\| < \frac{\varepsilon}{2}$$

holds and this means that

$$\lim_{m\to\infty} S(U_m,D) = S(U,D).$$

Therefore for any δ–fine partition D of $[a,b]$ there is a positive integer m_0 such that for $m > m_0$ we have

$$\|S(U,D) - \int_a^b DU_m(\tau,t)\| < \varepsilon. \tag{1.26}$$

First we get from (1.26) that for all positive integers $m, l > m_0$ the inequality

$$\|\int_a^b DU_m(\tau, t) - \int_a^b DU_l(\tau, t)\| < 2\varepsilon$$

holds. This means that $(\int_a^b DU_m(\tau, t))_{m=1}^{\infty}$ is a Cauchy sequence in $\mathbb{R}^n$ and therefore has a limit

$$\lim_{m\to\infty} \int_a^b DU_m(\tau, t) = I \in \mathbb{R}^n. \tag{1.27}$$

Another consequence of (1.26) is the inequality

$$\|S(U, D) - I\| \leq \|S(U, D) - \int_a^b DU_m(\tau, t)\| +$$

$$\|\int_a^b DU_m(\tau, t) - I\| < \varepsilon + \|\int_a^b DU_m(\tau, t) - I\|.$$

By (1.27) we obtain immediately from this inequality that for every δ–fine partition D of $[a, b]$ we have

$$\|S(U, D) - I\| < \varepsilon$$

and this means that the integral $\int_a^b D(U\tau, t)$ exists and (1.25) is satisfied. □

1.26 Definition. A sequence of integrable functions U_m : $[a, b] \times [a, b] \to \mathbb{R}^n$ ($U_m \in \mathcal{K}([a, b])$), $m = 1, 2, \ldots$ is called *equi - integrable* if the condition (1.24) of Theorem 1.25 is satisfied.

1.27 *Remark.* Theorem 1.25 gives a sufficient condition for a sequence of integrable functions to tend to an integrable limit and for the integrals of the members of the sequence to tend to the integral of the limit function. The convergence of the functions U_m to U is given by (1.23) and the sufficient condition is the equi - integrability (1.24) of the sequence (U_m).

Clearly a convergence result for integrals of $\mathbb{R}^n$–valued functions holds if and only if it holds for every component of the functions (cf. Theorem 1.6). Therefore without loss of generality we can consider sequences of real–valued functions only.

The fundamental idea of Theorem 1.25 lies in viewing the concept of the integral as a certain limiting process and in the fact that two limits are interchangeable provided one of them is uniform with respect to the limiting variable of the second. In our situation equi - integrability stands for this uniformity. In this sense Theorem 1.25 is a transparent mathematical fact, nevertheless from the practical point of view it is not easy to check that a given sequence of integrable functions is equi - integrable. This forces us to use another condition instead of equi - integrability and motivates the results given in the sequel.

For a given $V : [a,b] \times [a,b] \to \mathbb{R}$ and a tagged interval (τ, J) with $\tau \in J = [\alpha, \beta] \subset [a,b]$ we will use the notation

$$V(\tau, J) = V(\tau, \alpha) - V(\tau, \beta) \tag{1.28}$$

for the point–interval function which corresponds to V.

1.28 Theorem. *Let $U, U_m : [a,b] \times [a,b] \to \mathbb{R}, m = 1,2,\ldots$ where $U_m \in \mathcal{K}([a,b])$ for $m = 1,2,\ldots$.*

> *Assume that there is a gauge ω on $[a,b]$ such that for every $\varepsilon > 0$ there exist a $p : [a,b] \to \mathbb{N}$*) and a positive superadditive interval function Φ defined for closed intervals $J \subset [a,b]$ with $\Phi([a,b]) < \varepsilon$ such that for every $\tau \in [a,b]$ we have*
>
> $$|U_m(\tau, J) - U(\tau, J)| < \Phi(J) \tag{1.29}$$

---

*)By $\mathbb{N}$ the set of positive integers is denoted.

provided $m > p(\tau)$ and (τ, J) is an ω–fine tagged interval with $\tau \in J \subset [a,b]$.

Let us further assume that the sequence (U_m) satisfies the following condition.

There exist a gauge θ on $[a,b]$ and real constants $B < C$ such that for all choices of functions m defined on $[a,b]$ taking positive integer values ($m : [a,b] \to \mathbb{N}$) the inequalities

$$B \leq \sum_{j=1}^{k} U_{m(\tau_j)}(\tau_j, J_j) \leq C \tag{1.30}$$

hold provided

$$D = \{(\tau_j, J_j), j = 1, 2, \ldots, k\}$$

is an arbitrary θ–fine partition of $[a,b]$.

Then the sequence (U_m) is equi - integrable, i.e. (1.24) holds.

1.28a Definition. Let $U, U_m : [a,b] \times [a,b] \to \mathbb{R}, m = 1, 2, \ldots,$ $U_m \in \mathcal{K}([a,b])$ for $m = 1, 2, \ldots$ be given such that the conditions (1.29) and (1.30) of Theorem 1.28 are satisfied.

For a given positive integer $p \in \mathbb{N}$ let $\mathcal{S}_p$ be the family of all functions $V : [a,b] \times [a,b] \to \mathbb{R}$ such that there is a division

$$a = \beta_0 < \beta_1 < \cdots < \beta_{l-1} < \beta_l = b$$

of $[a,b]$ such that for any tagged interval (τ, J), $\tau \in J \subset [a,b]$ we have

$$\begin{aligned} V(\tau, J) =& U_{m_j}(\tau, J) \text{ if } \tau \in (\beta_{j-1}, \beta_j), \\ V(\tau, J) =& U_{m_j}(\tau, J^-) + U_{m_{j+1}}(\tau, J^+) \text{ if } \tau = \beta_j, \\ j =& 1, \ldots, l, J^- = J \cap (-\infty, \beta_j], J^+ = J \cap [\beta_j, \infty) \end{aligned} \tag{1.31}$$

where $m_j \in \mathbb{N}$ with

$$m_j \geq p, j = 1, 2, \ldots, l.$$

Let us give a series of statements concerning the families $\mathcal{S}_p$ described by this definition.

1.28b Lemma.

a) *For every $p \in \mathbb{N}$ we have $U_p \in \mathcal{S}_p$.*
b) *If $p_1, p_2 \in \mathbb{N}$, $p_1 > p_2$ then $\mathcal{S}_{p_1} \subset \mathcal{S}_{p_2}$.*
c) *If $V \in \mathcal{S}_1$ then $V \in \mathcal{K}([a, b])$.*

Proof. a) If we set $\beta_0 = a$, $\beta_1 = b$ then clearly U_p is of the form (1.31).

b) is clear by the definition of $\mathcal{S}_p$.

c) Given $V \in \mathcal{S}_p$ then it is easy to check by (1.31) that for every partition D^j of $[\beta_{j-1}, \beta_j]$ we have

$$S(V, D^j) = S(U_{m_j}, D^j) \tag{1.32}$$

for the corresponding integral sums. Since U_{m_j} is integrable over $[a, b]$, it is also integrable over $[\beta_{j-1}, \beta_j]$ by Theorem 1.10 and (1.32) yields the integrability of V over $[\beta_{j-1}, \beta_j]$. This holds for every $j = 1, \ldots, l$ and therefore by Theorem 1.11 we obtain $V \in \mathcal{K}([a, b])$. □

1.28c Lemma. *If $V \in \mathcal{S}_1$ then*

$$B \leq \sum_{j=1}^{k} V(\tau_j, J_j) = S(V, D) \leq C \tag{1.33}$$

for an arbitrary θ–fine partition $D = \{(\tau_j, J_j), j = 1, 2, \ldots, k\}$ of $[a, b]$.

Proof. Let $\{\beta_0, \ldots, \beta_m\}$ be the finite sequence of points used for the definition of V in (1.31). By (1.31) we have $V(\tau, J) =$

$U_{m_j}(\tau, J)$ if $\tau \notin \{\beta_0, \dots, \beta_m\}$ and $V(\tau, J) = U_{m_j}(\tau, J^-) + U_{m_{j+1}}(\tau, J^+)$ when $\tau \in \{\beta_0, \dots, \beta_m\}$.

It is easy to see that

$$\{(\tau_j, J_j^-); j = 1, \dots, k, \tau_j \notin \{\beta_0, \dots, \beta_m\}\} \cup$$
$$\cup \{(\tau_j, J_j^-); j = 1, \dots, k, \tau_j \in \{\beta_0, \dots, \beta_m\}\} \cup$$
$$\cup \{(\tau_j, J_j^+); j = 1, \dots, k, \tau_j \in \{\beta_0, \dots, \beta_m\}\}$$

forms a θ–fine partition of $[a, b]$ and therefore (1.33) follows immediately from (1.30). □

1.28d Lemma. *If (τ, J) is an ω–fine tagged interval, $\tau \in J \subset [a, b]$, $p \geq p(\tau)$, then*

$$|V(\tau, J) - U(\tau, J)| < \Phi(J) \tag{1.34}$$

for every $V \in \mathcal{S}_p$.

Proof. If $\tau_j \notin \{\beta_0, \dots, \beta_m\}$, then $V(\tau, J) = U_{m_j}(\tau, J)$ by (1.31) and because $m_j \geq p \geq p(\tau)$ (1.34) holds by (1.29).

If $\tau_j \in \{\beta_0, \dots, \beta_m\}$, then $\tau = \beta_j$ for some $j = 1, \dots, m$ and

$$V(\tau, J) = U_{m_j}(\tau, J^-) + U_{m_{j+1}}(\tau, J^+)$$

by Definition 1.28a. If $J = [\alpha, \beta]$ then $J^- = [\alpha, \beta_j]$, $J^+ = [\beta_j, \beta]$ and

$$U(\tau, J) = U(\beta_j, \beta) - U(\beta_j, \alpha) =$$
$$U(\beta_j, \beta) - U(\beta_j, \beta_j) + U(\beta_j, \beta_j) - U(\beta_j, \alpha) = U(\tau, J^-) + U(\tau, J^+).$$

Therefore again by (1.29) we get

$$|V(\tau, J) - U(\tau, J)| =$$
$$|U_{m_j}(\tau, J^-) + U_{m_{j+1}}(\tau, J^+) - U(\tau, J^-) - U(\tau, J^+)| \leq$$
$$\leq |U_{m_j}(\tau, J^-) - U(\tau, J^-)| + |U_{m_{j+1}}(\tau, J^+) - U(\tau, J^+)| \leq$$
$$\leq \Phi(J^-) + \Phi(J^+) \leq \Phi(J).$$

□

1.28e Lemma. *Assume that $\varepsilon > 0$ is given. Then for every $p \in \mathbb{N}$ there exist $V_p, V^p \in \mathcal{S}_p$ such that*

$$\int_a^b DV_p(\tau,t) - \frac{\varepsilon}{2^p} < \inf\{\int_a^b DV(\tau,t); V \in \mathcal{S}_p\} \leq$$

$$\leq \sup\{\int_a^b DV(\tau,t); V \in \mathcal{S}_p\} < \int_a^b DV^p(\tau,t) + \frac{\varepsilon}{2^p}. \tag{1.35}$$

Proof. By 1.28c the supremum and infimum exist and

$$B \leq \inf\{\int_a^b DV(\tau,t); V \in \mathcal{S}_p\} \leq \sup\{\int_a^b DV(\tau,t); V \in \mathcal{S}_p\} \leq C.$$

The existence of V_p, V^p satisfying (1.35) comes immediately from the definition of the infimum and supremum. □

1.28f Lemma. *Assume that $V \in \mathcal{S}_p$. Let $I_j = [\gamma_{j-1}, \gamma_j]$, $j = 1, \dots, s$ be an arbitrary finite sequence of disjoint intervals in $[a,b]$. Then for a given $\varepsilon > 0$ we have*

$$\sum_{j=1}^s \int_{I_j} DV_p(\tau,t) - \frac{\varepsilon}{2^p} \leq \sum_{j=1}^s \int_{I_j} DV(\tau,t) \leq$$

$$\leq \sum_{j=1}^s \int_{I_j} DV^p(\tau,t) + \frac{\varepsilon}{2^p} \tag{1.36}$$

where $V_p, V^p \in \mathcal{S}_p$ are the functions corresponding to ε by Lemma 1.28e.

Proof. Assume for example that the second inequality in (1.36) is not satisfied. Then there is a $V^* \in \mathcal{S}_p$ such that

$$\sum_{j=1}^s \int_{I_j} DV^p(\tau,t) + \frac{\varepsilon}{2^p} < \sum_{j=1}^s \int_{I_j} DV^*(\tau,t).$$

Define

$$\begin{aligned}V_0(\tau,J) =& V^*(\tau,J) \text{ if } \tau\in(\gamma_{j-1},\gamma_j),\\ V_0(\tau,J) =& V^p(\tau,J) \text{ if } \tau\in[a,b]\setminus\cup_{j=1}^s I_j,\\ V_0(\tau,J) =& V^*(\tau,J^-)+V^p(\tau,J^+) \text{ if } \tau=\gamma_j,\\ j =& 1,\dots,s,\ J^-=J\cap(-\infty,\gamma_j],\ J^+=J\cap[\gamma_j,\infty).\end{aligned}$$

Denote by L_k, $k=1,\dots,r$ the system of closed intervals forming the components of the closure of $[a,b]\setminus\cup_{j=1}^s I_j$. Then evidently $V_0\in\mathcal{S}_p$ and we have the inequality

$$\int_a^b DV_0(\tau,t)=\sum_{j=1}^s\int_{I_j}DV^*(\tau,t)+\sum_{k=1}^r\int_{L_k}DV^p(\tau,t)>$$

$$>\sum_{j=1}^s\int_{I_j}DV^p(\tau,t)+\sum_{k=1}^r\int_{L_k}DV^p(\tau,t)+\frac{\varepsilon}{2^p}=\int_a^b DV^p(\tau,t)+\frac{\varepsilon}{2^p}$$

which contradicts Lemma 1.28e and thus proves the second inequality in (1.36). The first inequality in (1.36) holds by a similar argument. □

Proof of Theorem 1.28. Let $\varepsilon>0$ be given. By (1.29) and Lemma 1.28d for every $\tau\in[a,b]$ there is a $p(\tau)\in\mathbb{N}$ such that for all $m\in\mathbb{N}$, $m\ge p(\tau)$ we have

$$|V_{p(\tau)}(\tau,J)-U_m(\tau,J)|\le$$

$$|V_{p(\tau)}(\tau,J)-U(\tau,J)|+|U_m(\tau)(\tau,J)-U(\tau,J)|\le 2\Phi(J),\quad (1.37)$$

and similarly also

$$|V^{p(\tau)}(\tau,J)-U_m(\tau,J)|\le 2\Phi(J)\qquad (1.38)$$

if (τ, J) is an ω–fine tagged pair, $\tau \in J \subset [a,b]$, where Φ is a positive superadditive interval function with $\Phi([a,b]) < \varepsilon$.

For a given $p \in \mathbb{N}$ the functions U_p, V_p, V^p are integrable and therefore there is a gauge δ_p on $[a,b]$ such that

$$\begin{aligned} |S(U_p, D) - \int_a^b DU_p(\tau,t)| &< \frac{\varepsilon}{2^p}, \\ |S(V_p, D) - \int_a^b DV_p(\tau,t)| &< \frac{\varepsilon}{2^p}, \\ |S(V^p, D) - \int_a^b DV^p(\tau,t)| &< \frac{\varepsilon}{2^p} \end{aligned} \tag{1.39}$$

for any δ_p– fine partition D of $[a,b]$. For $\tau \in [a,b]$ let us choose $\delta(\tau) > 0$ such that

$$\delta(\tau) \le \min(\omega(\tau), \delta_1(\tau), \delta_2(\tau), \dots, \delta_{p(\tau)}(\tau))$$

(ω is the gauge from (1.29) and (1.37), $p : [a,b] \to \mathbb{N}$ is introduced in Lemma 1.28d and the gauges $\delta_1, \delta_2, \dots$ come from (1.39)).

Assume that $D = \{(\tau_i, J_i), i = 1, \dots, k\}$ is a δ–fine partition of $[a,b]$ and that $m \in \mathbb{N}$ is arbitrary. Then

$$S(U_m, D) = \sum_{\substack{i=1 \\ m \le p(\tau_i)}}^{k} U_m(\tau_i, J_i) + \sum_{\substack{i=1 \\ m > p(\tau_i)}}^{k} U_m(\tau_i, J_i). \tag{1.40}$$

The first term is the sum of those $U_m(\tau_i, J_i)$ for which $m \le p(\tau_i)$, and similarly for the second term.

If $m > p(\tau_i)$ then by (1.38)

$$-2\Phi(J) \le U_m(\tau_i, J_i) - V^{p(\tau_i)}(\tau_i, J_i) \le 2\Phi(J),$$

therefore

$$U_m(\tau_i, J_i) \geq V^{p(\tau_i)}(\tau_i, J_i) - 2\Phi(J)$$

and for the second sum in (1.40) we have

$$\sum_{\substack{i=1 \\ m>p(\tau_i)}}^{k} U_m(\tau_i, J_i) \geq \sum_{\substack{i=1 \\ m>p(\tau_i)}}^{k} [V^{p(\tau_i)}(\tau_i, J_i) - 2\Phi(J)].$$

Therefore

$$S(U_m, D) \geq$$

$$\geq \sum_{\substack{i=1 \\ m\leq p(\tau_i)}}^{k} U_m(\tau_i, J_i) + \sum_{\substack{i=1 \\ m>p(\tau_i)}}^{k} V^{p(\tau_i)}(\tau_i, J_i) - 2 \sum_{\substack{i=1 \\ m>p(\tau_i)}}^{k} \Phi(J) =$$

$$= -2 \sum_{\substack{i=1 \\ m>p(\tau_i)}}^{k} \Phi(J) + \sum_{\substack{i=1 \\ m\leq p(\tau_i)}}^{k} U_m(\tau_i, J_i) + \sum_{l=1}^{m-1} \sum_{\substack{i=1 \\ p(\tau_i)=l}}^{k} V^{p(\tau_i)}(\tau_i, J_i).$$

From (1.39) and from the Saks–Henstock lemma 1.13 we obtain

$$| \sum_{\substack{i=1 \\ m\leq p(\tau_i)}}^{k} U_m(\tau_i, J_i) - \sum_{\substack{i=1 \\ m\leq p(\tau_i)}}^{k} \int_{J_i} DU_m(\tau, t)| < \frac{\varepsilon}{2^p},$$

$$| \sum_{\substack{i=1 \\ p(\tau_i)=l}}^{k} V^l(\tau_i, J_i) - \sum_{\substack{i=1 \\ p(\tau_i)=l}}^{k} \int_{J_i} DV^l(\tau, t)| < \frac{\varepsilon}{2^l},$$

i.e.

$$\sum_{\substack{i=1 \\ m\leq p(\tau_i)}}^{k} U_m(\tau_i, J_i) > \sum_{\substack{i=1 \\ m\leq p(\tau_i)}}^{k} \int_{J_i} DU_m(\tau, t) - \frac{\varepsilon}{2^m},$$

$$\sum_{\substack{i=1\\ p(\tau_i)=l}}^{k} V^l(\tau_i, J_i) > \sum_{\substack{i=1\\ p(\tau_i)=l}}^{k} \int_{J_i} DV^l(\tau,t) - \frac{\varepsilon}{2^l}$$

and

$$S(U_m, D) > -2 \sum_{\substack{i=1\\ m>p(\tau_i)}}^{k} \Phi(J) + \sum_{\substack{i=1\\ m\leq p(\tau_i)}}^{k} \int_{J_i} DU_m(\tau,t) - \frac{\varepsilon}{2^m} +$$

$$+ \sum_{l=1}^{m-1} \sum_{\substack{i=1\\ p(\tau_i)=l}}^{k} \int_{J_i} DV^l(\tau,t) - \sum_{l=1}^{m-1} \frac{\varepsilon}{2^l}.$$

Since $U_m \in \mathcal{S}_m$ we have $U_m \in \mathcal{S}_p$ for all $p = 1, 2. \dots, m-1$ by b) from Lemma 1.28b and by Lemma 1.28f we therefore get

$$\sum_{\substack{i=1\\ p(\tau_i)=l}}^{k} \int_{J_i} DV^l(\tau,t) \geq \sum_{\substack{i=1\\ p(\tau_i)=l}}^{k} \int_{J_i} DU_m(\tau,t) - \frac{\varepsilon}{2^l},$$

i.e.

$$\sum_{l=1}^{m-1} \sum_{\substack{i=1\\ p(\tau_i)=l}}^{k} \int_{J_i} DV^l(\tau,t) \geq \sum_{l=1}^{m-1} \sum_{\substack{i=1\\ p(\tau_i)=l}}^{k} \int_{J_i} DU_m(\tau,t) - \sum_{l=1}^{m-1} \frac{\varepsilon}{2^l} =$$

$$= \sum_{\substack{i=1\\ m>p(\tau_i)}}^{k} \int_{J_i} DU_m(\tau,t) - \sum_{l=1}^{m-1} \frac{\varepsilon}{2^l},$$

and

$$S(U_m, D) > -2 \sum_{\substack{i=1\\ m>p(\tau_i)}}^{k} \Phi(J) - \frac{\varepsilon}{2^m} - \sum_{l=1}^{m-1} \frac{\varepsilon}{2^l} - \sum_{l=1}^{m-1} \frac{\varepsilon}{2^l} +$$

$$+ \sum_{\substack{i=1 \\ m>p(\tau_i)}}^{k} \int_{J_i} DU_m(\tau,t) + \sum_{\substack{i=1 \\ m\le p(\tau_i)}}^{k} \int_{J_i} DU_m(\tau,t) \ge$$

$$\ge -2\Phi([a,b]) - \frac{\varepsilon}{2^m} - 2\sum_{l=1}^{m-1} \frac{\varepsilon}{2^l} + \int_a^b DU_m(\tau,t) \ge$$

$$\ge \int_a^b DU_m(\tau,t) - \varepsilon(2 + \frac{1}{2^m} + 2\sum_{l=1}^{m-1} \frac{1}{2^l}) > \int_a^b DU_m(\tau,t) - 4\varepsilon.$$

In a completely analogous way we can use V_p instead of V^p to show that

$$S(U_m, D) < \int_a^b DU_m(\tau,t) + 4\varepsilon,$$

i.e.

$$|S(U_m, D) - \int_a^b DU_m(\tau,t)| < 4\varepsilon$$

and the sequence (U_m) is equi - integrable, the gauge δ being independent of m. □

Using Theorems 1.28 and 1.25 the following new form of the convergence theorem for generalized Perron integral can be given.

1.29 Theorem. *Let $U, U_m : [a,b] \times [a,b] \to \mathbb{R}, m = 1,2,\dots$ where $U_m \in \mathcal{K}([a,b])$ for $m = 1,2,\dots$.*

Let us further assume that the sequence (U_m) satisfies the following conditions.

> *There is a gauge ω on $[a,b]$ such that for every $\varepsilon > 0$ there exist a $p : [a,b] \to \mathbb{N}$ and a positive superadditive interval function Φ defined for closed intervals $J \subset [a,b]$ with $\Phi([a,b]) < \varepsilon$ such that for every $\tau \in [a,b]$ we have*

$$|U_m(\tau, J) - U(\tau, J)| < \Phi(J) \quad (1.29)$$

provided $m > p(\tau)$ and (τ, J) is an ω–fine tagged interval with $\tau \in J \subset [a,b]$.

There exist a gauge θ on $[a,b]$ and real constants $B < C$ such that for all choices of functions m defined on $[a,b]$ taking positive integer values ($m : [a,b] \to \mathbb{N}$) the inequalities

$$B \leq \sum_{j=1}^{k} U_{m(\tau_j)}(\tau_j, J_j) \leq C \tag{1.30}$$

hold provided

$$D = \{(\tau_j, J_j), j = 1, 2, \dots, k\}$$

is an arbitrary θ–fine partition of $[a,b]$.

Then $U \in \mathcal{K}([a,b))]$ and

$$\lim_{m\to\infty} \int_a^b DU_m(\tau,t) = \int_a^b DU(\tau,t).$$

Proof. If (1.29) is satisfied then for $\tau \in [a,b]$ and all $t_1, t_2 \in \mathbb{R}$ such that $t_1 \leq \tau \leq t_2$, $[t_1, t_2] \subset [\tau - \omega(\tau), \tau + \omega(\tau)]$ we clearly have

$$\lim_{m\to\infty} [U_m(\tau, t_2) - U_m(\tau, t_1)] = U(\tau, t_2) - U(\tau, t_1).$$

By Theorem 1.28 the sequence (U_m) is equi - integrable and therefore by Theorem 1.25 the conclusion of this theorem holds. □

1.30 *Remark.* It is easy to see that if $f_m : [a,b] \to \mathbb{R}$, $m = 1, 2, \dots$ and if $g : [a,b] \to \mathbb{R}$ is of bounded variation where

$$\lim_{m\to\infty} f_m(\tau) = f(\tau), \quad \tau \in [a,b]$$

then the sequence $U_m : [a,b] \times [a,b] \to \mathbb{R}$ given by

$$U_m(\tau,t) = f_m(\tau).g(t), \quad \tau,t \in [a,b]$$

for $m = 1,2,\dots$ satisfies (1.29) because

$$U_m(\tau,J) - U(\tau,J) = (f_m(\tau) - f(\tau)).(g(\beta) - g(\alpha))$$

for $J = [\alpha,\beta]$ and for a given $\varepsilon > 0$ the corresponding superadditive interval function Φ can be given in the form $\Phi(J) = \varepsilon \operatorname{var}_\alpha^\beta g$.

A result similar to Theorem 1.28 is given in the book [84, Lemma 5.4] where the condition (1.30) is replaced by the following one.

There exists a constant $K > 0$ such that for every division $a = \beta_0 < \beta_1 < \dots < \beta_l = b$ of $[a,b]$ and every finite sequence $m_1, m_2, \dots, m_l$ the inequality

$$\left|\sum_{j=1}^{l} \int_{\beta_{j-1}}^{\beta_j} DU_{m_j}(\tau,t)\right| < K$$

holds.

A condition of this form was used for the first time in 1979 in a common text of David Preiss and the author on elementary Perron integration which was published in a mimeographed form in Czech for internal use only as a preparation to a chapter of a calculus textbook which has never been finished.

Our condition (1.30) in Theorems 1.28 and 1.29 is motivated by the results of R. Henstock [40, Theorem 9.1]. Henstock shows directly that a condition of type (1.30) yields the convergence theorem for Perron integrals. Here we have shown that (1.30) yields the more general condition of equi - integrability and the convergence result is derived via Theorem 1.25.

Now we give a statement which plays the role of the dominated convergence theorem for generalized Perron integrals.

1.31 Corollary. *Let $U, U_m : [a,b] \times [a,b] \to \mathbb{R}, m = 1,2,\dots$ where $U_m \in \mathcal{K}([a,b])$ for $m = 1,2,\dots$.*

Let us assume further that the sequence (U_m) satisfies the following conditions.

There is a gauge ω on $[a,b]$ such that for every $\varepsilon > 0$ there exist a $p : [a,b] \to \mathbb{N}$ and a positive superadditive interval function Φ defined for closed intervals $J \subset [a,b]$ with $\Phi([a,b]) < \varepsilon$ such that for every $\tau \in [a,b]$ we have

$$|U_m(\tau, J) - U(\tau, J)| < \Phi(J)$$

provided $m > p(\tau)$ and (τ, J) is an ω–fine tagged interval with $\tau \in J \subset [a,b]$.

Assume further that two functions $V, W : [a,b] \times [a,b] \to \mathbb{R}$, $V, W \in \mathcal{K}([a,b])$ are given and that there is a gauge Θ on $[a,b]$ such that for all $m \in \mathbb{N}$, $\tau \in [a,b]$ we have

$$V(\tau, J) \le U_m(\tau, J) \le W(\tau, J) \tag{1.41}$$

for any Θ–fine tagged interval (τ, J).

Then $U \in \mathcal{K}([a,b)]$ and

$$\lim_{m\to\infty} \int_a^b DU_m(\tau, t) = \int_a^b DU(\tau, t).$$

Proof. Assume that θ is a gauge on $[a,b]$ such that $\theta(\tau) \le \Theta(\tau)$, $\tau \in [a,b]$ and

$$|S(V, D) - \int_a^b DV(\tau, t)| < 1, |S(W, D) - \int_a^b DW(\tau, t)| < 1$$

for every θ–fine partition $D = \{(\tau_j, J_j), j = 1,2,\dots,k\}$ of $[a,b]$.

Then evidently for every $m : [a,b] \to \mathbb{N}$ and any θ–fine partition $D = \{(\tau_j, J_j), j = 1,2,\dots,k\}$ we have by (1.41)

$$V(\tau_j, J_j) \le U_{m(\tau_j)}(\tau_j, J_j) \le W(\tau_j, J_j)$$

and also

$$\int_a^b DV(\tau,t) - 1 < S(V,D) \le \sum_{j=1}^{k} U_{m(\tau_j)}(\tau_j, J_j) \le$$

$$\le S(W,D) < \int_a^b DW(\tau,t) + 1.$$

Putting $\int_a^b DV(\tau,t) - 1 = B$ and $\int_a^b DW(\tau,t) + 1 = C$ we obtain the condition (1.30) of Theorem 1.29 and also the statement of this corollary. □

Let us now show the corresponding result for Stieltjes integrals.

1.32 Corollary. *Let $g : [a,b] \to \mathbb{R}$ be a nondecreasing function on $[a,b]$. Assume that $f_m : [a,b] \to \mathbb{R}$ are such functions that the integral $\int_a^b f_m(s)dg(s)$ exists for every $m \in \mathbb{N}$. Suppose that for $\tau \in [a,b]$ we have*

$$\lim_{m\to\infty} f_m(\tau) = f(\tau)$$

and that for $m \in \mathbb{N}$, $\tau \in [a,b]$ the inequalities

$$v(\tau) \le f_m(\tau) \le w(\tau)$$

hold where $v, w : [a,b] \to \mathbb{R}$ are such functions that the integrals $\int_a^b v(s)dg(s)$, $\int_a^b w(s)dg(s)$ exist.

Then the integral $\int_a^b f(s)dg(s)$ *exists and*

$$\int_a^b f(s)dg(s) = \lim_{m\to\infty} \int_a^b f_m(s)dg(s).$$

Proof. If we set $U_m(\tau,t) = f_m(\tau).g(t)$, $m \in \mathbb{N}$, $U(\tau,t) = f(\tau).g(t)$, $V(\tau,t) = v(\tau).g(t)$, $W(\tau,t) = w(\tau).g(t)$ for $\tau \in [a,b]$, $t \in [a,b]$ then (1.41) is satisfied and by Corollary 1.31 we obtain the result. □

Definition. A function $f : [a,b] \to \mathbb{R}$ is called a *finite step function* if there is a finite division $a = \beta_0 < \beta_1 < \cdots < \beta_m = b$ such that in every open interval (β_{i-1}, β_i), $i = 1, \dots, m$ the function f is identically equal to a constant $c_i \in \mathbb{R}$.

1.33 Proposition. *If* $f : [a,b] \to \mathbb{R}$ *is a finite step function (of the form given by the previous definition) and* $g : [a,b] \to \mathbb{R}$ *is of bounded variation on* $[a,b]$ *then the integral* $\int_a^b f(s)dg(s)$ *exists and*

$$\int_a^b f(s)dg(s) =$$

$$= f(a)(g(a+) - g(a)) + \sum_{j=1}^{m} f(\beta_j)(g(\beta_j+) - g(\beta_j-))+$$

$$+f(b)(g(b) - g(b-)) + \sum_{j=1}^{m} c_j(g(\beta_j-) - g(\beta_{j-1}+)). \tag{1.42}$$

(By $g(s+)$, $g(s-)$ the onesided right and left limits of g at the point $s \in [a,b]$ are denoted.)

Proof. The function g is of bounded variation and therefore the onesided limits of this function exist at every point in $[a,b]$. For a given $\varepsilon > 0$ let us choose a gauge δ on $[a,b]$ such that $\delta(\beta_j) <$

$\rho(\varepsilon)$, $j = 0, 1, \ldots, m$, $\delta(\tau) < dist(\tau, \{\beta_0, \beta_1, \ldots, \beta_m\})$ for $\tau \neq \beta_j$, $j = 0, 1, \ldots, m$. For a gauge δ satisfying these assumptions we know that every δ–fine partition D of $[a, b]$ contains all the points $\beta_0, \beta_1, \ldots, \beta_m$ as tags of its δ–fine point–interval pairs. If we choose $\rho(\varepsilon)$ sufficiently small we obtain the existence of the integral $\int_a^b f(s)dg(s)$ as well as the equality (1.42) for its value. □

Definition. A function $f : [a, b] \to \mathbb{R}$ is called *regulated on the interval* $[a, b]$ if the onesided limits $f(s-)$, $f(s+)$ exist for every $s \in (a, b]$, $s \in [a, b)$, respectively.

It is well known that every function which is regulated on $[a, b]$ is bounded on this interval and is the uniform limit of finite step functions (See e.g. [43] or [6].) Clearly also every function of bounded variation on $[a, b]$ is regulated on $[a, b]$.

1.34 Corollary. *If $f : [a, b] \to \mathbb{R}$ is a regulated function on the interval $[a, b]$ and $g : [a, b] \to \mathbb{R}$ is of bounded variation on $[a, b]$ then the integral $\int_a^b f(s)dg(s)$ exists.*

Proof. Since f is bounded on $[a, b]$ and there is a sequence of finite step functions f_m which converges to f uniformly on $[a, b]$, there is an index $m_0 \in \mathbb{N}$ such that for $m > m_0$ we have $|f_m(\tau)| \leq K$ where $K \geq 0$ is a constant. Since the integrals $\int_a^b f_m(s)dg(s)$ exist, $m = 1, 2, \ldots$ by Proposition 1.33 and the assumptions of Corollary 1.31 are satisfied with $v(\tau) = -K$, $w(\tau) = K$, Corollary 1.31 yields the result. □

Inequalites for the integral

1.35 Theorem. *Assume that a function $U : [a, b] \times [a, b] \to \mathbb{R}^n$ is given for which the integral $\int_a^b DU(\tau, t)$ exists. If $V : [a, b] \times [a, b] \to \mathbb{R}$ is such that the integral $\int_a^b DV(\tau, t)$ exists and if there* *is a gauge θ on $[a,b]$ such that*

$$|t-\tau|.\|U(\tau,t)-U(\tau,\tau)\| \le (t-\tau).(V(\tau,t)-V(\tau,\tau)) \quad (1.43)$$

for every $t \in [\tau-\theta(\tau), \tau+\theta(\tau)]$ then the inequality

$$\left\| \int_a^b DU(\tau,t) \right\| \le \int_a^b DV(\tau,t) \quad (1.44)$$

holds.

Proof. Assume that $\varepsilon > 0$ is given. Since the integrals $\int_a^b DU(\tau,t)$, $\int_a^b DV(\tau,t)$ exist there is a gauge δ on $[a,b]$ with $\delta(s) \le \theta(s)$ for $s \in [a,b]$ such that for every δ–fine partition

$$D = \{\alpha_0, \tau_1, \alpha_1, \dots, \alpha_{k-1}, \tau_k, \alpha_k\}$$

of $[a,b]$ we have

$$\left\| \sum_{j=1}^k [U(\tau_j,\alpha_j) - U(\tau_j,\alpha_{j-1})] - \int_a^b DU(\tau,t) \right\| < \varepsilon, \quad (1.45)$$

$$\left| \sum_{j=1}^k [V(\tau_j,\alpha_j) - (\tau_j,\alpha_{j-1})] - \int_a^b DV(\tau,t) \right| < \varepsilon. \quad (1.46)$$

It is easy to see that (1.43) implies $\|U(\tau_i,\alpha_i) - U(\tau_i,\tau_i)\| \le V(\tau_i,\alpha_i) - V(\tau_i,\tau_i)$ when $\alpha_i > \tau_i$ and $\|U(\tau_i,\alpha_i) - U(\tau_i,\tau_i)\| \le V(\tau_i,\tau_i) - V(\tau_i,\alpha_i)$ when $\alpha_i < \tau_i$. Hence for $i = 1,2,\dots,k$ we have

$$\|U(\tau_i,\alpha_i) - U(\tau_i,\alpha_{i-1})\| \le \|U(\tau_i,\alpha_i) - U(\tau_i,\tau_i)\| +$$

$$+\|U(\tau_i,\tau_i) - U(\tau_i,\alpha_{i-1})\| \le V(\tau_i,\alpha_i) - V(\tau_i,\alpha_{i-1}).$$

By (1.45) and (1.46) we get

$$\left\| \int_a^b DU(\tau,t) \right\| \le$$

$$\le \left\| \sum_{j=1}^k [U(\tau_j,\alpha_j) - U(\tau_j,\alpha_{j-1})] - \int_a^b DU(\tau,t) \right\| +$$

$$+ \left\| \sum_{j=1}^k [U(\tau_j,\alpha_j) - U(\tau_j,\alpha_{j-1})] \right\| <$$

$$< \varepsilon + \sum_{j=1}^k [V(\tau_j,\alpha_j) - V(\tau_j,\alpha_{j-1})] =$$

$$= \varepsilon + \sum_{j=1}^k [V(\tau_j,\alpha_j) - V(\tau_j,\alpha_{j-1})] - \int_a^b DV(\tau,t) + \int_a^b DV(\tau,t) <$$

$$< 2\varepsilon + \int_a^b DV(\tau,t).$$

Since $\varepsilon > 0$ was arbitrary, the inequality (1.44) is satisfied. □

This theorem gives an estimate of the integral $\int_a^b DU(\tau,t)$ by another integral of a real valued function. In some cases it is useful to have an estimate of this type with a Stieltjes type integral. This is given in the following statement.

1.36 Corollary. *Assume that a function $U : [a,b] \times [a,b] \to \mathbb{R}^n$ is given for which the integral $\int_a^b DU(\tau,t)$ exists. If $V : [a,b] \times [a,b] \to \mathbb{R}$ is such that the integral $\int_a^b f(s)\,dg(s)$ exists for $f, g : [a,b] \to \mathbb{R}$ and if there is a gauge θ on $[a,b]$ such that*

$$|t-\tau|.\|U(\tau,t) - U(\tau,\tau)\| \le (t-\tau).f(\tau)(g(t) - g(\tau))$$

for every $t \in [\tau - \theta(\tau), \tau + \theta(\tau)]$ *then*

$$\left\| \int_a^b DU(\tau, t) \right\| \leq \int_a^b f(s)\, dg(s). \tag{1.47}$$

Proof. The result immediately follows from Theorem 1.34 for the case when $V(\tau, t) = f(\tau) g(t)$. □

1.37 *Remark.* Theorem 1.34 can be also used to deduce the following known result.

If $f : [a, b] \to \mathbb{R}$, $|f(s)| \leq c$ *for* $s \in [a, b]$ *where* c *is a constant,* $g : [a, b] \to \mathbb{R}$ *is of bounded variation on* $[a, b]$ *and the integral* $\int_a^b f(s)\, dg(s)$ *exists then*

$$\left| \int_a^b f(s)\, dg(s) \right| \leq c \operatorname{var}_a^b g.$$

Indeed, it is easy to see that

$$|t - \tau|.|f(\tau).g(t) - g(\tau)| \leq (t - \tau).c.(\operatorname{var}_a^t - \operatorname{var}_a^\tau)$$

for every $t, \tau \in [a, b]$ and the statement follows immediately from the known fact that the integral $\int_a^b c\, d(\operatorname{var}_a^s) = c. \operatorname{var}_a^b$ exists and from Theorem 1.34.

1.38 Lemma. *Let* $h : [a, b] \to \mathbb{R}$ *be a nonnegative nondecreasing function which is continuous from the left on* $(a, b]$. *Assume that* $f : [0, +\infty) \to [0, +\infty)$ *is a continuos nondecreasing function with the primitive* $F : [0, +\infty) \to \mathbb{R}$, *i.e.* $F'(s) = \dfrac{dF}{ds}(s) = f(s)$ *for* $s \in [0, +\infty)$.

Then the integral $\int_a^b f(h(s))\,dh(s)$ *exists and*

$$\int_a^b f(h(s))\,dh(s) \le F(h(b)) - F(h(a)). \tag{1.48}$$

Proof. The composition of functions f and h given by $f(h(s))$ for $s \in [a,b]$ is evidently nondecreasing on $[a,b]$. Therefore the integral $\int_a^b f(h(s))\,dh(s)$ exists by Corollary 1.34.

For proving (1.48) Theorem 1.34 will be used. Assume that $\varepsilon > 0$ is given. By the definition of the primitive F to f, for every $s \in [0,+\infty)$ there exists $\Theta(s) > 0$ such that for every η with $0 \le |\eta| < \Theta(s)$ we have

$$|F(s+\eta) - F(s) - f(s)\eta| \le \varepsilon|\eta|. \tag{1.49}$$

Since $\lim_{t\to\tau+} h(t) = h(\tau+)$ for $\tau \in [a,b)$, there is a $\delta^+(\tau) > 0$, $\delta^+(b) = 1$ such that for $t \in (\tau, \tau+\delta^+(\tau)] \cap [a,b]$ we have

$$0 \le h(t) - h(\tau+) \le \Theta(h(\tau+)).$$

Putting $s = h(\tau+)$ and $\eta = h(t) - h(\tau+)$ we obtain

$$f(h(\tau+))(h(t)-h(\tau+)) \le F(h(t)) - F(h(\tau+)) + \varepsilon(h(t)-h(\tau+))$$

for $t \in (\tau, \tau+\delta^+(\tau)] \cap [a,b]$. Further, we have

$$f(h(\tau+))(h(t)-h(\tau+)) - [F(h(t)) - F(h(\tau+))] =$$

$$= \int_{h(\tau)}^{h(\tau+)} [f(h(\tau)) - f(s)]\,ds \le 0$$

because $f(h(\tau)) \le f(s)$ for $s \in [h(\tau), h(\tau+)]$. Therefore

$$|f(h(\tau))(h(t) - h(\tau))| =$$

$$= f(h(\tau))(h(t) - h(\tau+)) + f(h(\tau))(h(\tau+) - h(\tau)) \leq$$
$$\leq f(h(\tau+))(h(t) - h(\tau+)) + F(h(\tau+)) - F(h(\tau)) \leq$$
$$\leq F(h(t)) - F(h(\tau)) + \varepsilon(h(t) - h(\tau))$$

provided $t \in (\tau, \tau + \delta^+(\tau)] \cap [a,b]$. From the inequality (1.49) and from the continuity from the left of the function h at the point $\tau \in (a,b]$ there is $\delta^-(\tau) > 0$, $\delta^-(a) = 1$ such that for $t \in [\tau - \delta^-(\tau), \tau] \cap [a,b]$ the inequality

$$|f(h(\tau))(h(t) - h(\tau))| \leq F(h(\tau)) - F(h(t)) + \varepsilon(h(\tau) - h(t))$$

is satisfied. Take $\delta(\tau) = \min(\delta^-(\tau), \delta^+(\tau))$ for $\tau \in [a,b]$. Then for $\tau \in [a,b]$ and $t \in [\tau - \delta(\tau), \tau + \delta(\tau) \cap [a,b]$ we obtain by the above inequalities the relation

$$|t - \tau|\,|f(h(\tau))h(t) - f(h(\tau))h(\tau))| \leq$$
$$(t - \tau)\,(F(h(t)) + \varepsilon h(t) - F(h(\tau)) - \varepsilon h(\tau))\,.$$

By Corollary 1.35 this inequality implies

$$\int_a^b f(h(s))\,dh(s) \leq \int_a^b d[F(h(s)) + \varepsilon h(s)] =$$
$$= F(h(b)) - F(h(a)) + \varepsilon(h(b) - h(a)).$$

This inequality yields (1.44) because $\varepsilon > 0$ can be chosen arbitrarily small. □

1.39 *Remark.* Using essentially the same technique as above it can be shown that if in Lemma 1.38 the function h is continuous from the right instead of being continuous from the left then the inequality

$$F(h(b)) - F(h(a)) \leq \int_a^b f(h(s))\,dh(s)$$

holds, where the integral $\int_a^b f(h(s))\,dh(s)$ exists for the same reason as was shown in the setting of Lemma 1.38.

If $\gamma \geq 0$ and $f(s) = s^\gamma$ for $s \in [0, +\infty)$ then $F(s) = \dfrac{1}{\gamma+1} s^{\gamma+1}$ is the primitive to f on $[0, +\infty)$. Using Lemma 1.38 for this special case ve obtain the inequality

$$\int_a^b h^\gamma(s)\,dh(s) \leq \frac{1}{\gamma+1}\left[h^{\gamma+1}(b) - h^{\gamma+1}(a)\right] \tag{1.50}$$

for an arbitrary nonnegative nondecreasing function $h : [a, b] \to \mathbb{R}$ which is continuous from the left.

It can be also shown that if $h : [a, b] \to \mathbb{R}$ is continuous from the right on $[a, b]$ then

$$\frac{1}{\gamma+1}\left[h^{\gamma+1}(b) - h^{\gamma+1}(a)\right] \leq \int_a^b h^\gamma(s)\,dh(s). \tag{1.51}$$

1.40 Theorem. *Let $\psi : [a, b] \to [0, +\infty)$, $h : [a, b] \to [a, +\infty)$ be given where ψ is bounded and h is nondecreasing and continuous from the left on the interval $[a, b]$. Suppose that the function $\omega : [0, +\infty) \to \mathbb{R}$ is continuous, nondecreasing, $\omega(0) = 0$, $\omega(r) > 0$ for $r > 0$.*

For $u > 0$ let us set

$$\Omega(u) = \int_{u_0}^u \frac{1}{\omega(r)}\,dr \tag{1.52}$$

with some $u_0 > 0$.

The function $\Omega : (0, +\infty) \to \mathbb{R}$ is increasing, $\Omega(u_0) = 0$ and $\lim_{u \to 0+} \Omega(u) = \alpha \geq -\infty$, $\lim_{u \to +\infty} \Omega(u) = \beta \leq +\infty$.

Assume that for $\xi \in [a, b]$ the inequality

$$\psi(\xi) \leq k + \int_a^\xi \omega(\psi(\tau))\,dh(\tau) \tag{1.53}$$

holds, where $k > 0$ is a constant.

If $\Omega(k) + h(b) - h(a) < \beta$ then for $\xi \in [a, b]$ we have

$$\psi(\xi) \leq \Omega^{-1}(\Omega(k) + h(\xi) - h(a)), \tag{1.54}$$

where $\Omega^{-1} : (\alpha, \beta) \to \mathbb{R}$ is the inverse function to the function Ω given by (1.52).

Proof. If we have $\Omega(l) + h(b) - h(a) < \beta$ for some $l > 0$ then for all $\tau \in [a, b]$ we have

$$\alpha < \Omega(l) + h(\tau) - h(a) < \beta.$$

Therefore the value of

$$\Omega(l) + h(\tau) - h(a)$$

belongs to the domain of Ω^{-1} provided $\tau \in [a, b]$, and for these τ we can define

$$w_l(\tau) = \Omega^{-1}(\Omega(l) + h(\tau) - h(a)).$$

Define further

$$\varphi(s) = \omega(\Omega^{-1}(\Omega(l) + s)) \text{ for } s \in [0, \beta - \Omega(l)]. \tag{1.55}$$

At $\Omega^{-1}(\Omega(l) + s)$ there exists a derivative Ω' of the function Ω and

$$\Omega'(\Omega^{-1}(\Omega(l) + s)) = \frac{1}{\omega(\Omega^{-1}(\Omega(l) + s))} \neq 0.$$

The well known formula for the derivative of the inverse function leads to

$$\frac{d}{ds}[\omega(\Omega^{-1}(\Omega(l) + s))] = \frac{1}{\Omega'(\Omega^{-1}(\Omega(l) + s))} = \tag{1.56}$$

$$= \omega(\Omega^{-1}(\Omega(l) + s)) = \varphi(s)$$

for $s \in [0, \beta - \Omega(l)]$. If now $\xi \in [a, b]$ is given then using the definition of the function φ from (1.55) we obtain

$$\int_a^\xi \omega(w_l(\tau))\, dh(\tau) = \int_a^\xi \omega(\Omega^{-1}(\Omega(l) + h(\tau) - h(a)))\, dh(\tau) =$$

$$= \int_a^\xi \varphi(h(\tau) - h(a))\, d(h(\tau) - h(a)).$$

This together with (1.56) and Lemma 1.38 implies

$$\int_a^\xi \omega(w_l(\tau))\, dh(\tau) \le \Omega^{-1}(\Omega(l) + h(\xi) - h(a)) - \Omega^{-1}(\Omega(l)) =$$

$$= w_l(\xi) - l,$$

and consequently for $\xi \in [a, b]$ we have the inequality

$$l + \int_a^\xi \omega(w_l(\tau))\, dh(\tau) \le w_l(\xi).$$

Assume that $\varepsilon_0 > 0$ is such that $\Omega(k + \varepsilon_0) + h(b) - h(a) < \beta$. Let us take an arbitrary $\varepsilon \in (0, \varepsilon_0)$ and set $l = k + \varepsilon$. For this case the last inequality reads

$$k + \varepsilon + \int_a^\xi \omega(w_{k+\varepsilon}(\tau))\, dh(\tau) \le w_{k+\varepsilon}(\xi),$$

and taking into account the relation (1.53) for every $\xi \in [a, b]$ we get

$$\psi(\xi) - w_{k+\varepsilon}(\xi) \le$$

$$\le k + \int_a^\xi \omega(\psi(\tau))\, dh(\tau) - k - \varepsilon - \int_a^\xi \omega(w_{k+\varepsilon}(\tau))\, dh(\tau) =$$

$$= -\varepsilon + \int_a^\xi [\omega(\psi(\tau)) - \omega(w_{k+\varepsilon}(\tau))]\, dh(\tau). \tag{1.57}$$

Hence $\psi(a) - w_{k+\varepsilon}(a) \leq -\varepsilon$ and also $\omega(\psi(a)) - \omega(w_{k+\varepsilon}(a)) \leq 0$ because the function ω is assumed to be nondecreasing. The functions ψ and $w_{k+\varepsilon}$ are bounded and therefore there is a constant $K > 0$ such that

$$|\omega(\psi(\tau)) - \omega(w_{k+\varepsilon}(\tau))| < K$$

for $\tau \in [a, b]$.

Using Theorem 1.16 and the estimate given in Remark 1.37 we get from the last two displayed inequalities

$$\psi(\xi) - w_{k+\varepsilon}(\xi) \leq -\varepsilon + [\omega(\psi(a)) - \omega(w_{k+\varepsilon}(a))](h(a+) - h(a)) +$$

$$+ \lim_{\delta \to 0+} \int_{a+\delta}^\xi [\omega(\psi(\tau)) - \omega(w_{k+\varepsilon}(\tau))]\, dh(\tau) \leq$$

$$\leq -\varepsilon + K \lim_{\delta \to 0+} [h(\xi) - h(a+\delta)] = -\varepsilon + K[h(\xi) - h(a+)].$$

Because $\lim_{\xi \to a+} h(\xi) = h(a+)$, an $\eta > 0$ can be found such that for $\xi \in (a, a+\eta)$ the inequality $h(\xi) - h(a+) < \dfrac{\varepsilon}{2K+1}$ holds and therefore also

$$\psi(\xi) - w_{k+\varepsilon}(\xi) < -\varepsilon + \frac{K\varepsilon}{2K+1} < -\frac{\varepsilon}{2} < 0$$

for $\xi \in (a, a+\eta)$.

Let us set

$$T = \sup \{t \in [a, b];\ \psi(\xi) - w_{k+\varepsilon}(\xi) < 0 \text{ for } \xi \in [a, t]\}.$$

As we have shown above, we have $T > a$ and for $\xi \in [a, T)$ the inequality $\psi(\xi) - w_{k+\varepsilon}(\xi) \leq 0$ and therefore also $\omega(\psi(\xi)) -$

$\omega(w_{k+\varepsilon}(\xi)) \le 0$ holds. The last conclusion is a consequence of the assumption that ω is nondecreasing. By (1.57) and Theorem 1.16 we have

$$\psi(T) - w_{k+\varepsilon}(T) \le$$

$$\le -\varepsilon + \lim_{\delta\to 0+} \int_a^{T-\delta} [\omega(\psi(\tau)) - \omega(w_{k+\varepsilon}(\tau))]\, dh(\tau) +$$

$$+[\omega(\psi(T)) - \omega(w_{k+\varepsilon}(T))](h(T) - h(T-)) \le -\varepsilon < 0$$

because $h(T) - h(T-) = h(T) - \lim_{\tau\to T-} h(\tau) = 0$ and

$$\lim_{\delta\to 0+} \int_a^{T-\delta} [\omega(\psi(\tau)) - \omega(w_{k+\varepsilon}(\tau))]\, dh(\tau) \le 0.$$

If we assume that $T < b$ then we can repeatthis procedure for $\xi > T$ by virtue of the inequality

$$\psi(\xi) - w_{k+\varepsilon}(\xi) \le -\varepsilon + \int_T^{\xi} [\omega(\psi(\tau)) - \omega(w_{k+\varepsilon}(\tau))]\, dh(\tau),$$

thus obtaining $\psi(\xi) - w_{k+\varepsilon}(\xi) < 0$ for $\xi \in [T, T + \eta]$ for some $\eta > 0$. Hence $T = b$ and

$$\psi(\xi) < w_{k+\varepsilon}(\xi) = \Omega^{-1}(\Omega(k+\varepsilon) + h(\xi) - h(a))$$

for $\xi \in [a, b]$. Since the function Ω is continuous and the last inequality holds for every sufficiently small $\varepsilon > 0$, we obtain the inequality (1.53). □

1.41 *Remark.* Theorem 1.40 represents a Bellman–Bihari–Győri type inequality (see e.g. [34]). Results of this type are especially useful for deriving uniqueness results for equations like ordinary differential equations.

1.42 Corollary. *Assume that the function $h : [a,b] \to [0,+\infty)$ is nondecreasing and continuous from the right. If the inequality (1.53) in Theorem 1.40 is replaced by*

$$\psi(\xi) \le k + \int_\xi^b \omega(\psi(\tau))\, dh(\tau) \tag{1.58}$$

where $k > 0$ is a constant and if $\Omega(k) + h(b) - h(a) < \beta$ then for $\xi \in [a,b]$ we have

$$\psi(\xi) \le \Omega^{-1}(\Omega(k) + h(b) - h(\xi)). \tag{1.59}$$

Proof. For $\sigma \in [-b,-a]$ define

$$\tilde{h}(\sigma) = h(b) - h(-\sigma).$$

It is easy to see that $\tilde{h} : [-b,-a] \to [0,+\infty)$ is continuous from the left and nondecreasing.

By the Substitution Theorem 1.18 we have

$$\int_\xi^b \omega(\psi(\tau))\, dh(\tau) = \int_{-\xi}^{-b} \omega(\psi(-\tau))\, dh(-\tau) =$$

$$= - \int_{-b}^{-\xi} \omega(\psi(-\tau))\, dh(-\tau) = \int_{-b}^{-\xi} \omega(\psi(-\tau))\, d\tilde{h}(\tau)$$

and we can write the inequality (1.58) in the form

$$\psi(\xi) \le k + \int_{-b}^{-\xi} \omega(\psi(-\tau))\, d\tilde{h}(\tau)$$

for $\xi \in [a,b]$, i.e.

$$\tilde{\psi}(\xi) \le k + \int_{-b}^{\xi} \omega(\tilde{\psi}(\tau))\, d\tilde{h}(\tau)$$

for $\xi \in [-b, -a]$, where $\tilde{\psi}(\sigma) = \psi(-\sigma)$ for $\sigma \in [-b, -a]$. Therefore Theorem 1.40 can be applied to obtain the inequality

$$\tilde{\psi}(\sigma) \le \Omega^{-1}(\Omega(k) + \tilde{h}(\sigma) - \tilde{h}(-b)) = \Omega^{-1}(\Omega(k) + h(b) - h(-\sigma))$$

for $\sigma \in [-b, -a]$. Hence for $\xi \in [a, b]$ we obtain the inequality

$$\psi(\xi) \le \Omega^{-1}(\Omega(k) + h(b) - h(\xi))$$

and the corollary is proved. □

1.43 Corollary. *If ψ, h and k satisfy the assumptions of Theorem 1.40 and if for $\xi \in [a, b]$ the inequality*

$$\psi(\xi) \le k + L \int_a^\xi \psi(\tau)\, dh(\tau), \quad \xi \in [a, b]$$

holds with a constant $L \ge 0$ instead of (1.53), then for every $\xi \in [a, b]$ the inequality

$$\psi(\xi) \le k e^{L(h(\xi) - h(a))}$$

is satisfied.

Proof. This result is an immediate consequence of Theorem 1.40 if we set $\omega(r) = Lr$ for $r \ge 0$. In this case we have

$$\Omega(u) = \frac{1}{L} \int_{u_0}^{u} \frac{dr}{r} = \frac{1}{L}(\ln u - \ln u_0) = \frac{1}{L} \ln \frac{u}{u_0}$$

and $\Omega^{-1}(u) = u_0 e^{Lu}$. The final inequality of the statement now easily follows from (1.54). □

In a completely analogous way also the following statement can be proved using Corollary 1.42.

1.44 Corollary. *If $h : [a,b] \to [0,+\infty)$ is nondecreasing and continuous from the right and if ψ, k satisfy the assumptions of Theorem 1.40 where for $\xi \in [a,b]$ the inequality*

$$\psi(\xi) \leq k + L\int_{\xi}^{b} \psi(\tau)\,dh(\tau), \quad \xi \in [a,b]$$

holds with a constant $L \geq 0$ instead of (1.53), then for every $\xi \in [a,b]$ the inequality

$$\psi(\xi) \leq ke^{L(h(b)-h(\xi))}$$

is satisfied.

A Tonelli -type theorem

1.45 Theorem. *Let $[a,b]$, $[c,d] \subset \mathbb{R}$, $-\infty < a < b < \infty, -\infty < c < d < \infty$ be given intervals. Assume that $\varphi : [a,b] \to \mathbb{R}$ is a function of bounded variation on $[a,b]$ and $w : [a,b]\times[c,d]\times[c,d] \to \mathbb{R}$ is a given function satisfying*

(A) *for every $(\sigma,s) \in [c,d]\times[c,d]$ the function $\varphi(t)w(\tau,\sigma,s)$ is integrable over $[a,b]$, i.e. the integral*

$$\int_a^b D_t[\varphi(t)w(\tau,\sigma,s)]$$

exists for every $(\sigma,s) \in [c,d]\times[c,d]$,

(B) *the system of functions $w(\tau,\cdot,\cdot)$, $\tau \in [a,b]$ is equi - integrable over $[c,d]$, i.e. for every $\eta > 0$ there is a gauge $\delta_0 : [c,d] \to (0,\infty)$ on $[c,d]$ such that*

$$|\sum_{j=1}^{m}[w(\tau,\sigma_j,\beta_j) - w(\tau,\sigma_j,\beta_{j-1})] - \int_c^d D_s w(\tau,\sigma,s)| < \eta$$

for every δ_0–fine partition

$$D_0 : c = \beta_0 \le \sigma_1 \le \beta_1 \le \cdots \le \beta_{m-1} \le \sigma_m \le \beta_m = d$$

of $[c, d]$ and any $\tau \in [a, b]$.

Then both the integrals $\int_a^b D_t(\int_c^d D_s[\varphi(t)w(\tau,\sigma,s)])$ and $\int_c^d (D_s \int_a^b D_t[\varphi(t)w(\tau,\sigma,s)])$ exist and have the same value, i.e.

$$\int_a^b D_t(\int_c^d D_s[\varphi(t)w(\tau,\sigma,s)]) = \int_c^d (D_s \int_a^b D_t[\varphi(t)w(\tau,\sigma,s)]) \tag{1.60}$$

Remark. By D_t, D_s the integration with respect to the couple $(\tau, t), (\sigma, s)$ is marked in the formulation of the theorem. This means that the symbol D_s corresponds to the integration over $[c, d]$ with $s, \sigma \in [c, d]$ where for tagged intervals (σ, K), $K \subset [c, d]$ the integral sums are determined by terms $h(\sigma, K) = V(\sigma, \gamma) - V(\sigma, \beta)$ for a function $V : [c, d] \times [c, d] \to \mathbb{R}$ and $K = [\beta, \gamma] \subset [c, d]$, and similarly for D_t and integration over $[a, b]$.

Proof. Let $\varepsilon > 0$ be given. By the assumption (B) there is a gauge δ_0 on $[c, d]$ such that

$$|\sum_{j=1}^{m}[w(\tau,\sigma_j,\beta_j) - w(\tau,\sigma_j,\beta_{j-1})] - \int_c^d D_s w(\tau,\sigma,s)| <$$

$$< \frac{\varepsilon}{2(1 + \operatorname{var}_a^b \varphi)} \tag{1.61}$$

for any $\tau \in [a, b]$ and for every δ_0–fine partition

$$D_0 = \{\beta_0, \sigma_1, \beta_1, \cdots, \beta_{m-1}, \sigma_m, \beta_m\}$$

of $[c, d]$ with σ_j being the tags in $[\beta_{j-1}, \beta_j]$.

First let us prove the existence of the integral

$$\int_a^b D_t(\int_c^d D_s[\varphi(t)w(\tau,\sigma,s)]) = \int_a^b D_t(\varphi(t)\int_c^d D_s w(\tau,\sigma,s)).$$

Assume that the δ_0–fine partition D_0 of $[c,d]$ is fixed. By (A) we find a gauge δ on $[a,b]$ such that for every δ–fine partition

$$D = \{\alpha_0, \tau_1, \alpha_1, \cdots, \alpha_{p-1}, \tau_p, \alpha_p\}$$

of $[a,b]$ we have

$$\sum_{j=1}^{m} |\sum_{k=1}^{p}(\varphi(\alpha_k) - \varphi(\alpha_{k-1}))w(\tau_k, \sigma_j, \beta_j) -$$

$$- \int_a^b D_t[\varphi(t)w(\tau,\sigma_j,\beta_j)]| < \frac{\varepsilon}{4} \tag{1.62}$$

and

$$\sum_{j=1}^{m} |\sum_{k=1}^{p}(\varphi(\alpha_k) - \varphi(\alpha_{k-1}))w(\tau_k, \sigma_j, \beta_{j-1}) -$$

$$- \int_a^b D_t[\varphi(t)w(\tau,\sigma_j,\beta_{j-1})]| < \frac{\varepsilon}{4}. \tag{1.63}$$

Denote $U(\tau,t) = \varphi(t)\int_c^d D_s w(\tau,\sigma,s)$ for $\tau, t \in [a,b]$. We show that for every δ–fine partition D of the interval $[a,b]$ the corresponding integral sum $S(U,D)$ is ε–close to the value of

$$\sum_{j=1}^{m} \int_a^b D_t[\varphi(t)(w(\tau,\sigma_j,\beta_j) - w(\tau,\sigma_j,\beta_{j-1}))].$$

Using the inequalities (1.61), (1.62) and (1.63) we obtain for every δ–fine partition D of $[a,b]$ the following estimate.

$$|S(U,D) - \sum_{j=1}^{m} \int_a^b D_t[\varphi(t)(w(\tau,\sigma_j,\beta_j) - w(\tau,\sigma_j,\beta_{j-1}))]| =$$

$$= |\sum_{k=1}^{p}(\varphi(\alpha_k) - \varphi(\alpha_{k-1})) \int_c^d D_s w(\tau_k,\sigma,s) -$$

$$-\sum_{j=1}^{m} \int_a^b D_t[\varphi(t)(w(\tau,\sigma_j,\beta_j) - w(\tau,\sigma_j,\beta_{j-1}))]| \leq$$

$$\leq |\sum_{k=1}^{p}(\varphi(\alpha_k) - \varphi(\alpha_{k-1})) \int_c^d D_s w(\tau_k,\sigma,s) -$$

$$-\sum_{k=1}^{p}\sum_{j=1}^{m}(\varphi(\alpha_k) - \varphi(\alpha_{k-1}))[w(\tau_k,\sigma_j,\beta_j) - w(\tau_k,\sigma_j,\beta_{j-1})]| +$$

$$+|\sum_{k=1}^{p}\sum_{j=1}^{m}(\varphi(\alpha_k) - \varphi(\alpha_{k-1}))[w(\tau_k,\sigma_j,\beta_j) - w(\tau_k,\sigma_j,\beta_{j-1})] -$$

$$-\sum_{j=1}^{m} \int_a^b D_t[\varphi(t)(w(\tau,\sigma_j,\beta_j) - w(\tau,\sigma_j,\beta_{j-1}))]| \leq$$

$$\leq \sum_{k=1}^{p} |\varphi(\alpha_k) - \varphi(\alpha_{k-1})| \quad |\int_c^d D_s w(\tau_k,\sigma,s) -$$

$$-\sum_{j=1}^{m}(\varphi(\alpha_k) - \varphi(\alpha_{k-1}))[w(\tau_k,\sigma_j,\beta_j) - w(\tau_k,\sigma_j,\beta_{j-1})]| +$$

$$+\sum_{j=1}^{m} |\sum_{k=1}^{p}(\varphi(\alpha_k) - \varphi(\alpha_{k-1}))w(\tau_k,\sigma_j,\beta_j) -$$

$$-\int_a^b D_t[\varphi(t)w(\tau,\sigma_j,\beta_j)]|+$$

$$+\sum_{j=1}^m|\sum_{k=1}^p(\varphi(\alpha_k)-\varphi(\alpha_{k-1}))w(\tau_k,\sigma_j,\beta_{j-1})-$$

$$-\int_a^b D_t[\varphi(t)w(\tau,\sigma_j,\beta_{j-1})]|\leq$$

$$\leq\frac{\varepsilon}{2(1+\operatorname{var}_a^b\varphi)}\sum_{k=1}^p|\varphi(\alpha_k)-\varphi(\alpha_{k-1})|+\frac{\varepsilon}{4}+\frac{\varepsilon}{4}<$$

$$<\frac{\varepsilon}{2(1+\operatorname{var}_a^b\varphi)}\operatorname{var}_a^b\varphi+\frac{\varepsilon}{2}<\varepsilon.$$

Therefore for every two δ–fine partitions D_1, D_2 of the interval $[a,b]$ we clearly have

$$|S(U,D_1)-S(U,D_2)|<2\varepsilon$$

and Theorem 1.7 yields the existence of the integral

$$\int_a^b D_tU(\tau,t)=\int_a^b D_t[\varphi(t)\int_c^d D_sw(\tau,\sigma,s)].$$

The inequality (1.61) further implies

$$|t-\tau||(\varphi(t)-\varphi(\tau))\sum_{j=1}^m[w(\tau,\sigma_j,\beta_j)-w(\tau,\sigma_j,\beta_{j-1})]-$$

$$-\int_c^d D_sw(\tau,\sigma,s)|\leq$$

$$\leq|t-\tau||\varphi(t)-\varphi(\tau)|\frac{\varepsilon}{2(1+\operatorname{var}_a^b\varphi)}\leq$$

$$\le |t-\tau| |\operatorname{var}_a^t \varphi - \operatorname{var}_a^\tau \varphi| \frac{\varepsilon}{2(1+\operatorname{var}_a^b \varphi)} =$$
$$= (t-\tau)(\operatorname{var}_a^t \varphi - \operatorname{var}_a^\tau \varphi) \frac{\varepsilon}{2(1+\operatorname{var}_a^b \varphi)}.$$

Since the integral $\int_a^b D_t[\varphi(t)\int_c^d D_s w(\tau,\sigma,s)]$ exists and by the assumption (A) also the integral

$$\int_a^b D_t[\varphi(t)\sum_{j=1}^m (w(\tau,\sigma_j,\beta_j) - w(\tau,\sigma_j,\beta_{j-1}))] =$$
$$= \sum_{j=1}^m \int_a^b D_t[\varphi(t)w(\tau,\sigma_j,\beta_j)] - \sum_{j=1}^m \int_a^b D_t[\varphi(t)w(\tau,\sigma_j,\beta_{j-1})]$$

exists we have by Corollary 1.36

$$|\int_a^b D_t[\varphi(t)\sum_{j=1}^m (w(\tau,\sigma_j,\beta_j) - w(\tau,\sigma_j,\beta_{j-1}))] -$$
$$- \int_a^b D_t[\varphi(t)\int_c^d D_s w(\tau,\sigma,s)]| \le$$
$$\le \int_a^b d(\operatorname{var}_a^\tau) \frac{\varepsilon}{2(1+\operatorname{var}_a^b \varphi)} = \frac{\varepsilon \operatorname{var}_a^b \varphi}{2(1+\operatorname{var}_a^b \varphi)} < \frac{\varepsilon}{2}.$$

Hence

$$|\sum_{j=1}^m [\int_a^b D_t[\varphi(t)w(\tau,\sigma_j,\beta_j)] - \int_a^b D_t[\varphi(t)w(\tau,\sigma_j,\beta_{j-1})]] -$$
$$- \int_a^b D_t[\varphi(t)\int_c^d D_s w(\tau,\sigma,s)]| < \varepsilon$$

for every $\varepsilon > 0$ and every δ_0-fine partition

$$D_0 = \{\beta_0, \sigma_1, \beta_1, \cdots, \beta_{m-1}, \sigma_m, \beta_m\}$$

of $[c,d]$. Consequently, the integral $\int_c^d (D_s \int_a^b D_t[\varphi(t)w(\tau,\sigma,s)])$ exists and by definition equals to $\int_a^b D_t[\varphi(t)\int_c^d D_s w(\tau,\sigma,s)]$. □

1.46 Corollary. *Let* $[a,b]$, $[c,d] \subset \mathbb{R}$, $-\infty < a < b < \infty$, $-\infty < c < d < \infty$ *be given intervals. Assume that* $\varphi : [a,b] \to \mathbb{R}$, $\psi : [c,d] \to \mathbb{R}$ *are functions of bounded variation on* $[a,b]$, $[c,d]$, *respectively, and that* $f : [a,b] \times [c,d] \to \mathbb{R}$ *is a given function such that*

for every $\sigma \in [c,d]$ *the integral* $\int_a^b f(\tau,\sigma)d\varphi(\tau)$ *exists, the system of functions* $f(\tau,\cdot)\psi(\cdot)$, $\tau \in [a,b]$ *is equi - integrable over* $[c,d]$, *i.e. for every* $\eta > 0$ *there is a gauge* $\delta_0 : [c,d] \to (0,\infty)$ *on* $[c,d]$ *such that*

$$|\sum_{j=1}^{m}[f(\tau,\sigma_j)(\psi(\beta_j) - \psi(\beta_{j-1})] - \int_c^d f(\tau,\sigma)\, d\psi(\sigma)| < \eta$$

for every δ_0*–fine partition*

$$D_0 : c = \beta_0 \le \sigma_1 \le \beta_1 \le \cdots \le \beta_{m-1} \le \sigma_m \le \beta_m = d$$

of $[c,d]$ *and any* $\tau \in [a,b]$.

Then both the integrals $\int_a^b(\int_c^d f(\tau,\sigma)\, d\psi(\sigma))\, d\varphi(\tau)$ *and* $\int_c^d(\int_a^b f(\tau,\sigma)\, d\varphi(\tau))d\psi(\sigma)$ *exist and have the same value, i.e.*

$$\int_a^b(\int_c^d f(\tau,\sigma)\, d\psi(\sigma))\, d\varphi(\tau) = \int_c^d(\int_a^b f(\tau,\sigma)\, d\varphi(\tau))d\psi(\sigma).$$

Proof. The convention described in Remark 1.5 for (Perron–) Stieltjes integrals is used here. Let us set $w(\tau,\sigma,s) = f(\tau,\sigma)\psi(s)$. For this function $w : [a,b] \times [c,d] \times [c,d] \to \mathbb{R}$ the assumptions of Theorem 1.45 are satisfied and this proves the corollary. □

1.47 *Remark.* The results of Theorem 1.45 remain valid if we assume that $\varphi : [a,b] \to L(\mathbb{R}^n)$ is an $n \times n$–matrix valued function of bounded variation on $[a,b]$ and the function w :

$[a, b] \times [c, d] \times [c, d] \to \mathbb{R}^n$ is n–vector valued. This can be shown using Theorem 1.45 for the components of the product $\varphi(t)w(\tau, \sigma, s)$ where Theorem 1.6 for $\mathbb{R}^n$– valued functions has to be taken into account.

1.48 *Remark.* Theorem 1.45 and Corollary 1.46 represent Tonelli – type results on the reversal of order of double integrals. More sophisticated results of this kind as well as Fubini – type theorems for the Perron integral can be found e.g. in [40], [46], [76].

It should be mentioned that in general it is not easy to check the equi - integrability condition (B) in Theorem 1.45. We will use the result in Chap. VI for generalized linear differential equations where fortunately it is not difficult to show that the condition (B) is satisfied.

CHAPTER II

ORDINARY DIFFERENTIAL EQUATIONS AND THE PERRON INTEGRAL

The present approach to the concept of an ordinary differential equation goes back to C. Carathéodory, in particular to his book [14] published in 1918. In this work C. Carathéodory accomplished the construction of an analysis (calculus) course based purely on the concept of the Lebesgue integral.

Given an ordinary differential equation of the form

$$\dot{x} = f(x,t) \tag{2.1}$$

with $f : B \times [a,b] \to \mathbb{R}^n$ where $B \subset \mathbb{R}^n$ is an open set (e. g. $B = B_c = \{x \in \mathbb{R}^n; \|x\| \le c\}$).To solve it, in the classical setting means:

find (if possible, all) functions $x : J \to \mathbb{R}^n$ defined on a nondegenerate interval $J \subset [a,b]$ such that

$$x(t) \in B \text{ for } t \in J \tag{2.2}$$

$$x \text{ is differentiable everywhere in } J, \tag{2.3}$$

i.e. the derivative $\dot{x}(t)$ exists for every $t \in J$ and

$$\dot{x}(t) = f(x(t),t) \text{ for every } t \in J. \tag{2.4}$$

A function $x : J \to \mathbb{R}^n$ satisfying (2.2), (2.3) and (2.4) is called a *solution* of (2.1) and of course the properties are satisfied componentwise, i.e. if $x = (x_1, \ldots, x_n)$ then (1.3) means that all x_k, $k = 1, \ldots, n$ are differentiable and (1.4) reads

$$\dot{x}_k(t) = f_m(x_1(t), \ldots, x_n(t), t) \text{ for } t \in J \text{ and } k = 1, \ldots, n,$$

where f_m are the components of $f = (f_1, \ldots, f_n)$.

Given a solution $x : J \to \mathbb{R}^n$ then

$$x(t) = x(\alpha) + \int_\alpha^t f(x(s), s)\, ds \tag{2.5}$$

provided $\alpha, t \in J$. On the other hand, if $x : J \to \mathbb{R}^n$ is a function satisfying (2.2) such that for some $\alpha \in J$ the equality (2.5) holds for $t \in J$, then x is a solution of (2.1) provided $f : B \times [a, b] \to \mathbb{R}^n$ satisfies some additional conditions. If e.g. f is continuous then this conclusion holds, provided the integral in (2.5) is the Riemann integral.

The starting point for Carathéodory's generalized approach to ordinary differential equations of the form (2.1) is the integral equation (2.5) where the Lebesgue integral is involved in (2.5). The fundamental question of existence of a solution of the ordinary differential equation (1.1) is treated by C. Carathéodory as the question of existence of a solution of the integral equation (2.5) with the Lebesgue integral on the right hand side. By the properties of the Lebesgue integral a function $x : J \to \mathbb{R}^n$ satisfying (2.5) is necessarily absolutely continuous in its interval of definition because the indefinite Lebesgue integral has this property. Therefore it cannot be expected that a solution of (2.1) in the sense of Carathéodory possesses a derivative everywhere in its domain of definition. Generalized solutions to (2.1) are absolutely continuous functions for which their derivative exists almost everywhere with respect to the Lebesgue measure.

More precisely, a function $x : J \to \mathbb{R}^n$, $J \subset [a, b]$ being an interval, is called *a solution of* (2.1) *in the Carathéodory sense* if

$$x(t) \in B \text{ for almost all } t \in J,$$

$$x \text{ is absolutely continuous on } J \qquad (x \in AC(J))$$

and

$$\dot{x}(t) = f(x(t), t) \text{ for almost all } t \in J,$$

or equivalently, if $x(t) \in B$ for almost all $t \in J$ and

$$x(t_2) - x(t_1) = (L)\int_{t_1}^{t_2} f(x(s), s)\, ds$$

for all $t_1, t_2 \in J$.*

Carathéodory's proof of existence of a solution to the initial value problem

$$\dot{x} = f(x, t),\ x(\alpha) = v \in B$$

makes use of successive approximations. Assume that $J = [\alpha, \alpha + \eta] \subset [a, b]$ where $\eta > 0$. Define the approximations as follows:

$$\begin{aligned} &\varphi_j(t) = v \text{ for } t \in [\alpha, \alpha + \frac{\eta}{j}], \\ &\varphi_j(t) = v + \int_\alpha^{t-\frac{\eta}{j}} f(\varphi_j(s), s)\, ds \text{ for } t \in (\alpha + \frac{\eta}{j}, \alpha + \eta] \\ &\qquad j = 1, 2, \ldots \end{aligned} \tag{2.6}$$

Using this definition a sequence of continuous functions $\varphi_j : [\alpha, \alpha + \eta] \to \mathbb{R}^n$ is defined. The idea of C. Carathéodory is to take a subsequence of (φ_j) (denote it again by (φ_j)) which for $j \to \infty$ tends to a limit $y(t)$ for every $t \in [\alpha, \alpha + \eta]$. This limit function y is a solution of (2.5) provided also

$$\lim_{j\to\infty} \int_\alpha^{t-\frac{\eta}{j}} f(\varphi_j(s), s)\, ds = \int_\alpha^t f(y(s), s)\, ds$$

*(L) in front of the integral indicates that the Lebesgue integral is considered.

is satisfied. To this end a convergence result for the Lebesgue integral has to be used and the Lebesgue dominated convergence theorem is the right and sufficiently powerful tool for this case . The Carathéodory assumptions are the following.

(C1) $f(\cdot, t)$ is continuous for almost all $t \in [a, b]$,

(C2) $f(x, \cdot)$ is measurable for $x \in B$,

(C3) there is a Lebesgue integrable function $m(t) \geq 0$, such that for $x \in B$ and almost all $t \in [\alpha, \alpha + \eta]$ the estimate

$$\|f(x,t)\| \leq m(t)$$

holds.

The local version of the existence theorem for a solution of (2.1) or (2.5) in the Carathéodory setting can be found in [18], [86] or [23]. For local existence theorems the requirement $x(t) \in B$ for t in the domain of definition of a solution x plays a role.

In this context a very natural problem arises. It is connected with the possibility of using Perron's concept of the nonabsolutely convergent integral in the integral equation (2.5). The question is what concept of a generalized solution of (2.1) will be the result of this approach. The fundamental problem of existence of a solution of

$$x(t) = x(\alpha) + (P) \int_{\alpha}^{t} f(x(s), s)\, ds,$$

$$\text{for } t \in [\alpha, \alpha + \eta] \subset [a, b],\ \eta > 0 \tag{2.5}$$

should give the first information about the properties of a solution and about the possibly most general right hand sides of (2.1).*

*We write (P) in front of the integral to emphasize that the Perron integration is used.

First, it is clear that when we are looking for a solution of (2.5) then the Perron integral $(P)\int_\alpha^t f(x(s),s)\,ds$ should exist for every $t \in [\alpha, \alpha+\eta]$ and therefore any function satisfying (2.5) behaves like the indefinite integral of a Perron integrable function. Hence a solution of (2.5) (or of (2.1) in the Perron – Henstock sense) is a function $x : J \to \mathbb{R}^n$ defined on some interval J on which it is ACG^* and has almost everywhere in J a derivative for which $\dot{x}(t) = f(x(t),t)$ a.e. in J, because this are the properties of the indefinite Perron integral.

In the book [40, Chap. 7, Sec. 19] R. Henstock is following the lines of C. Carathéodory in deriving an existence result for the integral equation (2.5). For the same reason as mentioned above (for the case studied by C. Carathéodory with the Lebesgue integral), convergence results giving conditions for the possibility of interchanging the limit and integral play an essential role. Therefore let us quote a convergence result from [40, Theorem 9.1] in a simplified version suitable for the purposes of differential equations presented here.

2.1 Theorem. *Let $g_j : [\alpha,\beta] \to \mathbb{R}$, $j = 1,2,\ldots$ be a sequence of functions Perron integrable over $[\alpha,\beta]$. Let $\delta : [\alpha,\beta] \to (0,\infty)$ be a gauge on $[\alpha,\beta]$ and*

$$D = \{\alpha_0, \tau_1, \alpha_1, \ldots, \alpha_{k-1}, \tau_k, \alpha_k\}$$

an arbitrary δ–fine partition of $[\alpha,\beta]$.

If

$$\lim_{j\to\infty} g_j(t) = g(t)$$

almost everywhere in $[\alpha,\beta]$ and

$$B \le \sum_{i=1}^{k} g_{j(\tau_i)}(\tau_i)(\alpha_i - \alpha_{i-1}) \le C$$

for some real numbers $B < C$, all δ–fine partitions D of $[\alpha, \beta]$ and for all choices of positive integer–valued functions $j(t)$, $t \in [\alpha, \beta]$ then g is Perron integrable over $[\alpha, \beta]$ and

$$(P)\int_\alpha^\beta g(s)\,ds = \lim_{j\to\infty}(P)\int_\alpha^\beta g_j(s)\,ds.$$

This convergence theorem for Perron integral is the basis for the following existence result for the equation (2.5) given by R. Henstock in [40, Theorem 19.1].

2.2 Theorem. *Assume that $f : \mathbb{R}^n \times [a, b] \to \mathbb{R}^n$ satisfies the following conditions:*

(H1) *$f(\cdot, t)$ is continuous for almost all $t \in [a, b]$,*

(H2) *the Perron integral$(P)\int_a^b f(z, s)\,ds$ exists for every $z \in \mathbb{R}^n$,*

(H3) *there is a compact set $S \subset \mathbb{R}^n$ and a gauge δ on $[a, b]$ such that for all δ–fine partitions $D = \{\alpha_0, \tau_1, \alpha_1, \ldots, \tau_k, \alpha_k\}$ and all functions $w : [a, b] \to \mathbb{R}^n$ we have*

$$\sum_{i=1}^k f(w(\tau_i), \tau_i)(\alpha_i - \alpha_{i-1}) \in S$$

Then for every $v \in \mathbb{R}_n$, $\alpha \in [a, b]$ there is a function $y : [a, b] \to \mathbb{R}^n$ such that

$$y(t) = v + (P)\int_\alpha^t f(y(s), s)\,ds$$

for $t \in [a, b]$.

The proof of this theorem is postponed at this moment. It should be mentioned that we assume here the function $f(x, t)$ being defined for all $t \in [a, b]$ and $x \in \mathbb{R}^n$. Therefore the existence of a solution on the whole interval $[a, b]$ is asserted in the theorem.

Let us pay attention to Henstock's conditions (H1) – (H3) for the existence of a solution of (2.5). Clearly (H1) is the same as Carathéodory's condition (C1). Condition (H2) is not mentioned explicitly in [40], nevertheless it is evidently used in the proof of the result and its role is similar to that of Carathéodory's (C2). In view of Theorem 2.1 (H3) is the condition that makes it possible to interchange the order of limit and integral in the form

$$\lim_{j\to\infty}(P)\int_\alpha^\beta f(\varphi_j(s),s)\,ds = (P)\int_\alpha^\beta f(\lim_{j\to\infty}\varphi_j(s),s)\,ds =$$

$$= (P)\int_\alpha^\beta f(\varphi(s),s)\,ds \qquad (2.6)$$

provided the sequence of functions $(\varphi_j(s))$ converges with $j\to\infty$ pointwise to $\varphi(s)$ for $s\in[\alpha,\beta]$ and (H1) = (C1) holds. This means that the role of (H3) in Theorem 2.2 is the same as that of Carathéodory's condition (C3) that guarantees (2.6) with the Lebesgue integral by the Lebesgue Dominated Convergence Theorem.

Now we will have a closer look at Henstock's conditions (H1) – (H3) and at the functions $f:[a,b]\times\mathbb{R}^n\to\mathbb{R}^n$ for which these conditions are satisfied.

2.3 Lemma. *If $f:\mathbb{R}^n\times[a,b]\to\mathbb{R}^n$ satisfies (H3) from Theorem 2.2 and $v\in\mathbb{R}^n$ is given then there exists a constant $A>0$ such that*

$$\sum_{i=1}^k \|f(w(\tau_i),\tau_i)-f(v,\tau_i)\|(\alpha_i-\alpha_{i-1})\le 4nA \qquad (2.7)$$

for every $w:[a,b]\to\mathbb{R}^n$ and all δ–fine partitions

$$D=\{\alpha_0,\tau_1,\alpha_1,\ldots,\alpha_{k-1},\tau_k,\alpha_k\}$$

of $[a, b]$ where the gauge δ is given in (H3).

Proof. Since S from (H3) is a compact set in $\mathbb{R}^n$ there is an $A > 0$ such that $S \subset [-A, A]^n \subset \mathbb{R}^n$ where $[-A, A]^n$ is the n–dimensional cube centered at the origin in $\mathbb{R}^n$ with the edge length $2A$. Let

$$D = \{\alpha_0, \tau_1, \alpha_1, \dots, \alpha_{k-1}, \tau_k, \alpha_k\}$$

be an arbitrary δ–fine partition of $[a, b]$. By (H3) we have

$$-A \le \sum_{i=1}^{k} f_m(w(\tau_i), \tau_i)(\alpha_i - \alpha_{i-1}) \le A$$

for every function $w : [a, b] \to \mathbb{R}^n$ and every $m = 1, \dots, n$ where f_m is the m-th component of the function f. For the special case of the constant function $w(t) = v$, $t \in [a, b]$ we have

$$-A \le \sum_{i=1}^{k} f_m(v, \tau_i)(\alpha_i - \alpha_{i-1}) \le A$$

and

$$-2A \le \sum_{i=1}^{k} [f_m(w(\tau_i), \tau_i) - f_m(v, \tau_i)](\alpha_i - \alpha_{i-1}) \le 2A.$$

Denote

$$[f_m(w(\tau_i), \tau_i) - f_m(v, \tau_i)]^+ = \max\{f_m(w(\tau_i), \tau_i) - f_m(v, \tau_i), 0\}$$

and

$$[f_m(w(\tau_i), \tau_i) - f_m(v, \tau_i)]^- = \min\{f_m(w(\tau_i), \tau_i) - f_m(v, \tau_i), 0\}$$

If $f_m(w(\tau_i),\tau_i) - f_m(v,\tau_i) \geq 0$ then define $z(\tau) = w(\tau)$ for $\tau \in [\alpha_{i-1},\alpha_i]$ and if $f_m(w(\tau_i),\tau_i) - f_m(v,\tau_i) < 0$ then we set $z(\tau) = v$ for $\tau \in [\alpha_{i-1},\alpha_i]$. Evidently $z : [a,b] \to \mathbb{R}^n$ and

$$f_m(z(\tau_i),\tau_i) - f_m(v,\tau_i) = [f_m(w(\tau_i),\tau_i) - f_m(v,\tau_i)]^+ \geq 0.$$

Therefore

$$0 \leq \sum_{i=1}^{k} [f_m(z(\tau_i),\tau_i) - f_m(v,\tau_i)](\alpha_i - \alpha_{i-1}) =$$
$$= \sum_{i=1}^{k} [f_m(w(\tau_i),\tau_i) - f_m(v,\tau_i)]^+(\alpha_i - \alpha_{i-1}) \leq 2A.$$

Similarly it can be shown that

$$-2A \leq \sum_{i=1}^{k} [f_m(w(\tau_i),\tau_i) - f_m(v,\tau_i)]^-(\alpha_i - \alpha_{i-1}) \leq 0.$$

Since evidently

$$|f_m(w(\tau_i),\tau_i) - f_m(v,\tau_i)| =$$
$$= [f_m(w(\tau_i),\tau_i) - f_m(v,\tau_i)]^+ - [f_m(w(\tau_i),\tau_i) - f_m(v,\tau_i)]^-$$

we obtain

$$\sum_{i=1}^{k} |f_m(w(\tau_i),\tau_i) - f_m(v,\tau_i)|(\alpha_i - \alpha_{i-1}) =$$
$$= \sum_{i=1}^{k} [f_m(w(\tau_i),\tau_i) - f_m(v,\tau_i)]^+(\alpha_i - \alpha_{i-1}) -$$
$$- \sum_{i=1}^{k} [f_m(w(\tau_i),\tau_i) - f_m(v,\tau_i)]^-(\alpha_i - \alpha_{i-1}) \leq 4A$$

for every $m = 1, \dots, n$ and therefore (2.7) is satisfied. □

2.4 Theorem. *A function $f : \mathbb{R}^n \times [a, b] \to \mathbb{R}^n$ fulfils (H1) – (H3) from Theorem 2.2 if and only if*

$$f(x,t) = g(t) + h(x,t) \tag{2.8}$$

for $(x, t) \in \mathbb{R}^n \times [a, b]$ where

$$g : [a, b] \to \mathbb{R}^n \text{ is Perron integrable over } [a, b] \tag{2.9}$$

and $h : \mathbb{R}^n \times [a, b] \to \mathbb{R}^n$ fulfils

(C1) *$h(\cdot, t)$ is continuous for almost all $t \in [a, b]$,*
(C2) *$h(x, \cdot)$ is measurable for $x \in \mathbb{R}^n$,*
(C3) *there exists $m : [a, b] \to [0, \infty)$ Lebesgue integrable over $[a, b]$ such that*

$$\|h(x,t)\| \le m(t)$$

for $x \in \mathbb{R}^n$ and almost all $t \in [a, b]$.

Proof. If f is of the form (2.8) with g and h satisfying (2.8) and (C1) – (C3) then evidently f satisfies (H1) and (H2) from Theorem 2.2.

Since the integrals $(P)\int_a^b g(s)ds$, $(L)\int_a^b m(s)ds$ exist, there is a gauge $\delta : [a, b] \to (0, \infty)$ on $[a, b]$ such that

$$\|\sum_{i=1}^{k} g(\tau_i)(\alpha_i - \alpha_{i-1}) - (P)\int_a^b g(s)\,ds\| < 1, \tag{2.10}$$

$$\|\sum_{i=1}^{k} m(\tau_i)(\alpha_i - \alpha_{i-1}) - (L)\int_a^b m(s)\,ds\| < 1 \tag{2.11}$$

provided $D = \{\alpha_0, \tau_1, \alpha_1, \dots, \alpha_{k-1}, \tau_k, \alpha_k\}$ is an arbitrary δ–fine partition of $[a,b]$. For such a partition and an arbitrary $w : [a,b] \to \mathbb{R}^n$ we have by (2.10), (2.11) and (C3) the inequality

$$\begin{aligned}
&\|\sum_{i=1}^{k} f(w(\tau_i), \tau_i)(\alpha_i - \alpha_{i-1})\| \le \\
\le \quad & \|\sum_{i=1}^{k} g(\tau_i)(\alpha_i - \alpha_{i-1})\| + \\
&+ \|\sum_{i=1}^{k} h(w(\tau_i), \tau_i)(\alpha_i - \alpha_{i-1})\| \le \\
\le \quad & \|\sum_{i=1}^{k} g(\tau_i)(\alpha_i - \alpha_{i-1}) - (P)\int_a^b g(s)\,ds\| + \\
+ \quad & \|(P)\int_a^b g(s)\,ds\| + \sum_{i=1}^{k} \|h(w(\tau_i), \tau_i)\|(\alpha_i - \alpha_{i-1}) \le \\
\le \quad & 1 + \|(P)\int_a^b g(s)\,ds\| + \sum_{i=1}^{k} m(\tau_i)(\alpha_i - \alpha_{i-1}) \le \\
\le \quad & \|(P)\int_a^b g(s)\,ds\| + \int_a^b m(s)\,ds + 2
\end{aligned}$$

and this means that the sum $\sum_{i=1}^{k} f(w(\tau_i), \text{Í}\tau_i)(\alpha_i - \alpha_{i-1})$ belongs to the compact ball

$$S = \{x \in \mathbb{R}^n; \|x\| \le \|(P)\int_a^b g(s)\,ds\| + \int_a^b m(s)\,ds + 2\}$$

in $\mathbb{R}^n$, i.e. (H3) is satisfied.

Let now $f : \mathbb{R}^n \times [a,b] \to \mathbb{R}^n$ fulfil (H1) – (H3) from Theorem 2.2. Assume that $v \in \mathbb{R}^n$ and set

$$g(t) = f(v,t) \text{ for } t \in [a,b],$$

$$h(x,t) = f(x,t) - g(t) = f(x,t) - f(v,t) \text{ for } t \in [a,b],\ x \in \mathbb{R}^n.$$

Then evidently (2.8) holds and by (H2) g is Perron integrable over $[a,b]$, i. e. (2.9) holds.

By Lemma 2.3 we have

$$\sum_{i=1}^{k} \|h(w(\tau_i),\tau_i)\|(\alpha_i - \alpha_{i-1}) \le C \tag{2.12}$$

with a constant $C > 0$ where $w : [a,b] \to \mathbb{R}^n$ and the δ–fine partition $D = \{\alpha_0, \tau_1, \alpha_1, \ldots, \alpha_{k-1}, \tau_k, \alpha_k\}$ of $[a,b]$ are arbitrary. The gauge δ is given in (H3).

Put $M(0) = 0$ and

$$M(s) = \sup\{\sum_{i=1}^{k} \|h(w(\sigma_i),\sigma_i)\|(\beta_i - \beta_{i-1})\}, \quad s \in (a,b]$$

where the supremum is taken over all δ–fine partitions

$$\{\beta_0, \sigma_1, \beta_1, \ldots, \beta_{m-1}, \sigma_m, \beta_m\}$$

of $[a,s]$ and all functions $w : [a,b] \to \mathbb{R}^n$.

Let us show that if $a \le s_1 < s_2 \le b$ then

$$M(s_1) \le M(s_2). \tag{2.13}$$

Assume that

$$\{\gamma_0, \rho_1, \gamma_1, \ldots, \gamma_{l-1}, \rho_l, \gamma_l\}$$

is an arbitrary δ–fine partition of $[s_1, s_2]$ and that

$$\{\beta_0, \sigma_1, \beta_1, \ldots, \beta_{m-1}, \sigma_m, \beta_m\}$$

is a δ–fine partition of $[a, s_1]$. Then

$$\{\beta_0, \sigma_1, \beta_1, \ldots, \beta_{m-1}, \sigma_m, \beta_m\} \cup \{\gamma_0, \rho_1, \gamma_1, \ldots, \gamma_{l-1}, \rho_l, \gamma_l\}$$

is a δ–fine partition of $[a, s_2]$ and

$$\sum_{i=1}^{k} \|h(w(\sigma_i), \sigma_i)\|(\beta_i - \beta_{i-1}) +$$

$$+ \sum_{j=1}^{l} \|h(w(\rho_j), \rho_j)\|(\gamma_j - \gamma_{j-1}) \le M(s_2)$$

Passing to the supremum over all δ–fine partitions of $[a, s_1]$ we get

$$M(s_1) + \sum_{j=1}^{l} \|h(w(\rho_j), \rho_j)\|(\gamma_j - \gamma_{j-1}) \le M(s_2).$$

Further evidently

$$M(s_1) \le M(s_1) + \sum_{j=1}^{l} \|h(w(\rho_j), \rho_j)\|(\gamma_j - \gamma_{j-1}) \le M(s_2)$$

and therefore (2.13) holds.

By (2.12) clearly

$$0 \le M(s) \le C, \quad s \in [a, b]. \tag{2.14}$$

Assume now that $t \in [a, b]$ and $a \le s_1 \le s_2 \le b$ where

$$t - \delta(t) < s_1 \le t \le s_2 < t + \delta(t). \tag{2.15}$$

Having an arbitrary δ–fine partition of $[a, s_1]$ we add to it the triple

$$\{s_1, t, s_2\}$$

and we obtain by (2.15) a δ–fine partition of $[a, s_2]$. Therefore

$$M(s_1) + \|h(x,t)\|(s_2 - s_1) \leq M(s_2) \text{ for any } x \in \mathbb{R}^n$$

and consequently

$$\|h(x,t)\|(s_2 - s_1) \leq M(s_2) - M(s_1), \quad x \in \mathbb{R}^n. \tag{2.16}$$

Since M is monotone by (2.13) the derivative $\dot{M}$ of M exists almost everywhere in $[a, b]$. Putting

$$m(t) = \dot{M}(t)$$

if the derivative $\dot{M}(t)$ exists and

$$m(t) = 0$$

otherwise, we get by (2.16) the inequality

$$\|h(x,t)\| \leq m(t)$$

for $x \in \mathbb{R}^n$ and almost all $t \in [a, b]$. Except a set of Lebesgue measure zero the function m is the derivative of the absolutely continuous part of the nondecreasing function M and is therefore Lebesgue integrable over $[a, b]$ with $(L)\int_a^b m(s)\,ds \leq M(b) \leq C$. Hence h satisfies (C3) and this concludes the proof. □

2.5 *Remark.* Theorem 2.4 represents a surprising result which states that Henstock's existence theorem 2.2 covers "only" the case of Carathéodory functions perturbed by a Perron integrable function, i.e. differential equations of the form

$$\dot{x} = h(x,t) + g(t) \tag{2.17}$$

where h is a Carathéodory function satisfying (C1) – (C3) and g is Perron integrable over the interval at which we are looking for solutions to (2.17) in the sense of the integral equation (2.5) , i.e. for functions $x : J \to \mathbb{R}^n$ such that

$$x(t) = x(\alpha) + (P)\int_\alpha^t h(x(s), s)\,ds + (P)\int_\alpha^t g(s)\,ds.$$

Theorem 2.4 was proved in [87].

2.6 Lemma. *If $f : \mathbb{R}^n \times [a,b] \to \mathbb{R}^n$ satisfies conditions (H1)–(H3) from Theorem 2.2 and if $\varphi : [a,b] \to \mathbb{R}^n$ is continuous then $(P)\int_\alpha^\beta f(\varphi(s),s)\,ds$ exists for every $a \le \alpha \le \beta \le b$.*

Proof. The function φ is continuous on $[a,b]$ and therefore it is the uniform limit of a sequence ψ_k of step functions. By (H2) the Perron integral $(P)\int_a^b f(\psi_k(s),s)\,ds$ exists for every $k \in \mathbb{N}$ and the continuity condition (H1) yields

$$\lim_{k\to\infty} f(\psi_k(s),s) = f(\varphi(s),s), \quad s \in [a,b].$$

Using (H3) makes it apparent that the conditions of Theorem 2.1 are satisfied for every component of the function f and therefore the integral $(P)\int_a^b f(\varphi_k(s),s)\,ds$ and also $(P)\int_\alpha^\beta f(\varphi(s),s)\,ds$ exists for every $a \le \alpha \le \beta \le b$. □

Before proceeding with the proof of the existence result stated in Theorem 2.2 we will prove one more lemma.

We define the Carathéodory approximations by

$$\varphi_j(t) = v \text{ for } t \in [\alpha, \alpha + \frac{\eta}{j}],$$
$$\varphi_j(t) = v + (P)\int_\alpha^{t-\frac{\eta}{j}} f(\varphi_j(s),s)\,ds \text{ for } t \in (\alpha + \frac{\eta}{j}, \alpha + \eta]$$
$$j = 1,2,\ldots \tag{2.18}$$

where f is the function from Theorem 2.2, $\alpha \in [a,b)$ and $\eta = b - \alpha$. For $t \in [\alpha, \alpha + \frac{\eta}{j}]$ the function φ_j is evidently continuous. Therefore by Lemma 2.6 the integral

$$(P)\int_\alpha^{t-\frac{\eta}{j}} f(\varphi_j(s),s)\,ds$$

exists and is continuous as a function of the variable t for $t \in [\alpha + \frac{\eta}{j}, \alpha + \frac{2\eta}{j}]$, i.e. φ_j is continuous on $[\alpha, \alpha + \frac{2\eta}{j}]$. In this way for every $j = 1, 2, \ldots$ the function φ_j is defined inductively on the whole interval $[\alpha, \alpha + \eta] = [\alpha, b]$, and is continuous in this interval.

2.7 Lemma. *If $f : [a, b] \times \mathbb{R}^n \to \mathbb{R}^n$ satisfies conditions (H1)–(H3) from Theorem 2.2 then the sequence of functions $(\varphi_j)_{j=1}^{\infty}$ defined on $[\alpha, b]$, $a \leq \alpha < b$ contains a pointwise convergent subsequence.*

Proof. Assume that

$$\alpha + \frac{\eta}{j} = c_0 < c_1 < \cdots < c_l = b$$

is an arbitrary division of $[\alpha + \frac{\eta}{j}, b]$. Using the formulas (2.18) for every $m = 1, \ldots, l$ we have

$$\|\varphi_j(c_m) - \varphi_j(c_{m-1}) - (P)\int_{c_{m-1}-\frac{\eta}{j}}^{c_m-\frac{\eta}{j}} f(v, s)\, ds\| =$$

$$= \|(P)\int_{\alpha}^{c_m-\frac{\eta}{j}} f(\varphi_j(s), s)\, ds - (P)\int_{\alpha}^{c_{m-1}-\frac{\eta}{j}} f(\varphi_j(s), s)\, ds -$$

$$-(P)\int_{c_{m-1}-\frac{\eta}{j}}^{c_m-\frac{\eta}{j}} f(v, s)\, ds\| =$$

$$= \|(P)\int_{c_{m-1}-\frac{\eta}{j}}^{c_m-\frac{\eta}{j}} (f(\varphi_j(s), s) - f(v, s))\, ds\|. \tag{2.19}$$

Let $\delta : [a, b] \to (0, \infty)$ be the gauge on $[a, b]$ given by the condition (H3) in Theorem 2.2. Since the integral

$$(P)\int_{c_{m-1}-\frac{\eta}{j}}^{c_m-\frac{\eta}{j}} (f(\varphi_j(s), s) - f(v, s))\, ds$$

exists, there is a gauge δ^* with $\delta^*(t) < \delta(t)$ such that if

$$D^m = \{\alpha_0^m, \tau_1^m, \alpha_1^m, \dots, \alpha_{k_m-1}^m, \tau_{k_m}^m, \alpha_{k_m}^m\}$$

is a δ^*–fine partition of $[c_{m-1} - \frac{\eta}{j}, c_m - \frac{\eta}{j}]$ then

$$\|(P)\int_{c_{m-1}-\frac{\eta}{j}}^{c_m-\frac{\eta}{j}} (f(\varphi_j(s),s) - f(v,s))\,ds-$$

$$-\sum_{i=1}^{k_m}[f(\tau_i^m,\varphi_j(\tau_i^m)) - f(v,\tau_i^m)](\alpha_i^m - \alpha_{i-1}^m)\| < \frac{1}{l}.$$

Hence

$$\|(P)\int_{c_{m-1}-\frac{\eta}{j}}^{c_m-\frac{\eta}{j}} (f(\varphi_j(s),s) - f(v,s))\,ds\| \le$$

$$\le \|(P)\int_{c_{m-1}-\frac{\eta}{j}}^{c_m-\frac{\eta}{j}} (f(\varphi_j(s),s) - f(v,s))\,ds-$$

$$-\sum_{i=1}^{k_m}[f(\varphi_j(\tau_i^m),\tau_i^m) - f(v,\tau_i^m)](\alpha_i^m - \alpha_{i-1}^m)\|+$$

$$+\|\sum_{i=1}^{k_m}[f(\varphi_j(\tau_i^m),\tau_i^m) - f(v,\tau_i^m)](\alpha_i^m - \alpha_{i-1}^m)\|$$

and using (2.19) we get

$$\|\varphi_j(c_m) - \varphi_j(c_{m-1}) - (P)\int_{c_{m-1}-\frac{\eta}{j}}^{c_m-\frac{\eta}{j}} f(v,s)\,ds\| \le$$

$$\le \frac{1}{l} + \|\sum_{i=1}^{k_m}[f(\varphi_j(\tau_i^m),\tau_i^m) - f(v,\tau_i^m)](\alpha_i^m - \alpha_{i-1}^m)\|$$

for every $m = 1, \dots, l$. Therefore

$$\sum_{m=1}^{l} \|\varphi_j(c_m) - \varphi_j(c_{m-1}) - (P)\int_{c_{m-1}-\frac{\eta}{j}}^{c_m - \frac{\eta}{j}} f(v,s)\,ds\| \le$$

$$\le 1 + \sum_{m=1}^{l}\sum_{i=1}^{k_m} \|f(\varphi_j(\tau_i^m),\tau_i^m) - f(v,\tau_i^m)\|(\alpha_i^m - \alpha_{i-1}^m).$$

Since the union of the above mentioned partitions D^m of the intervals $[c_{m-1} - \frac{\eta}{j}, c_m - \frac{\eta}{j}]$ is evidently a δ–fine partition of $[\alpha, b - \frac{\eta}{j}]$, Lemma 2.3 can be used for $w = \varphi_j$ and by the last inequality we obtain the estimate

$$\sum_{m=1}^{l} \|\varphi_j(c_m) - \varphi_j(c_{m-1}) - (P)\int_{c_{m-1}-\frac{\eta}{j}}^{c_m - \frac{\eta}{j}} f(v,s)\,ds\| \le 1 + 4nA \tag{2.20}$$

for every division $c_0 < c_1 < \dots < c_l$ of the interval $[\alpha + \frac{\eta}{j}, b]$. Hence

$$\operatorname{var}_{\alpha+\frac{\eta}{j}}^{b}(\varphi_j(t) - (P)\int_{\alpha}^{t-\frac{\eta}{j}} f(v,s)\,ds) \le 1 + 4nA$$

for every j. Put

$$\psi_j(t) = v \text{ for } t \in [\alpha, \alpha + \frac{\eta}{j}],$$

$$\psi_j(t) = \varphi_j(t) - (P)\int_{\alpha}^{t-\frac{\eta}{j}} f(v,s)\,ds \text{ for } t \in (\alpha + \frac{\eta}{j}, b]$$

$$j = 1, 2, \dots .$$

Then

$$\operatorname{var}_{\alpha}^{b} \psi_j = \operatorname{var}_{\alpha}^{\alpha+\frac{\eta}{j}} \psi_j + \operatorname{var}_{\alpha+\frac{\eta}{j}}^{b} \psi_j \le 1 + 4nA$$

by (2.20) and evidently also

$$\|\psi_j(t)\| \le \|\psi_j(\alpha)\| + \|\psi_j(t) - \psi_j(\alpha)\| \le \|v\| + \operatorname{var}_\alpha^b \psi_j \le$$
$$\le \|v\| + 1 + 4nA.$$

Since the sequence (ψ_j) is equibounded with uniformly bounded variation Helly' s Choice Theorem (see e.g. [104]) yields that there exists a subsequence of (ψ_j), we denote it again by (ψ_j), which is pointwise convergent, i.e.

$$\lim_{j\to\infty} \psi_j(t) = \lim_{j\to\infty} (\varphi_j(t) - (P)\int_\alpha^{t-\frac{\eta}{j}} f(v,s)\,ds), \quad t > \alpha$$

and

$$\lim_{j\to\infty} \psi_j(\alpha) = \lim_{j\to\infty} \varphi_j(\alpha) = v.$$

Since the indefinite Perron integral is continuous, we have

$$\lim_{j\to\infty} (P)\int_\alpha^{t-\frac{\eta}{j}} f(v,s)\,ds = (P)\int_\alpha^t f(v,s)\,ds$$

and therefore $\lim_{j\to\infty} \varphi_j(t)$ exists and equals some value $\varphi(t)$ for every $t \in [a,b]$. □

Proof of Theorem 2.2. Let $\alpha \in [a,b)$ be given and assume that the sequence of functions $\varphi_j : [\alpha, b] \to \mathbb{R}^n$,$j = 1,2,\ldots$ is defined by the relations (2.18). By Lemma 2.7 there is a subsequence of this sequence, which we denote again by (φ_j), such that

$$\lim_{j\to\infty} \varphi_j(t) = \varphi(t) \quad \text{for every } t \in [\alpha, b].$$

Since by (H1) $f(t,.)$ is continuous for almost all $t \in [a,b]$, we have

$$\lim_{j\to\infty} f(\varphi_j(t), t) = f(\varphi(t), t) \quad \text{almost everywhere in } [\alpha, b].$$

Further, the integral

$$(P)\int_{\alpha}^{b} f(\varphi_j(s),s)\,ds$$

exists for every $j=1,2,\dots$ by Lemma 2.6 because the functions $\varphi_j:[\alpha,b]\to\mathbb{R}^n$ are continuous. Finally, (H3) implies that for every $l=1,\dots,n$ we have

$$-A\le\sum_{i=1}^{k} f_l(\varphi_{j(\tau_i)}(\tau_i),\tau_i))(\alpha_i-\alpha_{i-1})\le A$$

for an arbitrary $j:[\alpha,b]\to\mathbb{N}$ and an arbitrary δ–fine partition

$$\{\alpha_0,\tau_1,\alpha_1,\dots,\alpha_{k-1},\tau_k,\alpha_k\}$$

of $[\alpha,t]$.

It remains to show that

$$\lim_{j\to\infty}(P)\int_{\alpha}^{t-\frac{\eta}{j}} f(s,\varphi_j(s))\,ds=(P)\int_{\alpha}^{t} f(\varphi(s),s)\,ds. \tag{2.21}$$

By Henstock' s Theorem 2.1 we have

$$\lim_{j\to\infty}(P)\int_{\alpha}^{t} f(\varphi_j(s),s)\,ds=(P)\int_{\alpha}^{t} f(\varphi(s),s)\,ds.$$

Since

$$(P)\int_{\alpha}^{t-\frac{\eta}{j}} f(\varphi_j(s),s)\,ds=$$

$$=(P)\int_{\alpha}^{t} f(\varphi_j(s),s)\,ds-(P)\int_{t-\frac{\eta}{j}}^{t} f(\varphi_j(s),s)\,ds$$

it is now sufficient to show that

$$\lim_{j\to\infty}(P)\int_{t-\frac{\eta}{j}}^{t} f(\varphi_j(s),s)\,ds = 0. \tag{2.22}$$

At this moment we use the result of Theorem 2.4 which states that under the given condition we have $f(\varphi_j(s),s) = g(s)+h(\varphi_j(s),s)$ where g is Perron integrable and h is a Carathéodory function with $\|h(x,s)\| \le m(s)$. Hence

$$(P)\int_{t-\frac{\eta}{j}}^{t} f(\varphi_j(s),s)\,ds =$$

$$= (P)\int_{t-\frac{\eta}{j}}^{t} g(s)\,ds + (P)\int_{t-\frac{\eta}{j}}^{t} h(\varphi_j(s),s)\,ds.$$

For this expression we have

$$\lim_{j\to\infty}(P)\int_{t-\frac{\eta}{j}}^{t} g(s)\,ds = 0$$

because the indefinite integral of a Perron integrable function is continuous, and

$$\|(P)\int_{t-\frac{\eta}{j}}^{t} h(\varphi_j(s),s)\,ds\| \le (P)\int_{t-\frac{\eta}{j}}^{t} \|h(\varphi_j(s),s)\|\,ds \le$$

$$\le (P)\int_{t-\frac{\eta}{j}}^{t} m(s)\,ds \to 0$$

for $j\to\infty$ for the same reason. Therefore (2.22) holds and (2.21) is true.

Now, taking the limit for $j \to \infty$ of the Carathéodory approximations given by (2.18) we obtain

$$\varphi(t) = v + (P)\int_\alpha^t f(\varphi(s), s)\, ds, \quad t \in [\alpha, b],$$

i.e. φ is by definition a solution of (2.1) in the Perron sense on $[\alpha, b]$.

It is straightforward to define the Carathéodory approximations for the interval $[a, \alpha]$ similarly as in (2.18) and repeat the above procedure for examining the existence of the desired solution on $[a, \alpha]$. Since both solutions coincide at α Theorem 2.2 is proved. □

2.8 *Remark.* The proof of Theorem 2.2 follows the lines of Henstock's proof from [40]. It should be pointed out that the relation (2.22) is used but not mentioned explicitly in Henstock's proof and it is not very clear how to derive it without using Theorem 2.4 to this end as we have done above.

2.9 *Remark.* Let us consider again the situation in which we are after having proved Theorem 2.2.

Denote by *Car* the class of functions $f : [a, b] \times \mathbb{R}^n \to \mathbb{R}^n$ satisfying conditions (C1), (C2) and (C3) and let $\mathcal{H}en$ be the class of functions $f : [a, b] \times \mathbb{R}^n \to \mathbb{R}^n$ for which (H1), (H2) and (H3) are satisfied.

A function $x : [c, d] \to \mathbb{R}^n$ is said to be a solution of the ordinary differential equation

$$\dot{x} = f(x, t) \tag{2.1}$$

in the *Carathéodory sense* on the interval $[c, d] \subset [a, b]$ if

$$x(t) = x(\alpha) + (L)\int_\alpha^t f(x(s), s)\, ds$$

for every $\alpha, t \in [c,d]$ and $x : [c,d] \to \mathbb{R}^n$ is called a solution of the ordinary differential equation (2.1) in the *Perron - Henstock sense* if

$$x(t) = x(\alpha) + (P)\int_\alpha^t f(x(s),s)\,ds$$

for every $\alpha, t \in [c,d]$.*)

By the well known relation between the Lebesgue and Perron integrals it is clear that if $x : [c,d] \to \mathbb{R}^n$ is a solution of (2.1) in the Carathéodory sense then it is a solution of (2.1) in the Perron – Henstock sense as well.

Moreover, Theorem 2.4 shows not only that $\mathcal{C}ar \subset \mathcal{H}en$ but it gives the formal relation

$$\mathcal{H}en = \mathcal{P} + \mathcal{C}ar$$

where by $\mathcal{P}$ the set of Perron integrable functions $g : [a,b] \to \mathbb{R}^n$ is denoted. This relation means that $f \in \mathcal{H}en$ if and only if $f = g + h$ where $g \in \mathcal{P}$ and $h \in \mathcal{C}ar$, and it contains the information that the Henstock – Perron concept of a solution of the ordinary differential equation (2.1) is more general than the concept of Carathéodory provided (H1) – (H3) is fulfilled. It shows also how far the generalization goes with respect to the possible choices of right hand sides of an ordinary differential equation.

It is worth mentioning here that Henstock's conditions (H1), (H2) and (H3) are not the only ones under which the Perron – Henstock notion of a solution of the ordinary differential equation (2.1) is studied. For example in [12] a result of P. Bullen and R. Výborný [13] is mentioned where the following conditions are used for $f : \mathbb{R} \times [a,b] \to \mathbb{R}$:

(B1) $f(\cdot,t)$ is continuous for almost all $t \in [a,b]$,

---

*)We write (L) , (P) in front of the integral to emphasize that the Lebesgue or Perron integration is used.

(B2) for all continuous ACG^* functions $\varphi : [a,b] \to \mathbb{R}$ the function $f(\varphi(t),t) : [a,b] \to \mathbb{R}$ is measurable ,

(B3) there exist continuous $M, \mu : [a,b] \to \mathbb{R}$, $M(\alpha) = \mu(\alpha) = 0$ for some $\alpha \in [a,b]$ such that for all continuous and ACG^* functions $\varphi : [a,b] \to \mathbb{R}$ we have

$$\underline{D}M(t) \geq f(\varphi(t),t) \geq \overline{D}\mu(t)$$

($\underline{D}$, $\overline{D}$ stand for the lower and upper derivative, respectively.)

It is clear that the conditions (B1) – (B3) have to be modified for $\mathbb{R}^n$– valued functions $f : \mathbb{R}^n \times [a,b] \to \mathbb{R}^n$ in assuming them for the components of the function f.

It turns out that if a function f satisfies (B1) – (B3) then $f \in \mathcal{P} + \mathcal{C}ar = \mathcal{H}en$. Indeed, define $g(t) = f(v,t)$ for some $v \in \mathbb{R}$ and $t \in [a,b]$ and put $h(x,t) = f(x,t) - g(t)$ for $x \in \mathbb{R}, \quad t \in [a,b]$. The function g is measurable by (B1) and since by (B3) we have

$$\underline{D}M(t) \geq g(t) \geq \overline{D}\mu(t)$$

the function g is Perron integrable by P. Bullen's result given in [12].

By (B3) we further have

$$\overline{D}\mu(t) - \underline{D}M(t) \leq f(x,t) - f(v,t) = h(x,t) \leq \underline{D}M(t) - \overline{D}\mu(t),$$

i.e.

$$0 \leq \|h(x,t)\| \leq \underline{D}M(t) - \overline{D}\mu(t) \leq \underline{D}[M(t) - \mu(t)], \quad t \in [a,b]. \tag{2.23}$$

Since M and μ are major and minor functions to g, respectively, the difference $M - \mu$ is nondecreasing in $[a,b]$ and therefore the derivative $\frac{d}{dt}(M - \mu)$ exists almost everywhere. Defining

$$m(t) = \frac{d}{dt}(M - \mu)$$

for $t \in [a,b]$ at which the derivative exists and

$$m(t) = 0$$

elsewhere, we have by (2.23) the inequality

$$h(x,t) \leq m(t) \text{ for } x \in \mathbb{R} \text{ and almost all } t \in [a,b].$$

The function m is nonnegative and integrable and therefore h satisfies Carathéodory's condition (C3). It is also clear that (B1) and (B2) imply respectively (C1) and (C2) for the function h and this yields that if f satisfies Bullen's conditions (B1) – (B3) then $f \in \mathcal{P} + \mathcal{C}ar = \mathcal{H}en$, i.e. Bullen's case includes the same class of ordinary differential equations as Henstock's. On the other hand, it is a matter of routine to show that f satisfies (B1) – (B3) whenever $f \in \mathcal{P} + \mathcal{C}ar$. This means that for the class $\mathcal{B}ul$ of functions f satisfying (B1) – (B3) we have

$$\mathcal{B}ul = \mathcal{P} + \mathcal{C}ar = \mathcal{H}en,$$

i.e. all the aproaches mentioned above concern the same class of equations.

Finally, let us turn our attention to another concept described by T.S. Chew and F. Flordeliza in [16]. They consider the class of functions $f : \mathbb{R} \times [a,b] \to \mathbb{R}$ such that

(C1) $f(\cdot, t)$ is continuous for almost all $t \in [a,b]$,

(C2) $f(x, \cdot)$ is measurable for every fixed x,

(C-F) there exist two Perron integrable functions g, r on $[a,b]$ such that for

$$g(t) \leq f(x,t) \leq r(t), \quad t \in [a,b],\ x \in \mathbb{R}.$$

Put in this case $h(x,t) = f(x,t) - g(t)$. Then h evidently satisfies (C1) and (C2) and by (C-F) we have

$$0 \le f(x,t) - g(t) \le r(t) - g(t)$$

i.e.

$$0 \le h(x,t) \le m(t) \tag{2.24}$$

where $m(t) = r(t) - g(t)$ is a nonnegative Perron integrable function and therefore m is also Lebesgue integrable. h is nonnegative and this yields by (2.24) that h satisfies Carathéodory's condition (C3). In this way we have again that conditions (C1), (C2) and (C-F) for f imply $f \in \mathcal{P} + \mathcal{C}ar$, i.e. the case described by T.S. Chew and F. Flordeliza in [16] leads to equations of the form (2.17).

The existence theorem 2.2 for Perron – Henstock solutions of (2.1) was presented in a global form by assuming that the right hand side f is defined for $t \in [a,b]$ and $x \in \mathbb{R}^n$. Now we present the local version of the existence theorem.

2.10 Theorem. *Assume that the function f is defined on a set*

$$Q = \{y \in \mathbb{R}^n; \|y - v\| \le\} \times [\alpha - A, \alpha + A]$$

with some $\varrho > 0$ and that $f(x,t) = g(t) + h(x,t)$ for $(x,t) \in Q$ where

$$g : [\alpha - A, \alpha + A] \to \mathbb{R}^n \quad \text{is Perron integrable,}$$

and

(1) *$h(\cdot,t)$ is continuous for almost all $t \in [\alpha - A, \alpha + A]$,*
(2) *$h(x,\cdot)$ is measurable for*

$$x \in B(v,\varrho) = \{y \in \mathbb{R}^n; \|y - v\| \le \varrho\}$$

(3) *there exists* $m : [\alpha - A, \alpha + A] \to \mathbb{R}$ *Lebesgue integrable over* $[\alpha - A, \alpha + A]$ *such that*

$$\|h(x,t)\| \le m(t)$$

for $x \in B(v, \varrho)$ *and almost all* $t \in [\alpha - A, \alpha + A]$.

Then there exists a $\delta > 0$, $\delta < \alpha$ *such that on the interval* $[\alpha - \delta, \alpha + \delta]$ *there exists a Perron – Henstock solution* $x(t)$ *of (2.1) such that* $x(\alpha) = v$.

Proof. The indefinite integral

$$G(t) = (P)\int_\alpha^t g(s)ds, \quad t \in [\alpha - A, \alpha + A]$$

being an ACG^* function is continuous at $t = \alpha$ and the same is true for the indefinite integral $M(t) = \int_\alpha^t m(s)ds$ because it is absolutely continuous. Hence there is a $\delta > 0$ such that $\|G(t)\| + |M(t)| < \varrho$ for every $t \in [\alpha - \delta, \alpha + \delta]$.

For $t \in [\alpha, \alpha + \delta]$ define the Carathéodory approximations

$$\begin{aligned} &\varphi_j(t) = v \text{ for } t \in [\alpha, \alpha + \frac{\delta}{j}], \\ &\varphi_j(t) = v + (P)\int_\alpha^{t-\frac{\delta}{j}} g(s)\,ds + \\ &\quad + (P)\int_\alpha^{t-\frac{\delta}{j}} h(\varphi_j(s), s)\,ds \text{ for } t \in (\alpha + \frac{\delta}{j}, \alpha + \delta] \\ &\quad j = 1, 2, \ldots \end{aligned} \tag{2.25}$$

as above in (2.18) for the proof of Theorem 2.2. Since by Theorem 2.4 the assumptions of Theorem 2.2 are satisfied and

$$\begin{aligned} \|\varphi_j(t) - v\| &\le \|G(t - \frac{\delta}{j}) - G(\alpha)\| + |M(t - \frac{\delta}{j}) - M(\alpha)| = \\ &= \|G(t - \frac{\delta}{j})\| + |M(t - \frac{\delta}{j})| < \varrho \end{aligned} \tag{2.26}$$

holds for every $t \in [\alpha, \alpha + \delta]$ we conclude from Theorem 2.2 and Lemma 2.7 that there exists a subsequence of (φ_j) (we denote it again by (φ_j)) which converges pointwise to a certain function $x(t)$ on $[\alpha, \alpha + \delta]$. Using the continuity of the function h in the second variable and the Lebesgue Dominated Convergence Theorem we obtain

$$\lim_{j \to \infty} (P) \int_{\alpha}^{t - \frac{\delta}{j}} h(\varphi_j(s), s)\, ds = (P) \int_{\alpha}^{t} h(\varphi(s), s)\, ds.$$

(The functions $h(s, \varphi_j(s))$ are dominated by m on $[\alpha, \alpha + \delta]$.) Passing to the limit for $j \to \infty$ in (2.25) we obtain

$$x(t) = v + (P) \int_{\alpha}^{t} g(s)\, ds + (P) \int_{\alpha}^{t} h(x(s), s)\, ds$$

for $t \in [\alpha, \alpha + \delta]$ and this means that $x : [\alpha, \alpha + \delta] \to \mathbb{R}^n$ is a solution of (2.1) in the Perron – Henstock sense because by (2.26) evidently

$$\|x(t) - v\| \le \varrho, \quad t \in [\alpha, \alpha + \delta],$$

i.e. $x(t) \in B(v, \varrho)$ for every $t \in [\alpha, \alpha + \delta]$.

For $t \in [\alpha - \delta, \alpha]$ the same can be proved by defining the appropriate Carathéodory approximations and this completes the proof. □

2.11 *Remark.* Theorem 2.10 gives the local existence of a Perron – Henstock solution of the ordinary differential equation (2.1) with the initial condition $x(\alpha) = v$. This result has the same form and essentially also the same proof as the well-known local existence theorem for Carathéodory solutions when the right hand side f of (2.1) satisfies conditions (C1),(C2) and (C3) locally (see e.g. Theorem 2.1 of Chap. II in [18]). The proof of Theorem 2.10 differs only slightly from the proof of the Carathéodory local existence theorem because of the special form $g(t)+h(x,t)$ of the right hand side of (2.1). Since we have the benefit of the result given in Theorem 2.4 we can claim that the local existence of a Perron – Henstock solution of an initial value problem for the equation (2.1) is guaranteed also in the general case of equations with right hand sides satisfying Henstock's conditions (H1), (H2) and (H3) given in Theorem 2.2.

If $f : G \to \mathbb{R}^n$ where $G \subset \mathbb{R}^n \times \mathbb{R} = \mathbb{R}^{n+1}$ is an open set then for every $(\alpha, v) \in G$ there exist $A, \varrho > 0$ such that

$$R_{A,\varrho} = \{y \in \mathbb{R}^n; \|y - v\| < \varrho\} \times [\alpha - A, \alpha + A] \subset G.$$

Assuming that the assumptions of Theorem 2.10 are satisfied for every such $R_{A,\varrho}$ we can use Theorem 2.10 to state that for every initial point $(t_0, \tilde{x}) \in G$ there exists locally a Perron–Henstock solution of (2.1).

Example. A typical equation which cannot be treated in the frame of Carathéodory's approach to ordinary differential equations is the following.

Define $F(t) = t^2 \sin \frac{1}{t^2}$ for $t \neq 0$, $F(0) = 0$. Then the derivative

$$\dot{F}(t) = f(t) = 2t \sin \tfrac{1}{t^2} - \tfrac{2}{t} \cos \tfrac{1}{t^2}, \quad t \neq 0,$$

$$\dot{F}(0) = f(0) = 0$$

exists. Since $f(t)$ is not Lebesgue integrable, for the differential equation

$$\dot{x} = f(t), \quad t \in [-1, 1]$$

the Carathéodory existence theorem does not hold. Nevertheless, $x(t) = F(t) + c$, where c is a constant, is evidently a Perron – Henstock solution of this equation. Even without knowing this, Theorem 2.10 applies to this case and gives the local existence of such a solution.

Another simple example based on this classical Perron integrable function f is the linear ordinary differential equation

$$\dot{x} = ax + f(t), \quad t \in [-1,1],\, x \in \mathbb{R}.$$

Its general solution has the form

$$x(t) = Ce^{at} + e^{at}\int_0^t e^{-as} f(s)\,ds, \quad t \in [-1,1]$$

where the integral on the right hand side exists when treated as the Perron integral.

2.12 *Remark.* In the book [23] the notion of a solution of the ordinary differential equation (2.1) was weakened to the following:

find a function $x : J \to \mathbb{R}^n$ defined on a nondegenerate interval $J \subset \mathbb{R}$ such that $(x(t),t)$ belongs to the domain G of the function f, x is continuous on J, almost everywhere differentiable in J and $\dot{x}(t) = f(x(t),t)$ for almost all $t \in J$.

This concept is more general than the concept of Carathéodory; absolute continuity is not required for the solution. In [23] it is pointed out that if x is a function satisfying

$$x(t) = x(\alpha) + (P)\int_\alpha^t f(x(s),s)\,ds$$

for $\alpha, t \in J$ then it is a solution of (2.1) in the above mentioned weakened sense. The fundamental existence and unicity results for such solutions of (2.1) are given in [23] for the case when f satisfies (H1) and (H2) from Theorem 2.2 provided the Lipschitz condition

$$\|f(x,t) - f(y,t)\| \le L(t)\|x - y\|$$

is satisfied locally in the domain of f with L integrable in the Lebesgue sense. The results in [23] concerning solutions in the weakened sense are reduced to the above mentioned case of solutions in the Perron–Henstock sense.

CHAPTER III

GENERALIZED ORDINARY DIFFERENTIAL EQUATIONS

Let an open set $G \subset \mathbb{R}^{n+1}$ be given. Assume that $F : G \to \mathbb{R}^{n+1}$ is a given $\mathbb{R}^{n+1}$–valued function $F(x,t)$ defined for $(x,t) \in G$, $x \in \mathbb{R}^n$, $t \in \mathbb{R}$.

3.1 Definition. A function $x : [\alpha, \beta] \to \mathbb{R}^n$ is called *a solution* of the *generalized ordinary differential equation*

$$\frac{dx}{d\tau} = DF(x,t) \tag{3.1}$$

on the interval $[\alpha, \beta] \subset \mathbb{R}$ if $(x(t),t) \in G$ for all $t \in [\alpha, \beta]$ and if

$$x(s_2) - x(s_1) = \int_{s_1}^{s_2} DF(x(\tau), t) \tag{3.2}$$

holds for every s_1, $s_2 \in [\alpha, \beta]$ the identity.

The integral on the right hand side of (3.2) has to be understood as the generalized Perron integral introduced by Definition 1.2n in Chapter I.

3.2 *Remark.* Let us mention that the notation (3.1) is symbolical only. The letter D indicates that (3.1) is a generalized differential equation, this concept being defined via its solutions. Even the symbol $\dfrac{dx}{d\tau}$ does not mean that the solution has a derivative.

For example if $r : [0,1] \to \mathbb{R}$ is a continuous function for which its derivative does not exist at any point of the interval $[0,1]$, then we can define $F(x,t) = r(t)$. In this case we evidently have

$$\int_{s_1}^{s_2} DF(x(\tau),t) = \int_{s_1}^{s_2} Dr(t) = r(s_2) - r(s_1)$$

by the definition of the integral, and this means that the function $x : [0,1] \to \mathbb{R}$ given by $x(s) = r(s)$ for $s \in [0,1]$ is a solution of the generalized differential equation $\dfrac{dx}{d\tau} = DF(x,t) = Dr(t)$ though it has no derivative at any point in $[0,1]$.

From this point of view a generalized ordinary differential equation is a formal equation–like object for which we have defined its solutions. The definition of the solution given by the integral relation (3.2) reminds us of the concept of a solution of an ordinary differential equation given for example in Chap. II .

3.3 *Remark.* Assuming that $F(x,\tau,t)$ is defined for $(x,\tau,t) \in H \subset \mathbb{R}^{n+2}$, where $H \subset \mathbb{R}^{n+2}$ is such that if $(x,s,s) \in H$ then there is a $\theta > 0$ such that $(x,s,t) \in H$ whenever $|t-s| < \theta$, we can introduce the solution of the generalized differential equation

$$\frac{dx}{d\tau} = DF(x,\tau,t) \tag{3.3}$$

similarly as in Definition 3.1.

3.4 Definition. A function $x : [\alpha,\beta] \to \mathbb{R}^n$ is called *a solution* of the *generalized ordinary differential equation* (3.3) on the interval $[\alpha,\beta] \subset \mathbb{R}$ if $(x(t),t,t) \in H$ for all $t \in [\alpha,\beta]$ and if for every s_1, $s_2 \in [\alpha,\beta]$ the identity

$$x(s_2) - x(s_1) = \int_{s_1}^{s_2} DF(x(\tau),\tau,t)$$

holds.

3.5 Proposition. *If a function $x : [\alpha, \beta] \to \mathbb{R}^n$ is a solution of the generalized ordinary differential equation (3.1) on $[\alpha, \beta]$ then for every fixed $\gamma \in [\alpha, \beta]$ we have*

$$x(s) = x(\gamma) + \int_{\gamma}^{s} DF(x(\tau), t), \quad s \in [\alpha, \beta]. \tag{3.4}$$

If a function $x : [\alpha, \beta] \to \mathbb{R}^n$ satisfies the integral equation (3.4) for some $\gamma \in [\alpha, \beta]$ and all $s \in [\alpha, \beta]$ then x is a solution of the generalized ordinary differential equation (3.1).

Proof. The first statement follows directly from the definition of a solution of (3.1) when we put $s_1 = \gamma$ and $s_2 = s$.

If $x : [\alpha, \beta] \to \mathbb{R}^n$ satisfies the integral equation (3.4) then by the additivity of the integral we have

$$x(s_2) - x(s_1) =$$

$$= x(\gamma) + \int_{\gamma}^{s_2} DF(x(\tau), t) - x(\gamma) - \int_{\gamma}^{s_1} DF(x(\tau), t) =$$

$$= \int_{s_1}^{s_2} DF(x(\tau), t)$$

for every $s_1, s_2 \in [\alpha, \beta]$, and x is a solution of (3.1). □

Definition 3.1 itself does not provide any information about the properties of the function $x : [\alpha, \beta] \to \mathbb{R}^n$ which is a solution of (3.1) in the sense of Definition 3.1. The only fact we know is that in this case the integral $\int_{s_1}^{s_2} DF(x(\tau), t)$ exists for every $s_1, s_2 \in [\alpha, \beta]$. By Proposition 3.5 we know that a solution x of (3.1) behaves like the indefinite generalized Perron integral of a function integrable in this sense. Using Theorem 1.16 from Chap. I the following result can be deduced.

3.6 Proposition. *If $x : [\alpha, \beta] \to \mathbb{R}^n$ is a solution of the generalized ordinary differential equation (3.1) on $[\alpha, \beta]$ then*

$$\lim_{s\to\sigma} [x(s) - F(x(\sigma), s) + F(x(\sigma), \sigma)] = x(\sigma) \tag{3.5}$$

for every $\sigma \in [\alpha, \beta]$.

Proof. Let $\sigma \in [\alpha, \beta]$ be fixed. Then by Proposition 3.5 we have

$$x(s) - \int_\sigma^s DF(x(\tau), t) = x(\sigma)$$

and therefore also

$$x(s) - F(x(\sigma), s) + F(x(\sigma), \sigma) -$$

$$- \int_\sigma^s DF(x(\tau), t) + F(x(\sigma), s) - F(x(\sigma), \sigma) - x(\sigma) = 0 \tag{3.6}$$

for every $s \in [\alpha, \beta]$.

By Theorem 1.16 we get

$$\lim_{s\to\sigma} \left[\int_\sigma^s DF(x(\tau), t) - F(x(\sigma), s) + F(x(\sigma), \sigma) \right] = 0$$

and this together with (3.6) yields the existence of the limit $\lim_{s\to\sigma} [x(s) - F(x(\sigma), s) + F(x(\sigma), \sigma) - x(\sigma)]$ as well as the relation

$$\lim_{s\to\sigma} [x(s) - F(x(\sigma), s) + F(x(\sigma), \sigma) - x(\sigma)] = 0$$

which is equivalent to (3.5). □

3.7 *Remark.* By virtue of Proposition 3.6 the following assertion is true.

If $x : [\alpha, \beta] \to \mathbb{R}^n$ is a solution of (3.1) then for every $\sigma \in [\alpha, \beta]$ the value of $x(s)$ can be approximated by $x(\sigma) + F(x(\sigma), s) - F(x(\sigma), \sigma)$ provided $s \in [\alpha, \beta]$ is sufficiently close to σ.

Now we introduce a class of functions $F : G \to \mathbb{R}^n$ for which it is possible to get more specific information about the solutions of (3.1).

Given $c > 0$, we denote

$$B_c = \{x \in \mathbb{R}^n;\ \|x\| < c\}.$$

Let $(a, b) \subset \mathbb{R}$ be an interval with $-\infty < a < b < \infty$ and let us set

$$G = B_c \times (a, b).$$

We will use the set $G \subset \mathbb{R}^{n+1}$ in our subsequent study of generalized differential equations (3.1). Assume that $h : [a, b] \to \mathbb{R}$ is a nondecreasing function defined on $[a, b]$ and that $\omega : [0, +\infty) \to \mathbb{R}$ is a continuous, increasing function with $\omega(0) = 0$.

3.8 Definition. A function $F : G \to \mathbb{R}^n$ belongs to the class $\mathcal{F}(G, h, \omega)$ if

$$\|F(x, t_2) - F(x, t_1)\| \le |h(t_2) - h(t_1)| \tag{3.7}$$

for all (x, t_2), $(x, t_1) \in G$ and

$$\|F(x, t_2) - F(x, t_1) - F(y, t_2) + F(y, t_1)\| \le$$

$$\le \omega(\|x - y\|)|h(t_2) - h(t_1)| \tag{3.8}$$

for all (x, t_2), (x, t_1), (y, t_2), $(y, t_1) \in G$.

3.9 Lemma. *Assume that $F : G \to \mathbb{R}^n$ satisfies the condition (3.7). If $[\alpha, \beta] \subset [a, b]$ and $x : [\alpha, \beta] \to \mathbb{R}^n$ is such that $(x(t), t) \in G$ for every $t \in [\alpha, \beta]$ and if the integral $\int_\alpha^\beta DF(x(\tau), t)$ exists, then for every $s_1, s_2 \in [\alpha, \beta]$ the inequality*

$$\left\| \int_{s_1}^{s_2} DF(x(\tau), t) \right\| \leq |h(s_2) - h(s_1)| \tag{3.9}$$

is satisfied.

Proof. Using (3.7) we obtain

$$|t - \tau|.\|F(x(\tau), t) - F(x(\tau), \tau)\| \leq (t - \tau)(h(t) - h(\tau))$$

for any $\tau, t \in [\alpha, \beta]$. The integral $\int_\alpha^\beta dh(t)$ exists and

$$\int_{s_1}^{s_2} dh(t) = h(s_2) - h(s_1)$$

for every $s_1, s_2 \in [\alpha, \beta]$. Therefore (3.9) is an immediate consequence of Corollary 1.35. □

3.10 Lemma. *Assume that $F : G \to \mathbb{R}^n$ satisfies the condition (3.7). If $[\alpha, \beta] \subset [a, b]$ and $x : [\alpha, \beta] \to \mathbb{R}^n$ is a solution of (3.1), then the inequality*

$$\|x(s_1) - x(s_2)\| \leq |h(s_2) - h(s_1)| \tag{3.10}$$

holds for every $s_1, s_2 \in [\alpha, \beta]$.

Proof. The result follows directly from Lemma 3.9 if we take into account that by definition we have

$$x(s_2) - x(s_1) = \int_{s_1}^{s_2} DF(x(\tau), t)$$

for every $s_1, s_2 \in [\alpha, \beta]$. □

3.11 Corollary. *Assume that $F : G \to \mathbb{R}^n$ satisfies the condition (3.7). If $[\alpha,\beta] \subset [a,b]$ and $x : [\alpha,\beta] \to \mathbb{R}^n$ is a solution of (3.1) then x is of bounded variation on $[\alpha,\beta]$ and*

$$\mathrm{var}_{\alpha}^{\beta} \leq h(\beta) - h(\alpha) < +\infty. \tag{3.11}$$

Moreover, every point in $[\alpha,\beta]$ at which the function h is continuous is a continuity point of the solution $x : [\alpha,\beta] \to \mathbb{R}^n$.

Proof. Let $\alpha = s_0 < s_1 < \cdots < s_k = \beta$ be an arbitrary division of the interval $[\alpha,\beta]$. By (3.10) we then have

$$\sum_{j=1}^{k} \|x(s_j) - x(s_{j-1}\| \leq \sum_{j=1}^{k} |h(s_j) - h(s_{j-1}| < h(\beta) - h(\alpha).$$

Passing to the supremum over all divisions of $[\alpha,\beta]$ we obtain (3.11). The second statement is a consequence of the inequality (3.10). □

3.12 Lemma. *If $x : [\alpha,\beta] \to \mathbb{R}^n$ is a solution of (3.1) and $F : G \to \mathbb{R}^n$ satisfies the condition (3.7) then*

$$x(s+) - x(s) = \lim_{\sigma \to s+} x(\sigma) - x(s) = F(x(s), s+) - F(x(s), s) \tag{3.12}$$

for $s \in [\alpha,\beta)$ and

$$x(s) - x(s-) = x(s) - \lim_{\sigma \to s-} x(\sigma) = F(x(s), s) - F(x(s), s-) \tag{3.13}$$

for $s \in (\alpha,\beta]$ where $F(x, s+) = \lim_{\sigma \to s+} F(x,\sigma)$ for $s \in [\alpha,\beta)$ and $F(x, s-) = \lim_{\sigma \to s-} F(x,\sigma)$ for $s \in (\alpha,\beta]$.

Proof. First of all note that the limits $F(x, s+)$ and $F(x, s-)$ exist because the function h, being a function of bounded variation, possesses onesided limits at every point of $[\alpha,\beta] \subset [a,b]$ and

(3.7) quarantees the Bolzano–Cauchy condition for the existence of the corresponding onesided limits of the function $F(x,\sigma)$: $[\alpha,\beta] \to \mathbb{R}^n$ for every $x \in B_c$.

For $\alpha \le s < \sigma \le \beta$ we have

$$x(\sigma) - x(s) = \int_s^\sigma DF(x(\tau),t)$$

by the definition of a solution of (3.1). By Theorem 1.16 we have

$$\lim_{\sigma \to s+} x(\sigma) - x(s) = \lim_{\sigma \to s+} \int_s^\sigma DF(x(\tau),t) =$$

$$= \lim_{\sigma \to s+} [F(x(s),\sigma) - F(x(s),s)]$$

and (3.12) holds. The relation (3.13) can be proved analogously. □

3.13 *Remark.* The above statements describe the properties of solutions of the generalized differential equation (3.1) whenever the "right hand side" F satisfies (3.7) from the definition of the class $\mathcal{F}(G,h,\omega)$. Of course this are merely preliminary results because at this moment we have not the fundamental information about the existence of a solution. Nevertheless, we have a description of the "candidates" for a solution.

Now we present a result concerning the existence of the integral involved in the definition of the solution of the generalized differential equation (3.1).

3.14 Theorem. *Assume that $F \in \mathcal{F}(G,h,\omega)$ is given and that $x : [\alpha,\beta] \to \mathbb{R}^n$, $[\alpha,\beta] \subset (a,b)$ is the pointwise limit of a sequence $(x_k)_{k\in\mathbb{N}}$ of functions $x_k : [\alpha,\beta] \to \mathbb{R}^n$ such that $(x(s),s) \in G$,$(x_k(s),s) \in G$ for every $k \in \mathbb{N}$ and $s \in [\alpha,\beta]$ and* *that* $\int_\alpha^\beta DF(x_k(\tau),t)$ *exists for every* $k \in \mathbb{N}$. *Then the integral* $\int_\alpha^\beta DF(x(\tau),t)$ *exists and*

$$\int_\alpha^\beta DF(x(\tau),t) = \lim_{k\to\infty} \int_\alpha^\beta DF(x_k(\tau),t).$$

Proof. According to Theorem 1.6 it can be assumed without any loss of generality that F is a real valued function. Assume that $\varepsilon > 0$ is given. By (3.8) we have

$$|F(x_k(\tau),t_2) - F(x_k(\tau),t_1) - F(x(\tau),t_2) + F(x(\tau),t_1)| \leq$$

$$\leq \omega(\|x_k(\tau) - x(\tau)\|)(h(t_2) - h(t_1)) \tag{3.14}$$

for every $\tau \in [\alpha,\beta]$, $t_1 \leq \tau \leq t_2$, $[t_1,t_2] \subset [\alpha,\beta]$. Let us set

$$\mu(t) = \frac{\varepsilon}{h(\beta) - h(\alpha) + 1} h(t) \text{ for } t \in [\alpha,\beta].$$

The function $\mu : [\alpha,\beta] \to \mathbb{R}$ is nondecreasing and $\mu(\beta) - \mu(\alpha) < \varepsilon$. Since

$$\lim_{k\to\infty} x_k(\tau) = x(\tau)$$

for every $\tau \in [\alpha,\beta]$ and the function ω is continuous at 0, there is a $p(\tau) \in \mathbb{N}$ such that for $k \geq p(\tau)$ we have

$$\omega(\|x_k(\tau) - x(\tau)\|) \leq \frac{\varepsilon}{h(\beta) - h(\alpha) + 1},$$

i.e. for $k \geq p(\tau)$ the inequality (3.14) can be rewritten to the form

$$|F(x_k(\tau),t_2) - F(x_k(\tau),t_1) - F(x(\tau),t_2) + F(x(\tau),t_1)| \leq$$

$$\leq \mu(t_2) - \mu(t_1).$$

By (3.7) we have

$$-h(t_2) + h(t_1) \leq F(x_k(\tau), t_2) - F(x_k(\tau), t_1) \leq$$
$$\leq h(t_2) - h(t_1)$$

for every $\tau \in [\alpha, \beta]$, $k \in \mathbb{N}$ and $\tau \in [\alpha, \beta]$, $t_1 \leq \tau \leq t_2$, $[t_1, t_2] \subset [\alpha, \beta]$. The integrals $\int_\alpha^\beta D[h(t)] = h(\beta) - h(\alpha)$, $\int_\alpha^\beta D[-h(t)] = h(\alpha) - h(\beta)$ obviously exist. Therefore by Corollary 1.31 the integral $\int_\alpha^\beta DF(x(\tau), t)$ exists and the conclusion of the theorem holds. □

3.15 Corollary. *Assume that $F \in \mathcal{F}(G, h, \omega)$ is given and that $x : [\alpha, \beta] \to \mathbb{R}^n$, $[\alpha, \beta] \subset (a, b)$ is the pointwise limit of a sequence of finite step functions such that $(x(s), s) \in G$ for every $s \in [\alpha, \beta]$. Then the integral $\int_\alpha^\beta DF(x(\tau), t)$ exists.*

Proof. By the previous Theorem 3.14 it is sufficient to prove that the integral $\int_\alpha^\beta DF(\varphi(\tau), t)$ exists for every finite step function $\varphi : [\alpha, \beta] \to \mathbb{R}^n$. If φ is a finite step function then there is a division $\alpha = s_0 < s_1 < \cdots < s_k = \beta$ of $[\alpha, \beta]$ such that $\varphi(s) = c_j \in \mathbb{R}^n$ for $s \in (s_{j-1}, s_j)$, $j = 1, \ldots, k$ where c_j, $j = 1, \ldots, k$ are constants. Assume that $(\varphi(s), s) \in G$ for every $s \in [\alpha, \beta]$.

By the definition of the integral it is easy to see that if $s_{j-1} < \sigma_1 < \sigma_2 < s_j$ then the integral $\int_{\sigma_1}^{\sigma_2} DF(\varphi(\tau), t)$ exists and

$$\int_{\sigma_1}^{\sigma_2} DF(\varphi(\tau), t) = F(c_j, \sigma_2) - F(c_j, \sigma_1).$$

Assume that $\sigma_0 \in (s_{j-1}, s_j)$ is given. We have

$$\lim_{s \to s_{j-1}+} [\int_s^{\sigma_0} DF(\varphi(\tau), t) +$$

$$+F(\varphi(s_{j-1}),s)-F(\varphi(s_{j-1}),s_{j-1})]=$$

$$=\lim_{s\to s_{j-1}+}[F(c_j,\sigma_0)-F(c_j,s)+F(\varphi(s_{j-1}),s)-$$

$$-F(\varphi(s_{j-1}),s_{j-1})]=$$

$$=F(c_j,\sigma_0)-F(c_j,s_{j-1}+)+$$

$$+F(\varphi(s_{j-1}),s_{j-1}+)-F(\varphi(s_{j-1}),s_{j-1}).$$

Hence by Theorem 1.14 (see Remark 1.15) the integral $\int_{s_{j-1}}^{\sigma_0} DF(\varphi(\tau),t)$ exists and equals the above computed limit.

Similarly it can be shown that also the integral $\int_{\sigma_0}^{s_j} DF(\varphi(\tau),t)$ exists and

$$\int_{\sigma_0}^{s_j} DF(\varphi(\tau),t)=F(c_j,s_j-)-F(c_j,\sigma_0)-$$

$$-F(\varphi(s_j),s_j-)+F(\varphi(s_j),s_j).$$

Therefore by Theorem 1.11 we obtain the existence of the integral $\int_{s_{j-1}}^{s_j} DF(\varphi(\tau),t)$ and

$$\int_{s_{j-1}}^{s_j} DF(\varphi(\tau),t)=$$

$$=F(c_j,s_j-)-F(c_j,\sigma_0)-F(\varphi(s_j),s_j-)+F(\varphi(s_j),s_j)+$$

$$+F(c_j,\sigma_0)-F(c_j,s_{j-1}+)+$$

$$+F(\varphi(s_{j-1}),s_{j-1}+)-F(\varphi(s_{j-1}),s_{j-1})=$$

$$=F(c_j,s_j-)-F(c_j,s_{j-1}+)+F(\varphi(s_{j-1}),s_{j-1}+)-$$

$$-F(\varphi(s_{j-1}),s_{j-1})-F(\varphi(s_j),s_j-)+F(\varphi(s_j),s_j).$$

Repeating this argument for every interval $[s_{j-1}, s_j]$, $j = 1, \dots, k$ and using again Theorem 1.11 we obtain the existence of the integral $\int_\alpha^\beta DF(\varphi(\tau), t)$ as well as the identity

$$\int_\alpha^\beta DF(\varphi(\tau), t) = \sum_{j=1}^{k} [F(c_j, s_j-) - F(c_j, s_{j-1}+)] +$$

$$+ \sum_{j=1}^{k} [F(\varphi(s_{j-1}), s_{j-1}+) - F(\varphi(s_{j-1}), s_{j-1}) -$$

$$- F(\varphi(s_j), s_j-) + F(\varphi(s_j), s_j)].$$

This completes the proof. □

3.16 Corollary. *If $F \in \mathcal{F}(G, h, \omega)$ is given and $x : [\alpha, \beta] \to \mathbb{R}^n$, $[\alpha, \beta] \subset (a, b)$ is a regulated function (in particular a function of bounded variation) on $[\alpha, \beta]$ such that $(x(s), s) \in G$ for every $s \in [\alpha, \beta]$ then the integral $\int_\alpha^\beta DF(x(\tau), t)$ exists.*

Proof. The result follows from Corollary 3.15 because every regulated function is the uniform limit of finite step functions (see e.g. [6]). □

3.17 *Remark.* Corollary 3.16 implies that the class of functions $x : [\alpha, \beta] \to \mathbb{R}^n$, $[\alpha, \beta] \subset (a, b)$ for which the integral $\int_\alpha^\beta DF(x(\tau), t)$ exists is sufficiently large provided $F \in \mathcal{F}(G, h, \omega)$. At least it is clear that $\int_\alpha^\beta DF(x(\tau), t)$ exists for every function which can be expected to be a solution of the generalized differential equation (3.1) with $F \in \mathcal{F}(G, h, \omega)$ (see Corollary 3.11).

3.18 *Remark.* In a more general setting it can be assumed that $G \subset \mathbb{R}^{n+1}$ is an open set and that the function $F : G \to \mathbb{R}^n$ locally satisfies the requirements given by the properties of the class $\mathcal{F}(G, h, \omega)$. More precisely, if $(\tilde{x}, \tilde{t}) \in G$ then there exists $\Delta > 0$ such that

$$\tilde{G} = \{x \in \mathbb{R}^n; \|x - \tilde{x}\| < \Delta\} \times (\tilde{t} - \Delta, \tilde{t} + \Delta) \subset G$$

and F satisfies on $\tilde{G}$ the conditons (3.7) and (3.8) from Definition 3.8 with some functions $\tilde{h}$ and $\tilde{\omega}$, and of course these functions can be specific for the set $\tilde{G}$.

CHAPTER IV

EXISTENCE AND UNIQUENESS OF SOLUTIONS OF GENERALIZED DIFFERENTIAL EQUATIONS

Let us consider the generalized ordinary differential equation

$$\frac{dx}{d\tau} = DF(x,t) \tag{4.1}$$

described in Definition 3.1 in the previous chapter for the case when $F : G \to \mathbb{R}^n$ belongs to the class $\mathcal{F}(G, h, \omega)$ (see Definition 3.8).

We assume here that $G = B_c \times (a,b)$ where $B_c = \{x \in \mathbb{R}^n; \|x\| < c\}$, $c > 0$, $(a,b) \subset \mathbb{R}$ $-\infty < a < b < +\infty$ and $h : [a,b] \to \mathbb{R}$ is a nondecreasing function defined and *continuous from the left* on $[a,b]$, $\omega : [0,+\infty) \to \mathbb{R}$ is a continuous, increasing function with $\omega(0) = 0$. Let us mention that in addition to the assumptions from Chap. III we now assume that the function h is continuous from the left. This assumption is made to avoid some technicalities which are not substantial for the problem itself.

Existence of solutions

Some of the fundamental properties of a solution of (4.1) are established in Lemma 3.10, Corollary 3.11 and Lemma 3.12. A solution of (4.1) has to be a function of bounded variation which in our case is continuous from the left and has discontinuities of the first kind given by Lemma 3.12, i.e. if for some $t_0 \in (a,b)$ the value of the solution x of (4.1) is $x(t_0) = \widetilde{x}$ then the right limit at the point t_0 fulfils

$$x(t_0+) = x(t_0) + F(x(t_0), t_0+) - F(x(t_0), t_0) =$$

$$= \widetilde{x} + F(\widetilde{x}, t_0+) - F(\widetilde{x}, t_0).$$

Because of the possible discontinuities of a solution it can happen that for some $\widetilde{x} \in B_c$, i.e. for some $(\widetilde{x}, t_0) \in G$, the value

$$\widetilde{x}+ = \widetilde{x} + F(\widetilde{x}, t_0+) - F(\widetilde{x}, t_0)$$

does not belong to B_c and this means that the corresponding solution x with $x(t_0) = \widetilde{x}$ jumps off the open set B_c at the moment t_0 and cannot be continued for $t > t_0$.

Therefore to prove a local existence theorem for a solution of (4.1) satisfying the initial condition $x(t_0) = \widetilde{x}$ we make the quite natural assumption

$$\widetilde{x}+ = \widetilde{x} + F(\widetilde{x}, t_0+) - F(\widetilde{x}, t_0) \in B_c. \tag{4.2}$$

4.1 *Remark.* The proof of the main result will be based on the Schauder–Tichonov fixed point theorem. We use the usual notation $BV[\alpha, \beta] = BV$ for the space of functions of bounded variation on the interval $[\alpha, \beta]$ with the usual norm given by

$$\|x\|_{BV} = \|x(\alpha)\| + \operatorname{var}_\alpha^\beta x$$

for $x : [\alpha, \beta] \to \mathbb{R}^n$. The classical monograph [21] is the basic reference for the space BV as well as for all fundamental facts from functional analysis used in this text.

4.2 Theorem. *Let $F : G \to \mathbb{R}^n$ belongs to the class $\mathcal{F}(G,h,\omega)$ and let $(\widetilde{x},t_0) \in G$ be such that (4.2) is satisfied. Then there exist $\Delta^-,\Delta^+ > 0$ such that on the interval $[t_0-\Delta^-,t_0+\Delta^+]$ there exists a solution $x : [t_0-\Delta^-,t_0+\Delta^+] \to \mathbb{R}^n$ of the generalized ordinary differential equation (4.2) for which $x(t_0) = \widetilde{x}$.*

Proof. The Schauder–Tichonov fixed point theorem will be used for proving this result (see [21]).

Since B_c is open there is a $\Delta^- > 0$ such that if $t \in [t_0-\Delta^-,t_0]$ and $x \in \mathbb{R}^n$ is such that $\|x-\widetilde{x}\| < |h(t)-h(t_0)|$ then $(x,t) \in G = B_c \times (a,b)$, and similarly there exists a $\Delta^+ > 0$ such that if $t \in [t_0,t_0+\Delta^+]$ and $x \in \mathbb{R}^n$ is such that $\|x-\widetilde{x}+\| < |h(t)-h(t_0+)|$ then $(x,t) \in G = B_c \times (a,b)$. This follows easily from (4.2) and from the properties of the function h.

Denote by $\mathcal{A}$ the set of all functions $z : [t_0-\Delta^-,t_0+\Delta^+] \to \mathbb{R}^n$ such that $z \in BV[t_0-\Delta^-,t_0+\Delta^+]$, $\|z(t)-\widetilde{x}\| < |h(t)-h(t_0)|$ for $t \in [t_0-\Delta^-,t_0)$ and $\|z(t)-\widetilde{x}+\| < |h(t)-h(t_0+)|$ for $t \in [t_0,t_0+\Delta^+]$.

It is easy to see that if $z \in \mathcal{A}$ then $z(t_0) = \widetilde{x}$ and $z(t_0+) = \widetilde{x}+$ by virtue of the inequalities in the definition of $\mathcal{A}$. It is also easy to check that the set $\mathcal{A} \subset BV[t_0-\Delta^-,t_0+\Delta^+]$ is convex, i.e. if $z_1,z_2 \in \mathcal{A}$, $\alpha \in [0,1]$ then $\alpha z_1 + (1-\alpha)z_2 \in \mathcal{A}$.

Next, let us show that $\mathcal{A}$ is a closed subset of $BV[t_0-\Delta^-,t_0+\Delta^+] = BV$. Let $z_k \in \mathcal{A}$, $k \in \mathbb{N}$ be a sequence which converges in BV to a function z. Since

$$\|z_k(t)-z(t)\| \le \|z_k-z\|_{BV} \tag{4.3}$$

for every $t \in [t_0-\Delta^-,t_0+\Delta^+]$, we have

$$\lim_{k\to\infty} \|z_k(t)-z(t)\| = 0$$

uniformly for $t \in [t_0-\Delta^-,t_0+\Delta^+]$. Therefore we have

$$\|z(t)-\widetilde{x}+\| \le \|z_k(t)-z(t)\| + \|z_k(t)-\widetilde{x}+\| < \varepsilon + |h(t)-h(t_0+)|$$

for any $\varepsilon > 0$ whenever $k \in \mathbb{N}$ is sufficiently large and $t \in [t_0, t_0 + \Delta^+]$. This yields

$$\|z(t) - \widetilde{x}+\| \leq h(t) - h(t_0+)$$

for $t \in [t_0, t_0 + \Delta^+]$. Similarly we can show that

$$\|z(t) - \widetilde{x}\| < h(t_0) - h(t)$$

for $t \in [t_0 - \Delta^-, t_0]$ and therefore for the limit z we have $z \in \mathcal{A}$ and $\mathcal{A}$ is closed.

For a given $z \in \mathcal{A}$ define the map

$$Tz(s) = \widetilde{x} + \int_{t_0}^{s} DF(z(\tau), t), \quad s \in [t_0 - \Delta^-, t_0 + \Delta^+]. \tag{4.4}$$

The map T is well–defined because by Corollary 3.16 the integral $\int_{t_0}^{s} DF(z(\tau), t)$ exists for every $s \in [t_0 - \Delta^-, t_0 + \Delta^+]$.

Using Lemma 3.9 and Theorem 1.14 (Remark 1.15) we get

$$\|Tz(s) - \widetilde{x}+\| = \|\widetilde{x} + \int_{t_0}^{s} DF(z(\tau), t) - \widetilde{x}\| =$$

$$= \| \int_{t_0}^{s} DF(z(\tau), t) - [F(\widetilde{x}, t_0+) - F(\widetilde{x}, t_0)]\| =$$

$$= \| \int_{t_0}^{s} DF(z(\tau), t) - [F(z(t_0), t_0+) - F(z(t_0), t_0+)]\| =$$

$$= \lim_{r \to t_0+} \| \int_{r}^{s} DF(z(\tau), t)\| \leq \lim_{r \to t_0+} |h(s) - h(r)| = h(s) - h(t_0+)$$

for $s \in [t_0, t_0 + \Delta^+]$. Analogously we have also

$$\|Tz(s) - \widetilde{x}\| \leq h(t_0) - h(s)$$

for every $s \in [t_0 - \Delta^-, t_0]$. Hence $Tz \subset \mathcal{A}$ for $z \in \mathcal{A}$, i.e. T maps the set $\mathcal{A}$ into itself.

Let us show that $T : \mathcal{A} \to \mathcal{A}$ is continuous. If $z, v \in \mathcal{A}$ then

$$\|Tz - Tv\|_{BV} =$$

$$= \|Tz(t_0 - \Delta^-) - Tv(t_0 - \Delta^-)\| + \operatorname{var}_{t_0-\Delta^-}^{t_0+\Delta^+}(Tz - Tv) \leq$$

$$\leq \|Tz(t_0) - Tv(t_0)\| + 2\operatorname{var}_{t_0-\Delta^-}^{t_0+\Delta^+}(Tz - Tv) =$$

$$= 2\operatorname{var}_{t_0-\Delta^-}^{t_0+\Delta^+}(Tz - Tv). \tag{4.5}$$

Further, by (3.8) from Definition 3.8 and by Theorem 1.35 we also obtain

$$\|Tz(t_2) - Tv(t_2) - Tz(t_1) + Tv(t_1)\| =$$

$$= \|\int_{t_1}^{t_2} D[F(z(\tau), t) - F(v(\tau), t)]\| \leq$$

$$\leq \int_{t_1}^{t_2} D[\omega(\|z(\tau) - v(\tau)\|)h(t)]$$

for $t_1, t_2 \in [t_0 - \Delta^-, t_0 + \Delta^+]$ and therefore

$$\operatorname{var}_{t_0-\Delta^-}^{t_0+\Delta^+}(Tz - Tv) \leq \int_{t_0-\Delta^-}^{t_0+\Delta^+} D[\omega(\|z(\tau) - v(\tau)\|)h(t)]. \tag{4.6}$$

Assume that $z, z_k \in \mathcal{A}$, $k \in \mathbb{N}$ and $\lim_{k\to\infty} \|z_k - z\|_{BV}$. Then by (4.3)

$$\lim_{k\to\infty} \|z_k(t) - z(t)\| = 0$$

uniformly for $t \in [t_0 - \Delta^-, t_0 + \Delta^+]$ and by Corollary 1.32 of the convergence theorem 1.29 we obtain

$$\lim_{k\to\infty} \int_{t_0-\Delta^-}^{t_0+\Delta^+} D[\omega(\|z_k(\tau) - z(\tau)\|)h(t)] = 0.$$

Therefore by (4.6)

$$\lim_{k\to\infty} \operatorname{var}_{t_0-\Delta^-}^{t_0+\Delta^+}(Tz_k - Tz) = 0$$

and (4.5) yields $\lim_{k\to\infty} \|Tz_k - Tz\|_{BV} = 0$, i.e. T is a continuous map.

Finally we show that $T(\mathcal{A}) \subset \mathcal{A}$ is sequentially compact in the Banach space $BV[t_0 - \Delta^-, t_0 + \Delta^+]$. Let $z_k \in \mathcal{A}$, $k \in \mathbb{N}$ be an arbitrary sequence in $\mathcal{A}$. The sequence (z_k) consists of equally bounded functions of equibounded variation and therefore Helly's Choice Theorem (see e.g. [104] or [21]) yields that this sequence contains a pointwise convergent subsequence which we again denote by (z_k). Hence we have $\lim_{k\to\infty} z_k(\tau) = z(\tau)$ for every $\tau \in [t_0 - \Delta^-, t_0 + \Delta^+]$, the values of z belong to B_c and $z \in BV[t_0 - \Delta^-, t_0 + \Delta^+]$. Put

$$y(s) = Tz(s) = \widetilde{x} + \int_{t_0}^{s} DF(z(\tau), t)$$

for $s \in [t_0 - \Delta^-, t_0 + \Delta^+]$. By Lemma 3.9 we have $y \in BV[t_0 - \Delta^-, t_0 + \Delta^+]$ and it is not difficult to show that

$$\lim_{k\to\infty} \|Tz_k - y\|_{BV[t_0-\Delta^-,t_0+\Delta^+]} = 0.$$

This immediately leads to the conclusion that every sequence in $T(\mathcal{A})$ contains a subsequence which converges in $BV[t_0 - \Delta^-, t_0 + \Delta^+]$ and consequently, $T(\mathcal{A})$ is sequentially compact.

All assumptions of the Schauder–Tichonov fixed point theorem being satisfied we can conclude that there exists at least one $x \in \mathcal{A}$ such that $x = Tx$, i.e.

$$x(s) = Tx(s) = \widetilde{x} + \int_{t_0}^{s} DF(x(\tau), t), \quad s \in [t_0 - \Delta^-, t_0 + \Delta^+],$$

and this implies (by the definition) that x is a solution of (4.1) on $[t_0 - \Delta^-, t_0 + \Delta^+]$, which concludes the proof. □

4.3 *Remark.* Theorem 4.2 represents the fundamental *local existence* result of a solution of the generalized differential equation (4.1) with a given initial value $\widetilde{x}$. It is easy to see that the proof can be repeated also for the general case when the left continuity of the function h is not assumed. In this case we define

$$\widetilde{x}- = \widetilde{x} + F(\widetilde{x}, t_0-) - F(\widetilde{x}, t_0) \in B_c$$

and set $\Delta^- > 0$ such that for $t \in [t_0 - \Delta^-, t_0]$ and $x \in \mathbb{R}^n$ with $\|x - \widetilde{x} - \| \leq |h(t) - h(t_0-)|$ we have $(x,t) \in G = B_c \times (a,b)$. Then also the set $\mathcal{A}$ in the proof Theorem 4.2 has to be modified as the set of all functions $z : [t_0 - \Delta^-, t_0 + \Delta^+] \to \mathbb{R}^n$ such that $z \in BV[t_0 - \Delta^-, t_0 + \Delta^+]$, $z(t_0) = \widetilde{x}$ $\|z(t) - \widetilde{x} - \| < |h(t) - h(t_0-)|$ for $t \in [t_0 - \Delta^-, t_0)$ and $\|z(t) - \widetilde{x} + \| < |h(t) - h(t_0+)|$ for $t \in (t_0, t_0 + \Delta^+]$.

A theorem similar to our Theorem 4.2 was proved in [69 , Theorem 2.1] and [70] by J. Kurzweil for increasing values of t, i.e. for $t \in [t_0, t_0 + \Delta^+]$ by the method of successive approximations with shifted arguments (see also Chap. II).

As we have mentioned in Remark 3.18 it can be assumed that $G \subset \mathbb{R}^{n+1}$ is an open set and that the function $F : G \to \mathbb{R}^n$ locally satisfies the requirements following from the properties of the class $\mathcal{F}(G, h, \omega)$, i.e. if $(\widehat{x}, \widehat{t}) \in G$ then there exists $\Delta > 0$ such that

$$\widehat{G} = \{x \in \mathbb{R}^n; \|x - \widehat{x}\| < \Delta\} \times (\widehat{t} - \Delta, \widehat{t} + \Delta) \subset G,$$

F satisfies on $\widehat{G}$ (3.7) and (3.8) from Definition 3.8 with some functions $\widehat{h}$ and $\widehat{\omega}$ and these functions can be specific for the set $\widehat{G}$. It is clear that if we are looking for a local existence result and if we require (4.2) then we can use Theorem 4.2 without any essential changes for showing that a solution of (4.2) locally exists. It is evident that if $x : [t_0 - \Delta^-, t_0 + \Delta^+] \to \mathbb{R}^n$ is a solution of (4.1) on $[t_0 - \Delta^-, t_0 + \Delta^+]$ given by Theorem 4.2 and if its value at the right endpoint of this interval satisfies the condition (4.2), i.e. if

$$x((t_0 + \Delta^+)+) =$$

$$= x(t_0+\Delta^+)+F(x(t_0+\Delta^+), (t_0+\Delta^+)+)-F(x(t_0+\Delta^+), t_0+\Delta^+)$$

and $(x((t_0 + \Delta^+)+), t_0 + \Delta^+) \in G$ then the local existence result can be applied to this point for continuing the solution to the right.

For this reason it is useful to introduce the set

$$G_F = \{(x,t) \in G;\ (x + F(x,t+) - F(x,t), t) \in G\}.$$

Then we can say e.g. that if $(x(t_0 + \Delta^+), t_0 + \Delta^+) \in G_F$ then the solution x can be continued to the right.

The background for the continuation of a solution of (4.1) is given by the following result.

4.4 Lemma. *If $x : [\alpha, \beta] \to \mathbb{R}^n$ and $y : [\beta, \gamma] \to \mathbb{R}^n$ are solutions of the equation (4.1) on $[\alpha, \beta]$ and $[\beta, \gamma]$, respectively, where $[\alpha, \gamma] \subset (a, b)$ and $x(\beta) = y(\beta)$ then the function $z : [\alpha, \gamma] \to \mathbb{R}^n$ defined by the relations $z(s) = x(s)$ for $s \in [\alpha, \beta]$, $z(s) = y(s)$ for $s \in [\beta, \gamma]$ is a solution of the equation (4.1).*

Proof. For $s_1, s_2 \in [\alpha, \gamma]$, $s_1 < \beta < s_2$ we have by Theorem 1.11

$$z(s_2) - z(s_1) = z(s_2) - z(\beta) + z(\beta) - z(s_1) =$$

$$= y(s_2) - y(\beta) + x(\beta) - x(s_1) =$$

$$= \int_\beta^{s_2} DF(y(\tau), t) + \int_{s_1}^\beta DF(x(\tau), t) =$$

$$= \int_{s_1}^{s_2} DF(z(\tau), t)$$

and this evidently yields the result. □

4.5 *Remark.* If $x : [t_0, t_1] \to \mathbb{R}^n$ is a solution of (4.1) on $[t_0, t_1]$ such that $(x(s), s) \in G_F$ for every $s \in [t_0, t_1]$, then it can be *continued to the right (for values $s > t_1$)* to a solution of (4.1). This procedure is well-known for the case of ordinary differential equations. In the present situation we can say that if $x : [t_0, t_1] \to \mathbb{R}^n$ is a solution of (4.1) which cannot be continued to the right, then either $(x(t_1), t_1) \notin G_F$ or $t_1 = b$.

A different situation occurs when we wish to continue a solution to the left. Given a solution $x : (t_1, t_0] \to \mathbb{R}^n$ of (4.1) on the interval $(t_1, t_0] \subset (a, b)$, $t_1 < t_0$, it can be continued to the left to the value t_1 if and only if there exists $x^\star$ such that

$$x(t_1+) = x^\star + F(x^\star, t_1+) - F(x^\star, t_1),$$

i.e. if the value $x(t_1+) = \lim_{s \to t_1+} x(s)$ belongs to the range of the map

$$x \in B_c \mapsto x + F(x, t_1+) - F(x, t_1) \in \mathbb{R}^n.$$

It is easy to see that this is true when the function h is continuous at the point t_1.

4.6 *Example.* Let us set

$$G = \{(x, t);\ |x| < 1, |t| < 1\} \subset \mathbb{R}^2$$

and define

$$F(x, t) = x \text{ for } |x| < 1,\ -1 < t \leq 0,$$

$$F(x, t) = 0 \text{ for } |x| < 1,\ 0 < t < 1.$$

Then $F \in \mathcal{F}(G, h, \omega)$ where $h(t) = 0$ for $-1 \leq t \leq 0$, $h(t) = 1$ for $0 < t \leq 1$, $\omega(r) = r$, $r \geq 0$.

For $t \neq 0$ we have $F(x, t+) - F(x, t) = 0$ and $F(x, 0+) - F(x, 0) = -x$. Therefore $x + F(x, 0+) - F(x, 0) = x - x = 0 \in B_1$, i.e. $G_F = G$ and every solution of the equation

$$\frac{dx}{d\tau} = DF(x, t) \tag{4.1}$$

with $x(0) = x$ jumps to the point 0 and satisfies $x(t) = 0$ for $t \in (0, 1)$.

It is not difficult to verify that if $(x^*, t_0) \in G$, $t_0 \leq 0$ then the solution of (4.1) with $x(t_0) = x^*$ is given by $x(t) = x^*$ for $-1 < t \leq 0$, $x(t) = 0$ for $0 < t < 1$. If $(x^*, t_0) \in G$, $t_0 > 0$ and $x^+ \neq 0$ then the function $x(t) = x^*$ is a solution of (4.1) with $x(t_0) = x^*$ on $(0, 1)$ and cannot be continued for $t \leq 0$ because the range of the mapping

$$x \in B_1 \mapsto x + F(x, 0+) - F(x, 0) = 0$$

consists of the point $0 \in \mathbb{R}$ and $x^* = \lim_{t \to 0+} x(t)$ is beyond of this range.

If $x^* = 0$, $t_0 > 0$ then $x(t) = 0$ is a solution of (4.1) on the whole interval $(-1, 1)$.

Uniqueness of solutions

4.7 Definition. A solution $x : [t_0, t_0 + \eta] \to \mathbb{R}^n$ of the generalized differential equation (4.1) is called *locally unique for increasing values of t (locally unique in the future)* if for any solution $y : [t_0, t_0 + \sigma] \to \mathbb{R}^n$ of (4.1) with $y(t_0) = x(t_0)$ there exists $\eta_1 > 0$ such that $x(t) = y(t)$ for $t \in [t_0, t_0 + \eta] \cap [t_0, t_0 + \sigma] \cap [t_0, t_0 + \eta_1]$.

A point $(\widetilde{x}, t_0) \in G$ is called a *point of local uniqueness in the future for the equation* (4.1) if every solution x of (4.1) such that $x(t_0) = \widetilde{x}$ is locally unique in the future.

We say that *the equation* (4.1) *has the local uniqueness property* in the future if every point $(\widetilde{x}, t_0) \in G$ is a point of local uniqueness in the future for the equation (4.1).

Similarly the *local uniqueness for decreasing values of t (local uniqueness in the past)* can be defined for a solution of (4.1).

4.8 Theorem. *Let us assume that $F \in \mathcal{F}(G, h, \omega)$ where the function h is nondecreasing and continuous from the left, $\omega : [0, +\infty) \to \mathbb{R}$ is continuous, nondecreasing, $\omega(r) > 0$ for $r > 0$, $\omega(0) = 0$ and*

$$\lim_{v \to 0+} \int_v^u \frac{1}{\omega(r)}\, dr = +\infty \tag{4.7}$$

for every $u > 0$. Then every solution x of (4.1) such that $x(t_0) = \widetilde{x}$ where $(\widetilde{x}, t_0) \in G_F$ is locally unique for increasing values of t.

Proof. Assume that $x, y : [t_0, t_0 + \eta] \to \mathbb{R}^n$ are solutions of (4.1) such that $x(t_0) = y(t_0) = \widetilde{x}$. Then

$$\|x(s) - y(s)\| = \| \int_{t_0}^{s} D[F(x(\tau), t) - F(y(\tau), t)]\| \leq$$

$$\leq \int_{t_0}^{s} \omega(\|x(\tau) - y(\tau)\|)\, dh(\tau) =$$

$$= \int_{t_0}^{t_0+\delta} \omega(\|x(\tau) - y(\tau)\|)\, dh(\tau) + \int_{t_0+\delta}^{s} \omega(\|x(\tau) - y(\tau)\|)\, dh(\tau)$$

where $0 < \delta < s - t_0$. We have

$$\int_{t_0}^{t_0+\delta} \omega(\|x(\tau) - y(\tau)\|)\, dh(\tau) =$$

$$= \omega(\|x(t_0) - y(t_0)\|)[h(t_0+) - h(t_0)] +$$

$$+ \lim_{t_1 \to t_0+} \int_{t_1}^{t_0+\delta} \omega(\|x(\tau) - y(\tau)\|)\, dh(\tau \leq$$

$$\leq \sup_{\tau \in (t_0, t_0+\delta]} \omega(\|x(\tau) - y(\tau)\|)[h(t_0+\delta) - h(t_0+)] = A(\delta)$$

because $\omega(\|x(t_0) - y(t_0)\|) = \omega(0) = 0$. Since the limit $h(t_0+)$ exists we have also $\lim_{\delta \to 0+} A(\delta) = 0$. Therefore

$$\|x(s) - y(s)\| \leq A(\delta) + \int_{t_0+\delta}^{s} \omega(\|x(\tau) - y(\tau)\|)\, dh(\tau)$$

for $s \in [t_0+\delta, t_0+\eta]$.

Take $u_0 > 0$ and set

$$\Omega(u) = \int_{u_0}^{u} \frac{1}{\omega(r)}\, dr.$$

Using Theorem 1.40 we get

$$\|x(s) - y(s)\| \leq \Omega^{-1}(\Omega(A(\delta)) + h(s) - h(t_0+\delta)) \tag{4.8}$$

for $s \in [t_0+\delta, t_0+\eta]$ provided $\Omega(A(\delta)) + h(t_0+\eta) - h(t_0+\delta) < \beta$ where $\beta = \lim_{u \to +\infty} \Omega(u) \leq +\infty$.

Evidently, we have

$$\Omega(A(\delta)) + h(t_0+\eta) - h(t_0+\delta) \leq \Omega(A(\delta)) + h(t_0+\eta) - h(t_0+)$$

and because $\lim_{\delta \to 0+} A(\delta) = 0$ and $\lim_{u \to 0+} \Omega(u) = -\infty$ we obtain

$$\lim_{\delta \to 0+} \Omega(A(\delta)) + h(t_0+\eta) - h(t_0+) = -\infty.$$

Hence there is a $\delta_0 > 0$ such that for $\delta \in (0, \delta_0)$ the inequality

$$\Omega(A(\delta)) + h(t_0+\eta) - h(t_0+) < \beta$$

holds. Applying now the map Ω to both sides of (4.8) we obtain

$$\Omega(\|x(s) - y(s)\|) \leq \Omega(A(\delta)) + h(s) - h(t_0+\delta),$$

and this yields

$$\Omega(\|x(s)-y(s)\|)-\Omega(A(\delta))\le h(s)-h(t_0+\delta)\le h(s)-h(t_0+).$$

From the definition of Ω we therefore have

$$\int_{A(\delta)}^{\|x(s)-y(s)\|}\frac{1}{\omega(r)}\,dr\le h(t_0+\eta)-h(t_0+)$$

for $s\in[t_0+\delta,t_0+\eta]$ and $\delta\in(0.\delta_0)$.

Assume now that $\|x(s^*)-y(s^*)\|=k>0$ for some $s^*\in(t_0,t_0+\eta]$. Then

$$\int_{A(\delta)}^{k}\frac{1}{\omega(r)}\,dr\le h(t_0+\eta)-h(t_0+)<+\infty$$

for every $\delta\in(0,\delta_0)$ such that $\delta<s^*-t_0$ (i.e. $s^*>t_0+\delta$). Now it is possible to take $\delta\to 0+$ for obtaining the inequality

$$\lim_{\delta\to 0+}\int_{A(\delta)}^{k}\frac{1}{\omega(r)}\,dr\le h(t_0+\eta)-h(t_0+)<+\infty,$$

which contradicts the assumption on the function ω. Therefore $\|x(s)-y(s)\|=0$ for $s\in(t_0,t_0+\eta]$ and the result is proved. □

4.9 *Remark.* Theorem 4.8 represents an Osgood type uniqueness theorem for the case of local uniqueness for increasing values of t.

4.10 Corollary. *If $F\in\mathcal{F}(G,h,\omega)$ with $\omega(r)=Lr$, $r\ge 0$, $L>0$ then every solution of the generalized differential equation (4.1) starting from a point $(\widetilde{x},t_0)\in G_F$ is locally unique for increasing values of t.*

Proof. For $u>0$ we have evidently

$$\lim_{v\to 0+}\int_{v}^{u}\frac{1}{\omega(r)}\,dr=\lim_{v\to 0+}\frac{1}{L}\ln\frac{u}{v}=+\infty$$

and the condition of Theorem 4.8 is satisfied. □

The local uniqueness for increasing values of t can be extended to the global uniqueness for increasing values of t in the same manner as this is done for the case of classical ordinary differential equations.

4.11 Theorem. *Assume that $F \in \mathcal{F}(G,h,\omega)$ and that $x : [\alpha_1,\beta_1] \to \mathbb{R}^n$, $y : [\alpha_2,\beta_2] \to \mathbb{R}^n$ are two solutions of the generalized differential equation (4.1).*

If the condition (4.7) is satisfied and if $x(s) = y(s)$ for some $s \in [\alpha_1,\beta_1] \cap [\alpha_2,\beta_2]$, then $x(t) = y(t)$ for all $t \in [\alpha_1,\beta_1] \cap [\alpha_2,\beta_2] \cap [s,b)$.

Proof. The intersection $[\alpha_1,\beta_1] \cap [\alpha_2,\beta_2] \cap [s,b)$ is a closed interval of the form $[s,c]$, where $c < b$. Denote

$$M = \{t \in [s,c];\ x(\sigma) = y(\sigma) \text{ for } \sigma \in [s,t]\}.$$

If $s = c$ then there is nothing to prove. Assume therefore that $s < c$ and put $\beta = \sup M$. We evidently have $\beta \le c$.

Because the solutions x nd y are continuous from the left in virtue of the assumption that the function h is continuous, we have $[s,\beta] \subset M$ and our goal is to show that $\beta = c$. If we had $\beta < c$ then Theorem 4.8 could be used to show that there is $\eta > 0$ such that $x(\sigma) = y(\sigma)$ for all $\sigma \in [\beta,\beta+\eta]$ because $\beta \in M$ and $x(\beta) = y(\beta)$. This contradicts the definition of β and consequently $M = [s,c]$. The theorem is proved. □

Theorems 4.8 and 4.11 give conditions for local, and consequently also global uniqueness in the future. A completely different situation arises when we consider the question of uniqueness in the past, i.e. for decreasing values of the independent variable.

4.12 *Example.* Let $B = \{x \in \mathbb{R};\ |x| \le 1\}$, $[a,b] = [-1,1]$. For $(x,t) \in G = B \times [-1,1]$ define

$$F(x,t) = a.g(t).x$$

where $a > 0$, $g(t) = t$ for $t \leq 0$, $g(t) = t - 1$ for $t > 0$.

Then $x(t) = 0$, $t \in [-1, 1]$ is a solution of the generalized differential equation

$$\frac{dx}{d\tau} = DF(x, t),$$

and also the function

$$y(t) = e^{at}, \quad t \in [-1, 0]$$

$$y(t) = 0, \quad t \in (0, 1]$$

is a solution of this equation on the interval $[-1, 1]$. We have $x(1) = y(1)$ but for $t \in [-1, 0]$ the solutions x, y are different in spite of the fact that the assumptions of Corollary 4.10 are evidently satisfied. The problem of nonuniqueness for decreasing values of the independent variable is specific for generalized differential equations wich can have solutions with right hand side discontinuities.

Maximal solutions

Let us assume that $F \in \mathcal{F}(G, h, \omega)$. In the situation considered in this chapter where the function h describing the class $\mathcal{F}(G, h, \omega)$ is assumed to be continuous from the left, a solution of the generalized differential equation (4.1) can be in general continued – if at all – only to the right,that is for increasing values of the independent variable.

If the local uniqueness of a solution for increasing values of the independent variable is ensured (for example by the conditions on the function ω given in Theorem 4.8) then a unique "forward" maximal solution of (4.1) can be defined when an initial condition $x(t_0) = \widetilde{x}$ is prescribed for some $t_0 \in (a, b)$ and $\widetilde{x} \in B_c$.

It is clear that a maximal forward solution can be defined only if $(\widetilde{x}, t_0) \in G_F$, i.e. if

$$\widetilde{x} + F(\widetilde{x}, t_0+) - F(\widetilde{x}, t_0) \in B_c$$

because otherwise for the possible solution x it can happen that $x(t) \notin B_c$ for $t > t_0$ and this would contradict the definition of a solution.

Assume therefore that $G_f = G$, i.e.

$$x + F(x, t+) - F(x, t) \in B_c$$

for every $x \in B_c$, $t \in (a, b)$. This means that there are no points in G from which the solution of (4.1) can jump off the set B_c.

Let $x : [t_0, t_0 + \eta] \to \mathbb{R}^n$, $\eta > 0$ be a solution of (4.1) on $[t_0, t_0 + \eta]$. The solution $y : I \to \mathbb{R}^n$ of (4.1) where $I = [t_0, t_0 + \sigma]$ or $I = [t_0, t_0 + \sigma)$ $\sigma > 0$ is called a *prolongation* of x if $[t_0, t_0 + \eta] \subset I$ and $x(t) = y(t)$ for $t \in [t_0, t_0 + \eta]$. If $[t_0, t_0 + \eta] \subsetneqq I$, i.e. $\sigma > \eta$ then I is called a *proper prolongation* of x to the right.

If $(\widetilde{x}, t_0) \in G$ then a solution x of (4.1) with $x(t_0) = \widetilde{x}$ defined for $t \geq t_0$ is *maximal* if there is a value $b(\widetilde{x}, t_0) > t_0$ such that x exists on $[t_0, b(\widetilde{x}, t_0))$ and cannot be prolonged to a larger interval of the form $[t_0, \beta]$ where $\beta \geq b(\widetilde{x}, t_0)$, or in other words there is no proper prolongation to the right of the solution $x : [t_0, b(\widetilde{x}, t_0)) \to \mathbb{R}^n$ of (4.1).

Let us give a few results concerning maximal solutions of (4.1).

4.13 Proposition. *Let $F \in \mathcal{F}(G, h, \omega)$ and $(\widetilde{x}, t_0) \in G$. If the equation (4.1) has the local uniqueness property in the future then there exists an interval J with the left endpoint t_0 and a function $x : J \to \mathbb{R}^n$ such that $t_0 \in J$, $x(t_0) = \widetilde{x}$ and $x : J \to \mathbb{R}^n$ is a maximal solution of (4.1). The interval J and the function x are uniquely defined by the initial condition $x(t_0) = \widetilde{x}$ and the maximality property of the solution.*

Proof. Assume that $x^1 : J^1 \to \mathbb{R}^n$, $x^2 : J^2 \to \mathbb{R}^n$ are two maximal solutions of (4.1) with $x^1(t_0) = x^2(t_0) = \widetilde{x}$. The local uniqueness property implies $x^1(t) = x^2(t)$ for every $t \in J^1 \cap J^2 \cap [t_0, +\infty)$. Define $x(t) = x^1(t)$ for $t \in J^1$ and $x(t) = x^2(t)$ for $t \in J^2$. Then it is easy to show that $x : J^1 \cup J^2 \to \mathbb{R}^n$ is a solution of (4.1) on $x : J^1 \cup J^2$. Since the solutions x^1, x^2 are assumed to be maximal, we have $J^1 = J^2 = J$ and $x^1(t) = x^2(t)$ for $t \in J$. Hence the maximal solution x is unique.

Now let us show that the solution $x : J \to \mathbb{R}^n$ exists. Denote by $\mathcal{S}$ the set of all solutions $x : J_x \to \mathbb{R}^n$ of (4.1) with $x(t_0) = \widetilde{x}$ and the interval of definition J_x for which t_0 is the left endpoint of J_x. The set $\mathcal{S}$ is nonempty by the local existence of a solution given in Theorem 4.2. Define

$$J = \bigcap_{x \in \mathcal{S}} J_x.$$

If $t \in J_z \cap J_y$ where $y, z \in \mathcal{S}$ then $z(t) = y(t)$ by the assumption of the uniqueness. Hence if we define $x : J \to \mathbb{R}^n$ by the relation $x(t) = y(t)$ where $y \in \mathcal{S}$ and $t \in J_y$ we obtain a solution of (4.1) defined on J which satisfies the initial condition $x(t_0) = \widetilde{x}$. Looking at the definition of J we can immediately see that $x : J \to \mathbb{R}^n$ is a solution of (4.1). □

4.14 Proposition. *Let $F \in \mathcal{F}(G, h, \omega)$ and $(\widetilde{x}, t_0) \in G$. Assume that the equation (4.1) has the local uniqueness property in the future. Let $x : J \to \mathbb{R}^n$ be th maximal solution of (4.1) with $x(t_0) = \widetilde{x}$, where $t_0 \in J$ is the left endpoint of the interval J. Then $J = [t_0, \beta) \cap (a, b)$, $t_0 < \beta \leq +\infty$.*

Proof. It is clear that for the maximal interval J we have $J \in (a, b)$. Let $t^* \in J$. Take $y = x(t^*) \in B_c$. Theorem 4.2 yields the existence of a $\delta > 0$ such that on $[t^*, t^* + \delta]$ there is a solution $v : [t^*, t^* + \delta] \to \mathbb{R}^n$ of (4.1) such that $v(t^*) = y$. The point (y, t^*) is a point of local uniqueness in the future and therefore x is a prolongation of v and $[t^*, t^* + \delta] \subset J$. This means that relatively to (a,b) the interval J is open at its right endpoint and the statement holds true. $\square$

4.15 Proposition. *Let $F \in \mathcal{F}(G,h,\omega)$ and $(\widetilde{x},t_0) \in G$. Assume that the equation (4.1) has the local uniqueness property in the future. Let $x : [t_0,\beta) \to \mathbb{R}^n$ be a maximal solution of (4.1) and let $M \subset G = B_c \times (a,b) \subset \mathbb{R}^{n+1}$ be a compact set. Then there exists $c \in [t_0,\beta)$ such that $(x(t),t) \notin M$ for $t \in (c,\beta)$.*

Proof. Assume to the contrary that the statement does not hold. Then there is a sequence $t_k \in [t_0,\beta)$, $k \in \mathbb{N}$ such that $\lim_{k\to\infty} t_k = \beta$ and $(x(t_k),t_k) \in M$, $k \in \mathbb{N}$. Since M is assumed to be compact and $b < +\infty$, the sequence $(x(t_k),t_k)_{k\in\mathbb{N}}$ contains a convergent subsequence which we denote again by $(x(t_k),t_k)_{k\in\mathbb{N}}$. Then $\lim_{k\to\infty} x(t_k) = y$ and $(y,\beta) \in M \subset G$.

By Theorem 4.2 there exists a $\delta > 0$ such that on $[\beta,\beta+\delta]$ there is a solution v of (4.1) with $v(\beta) = y$. Define $u : [t_0,\beta+\delta] \to \mathbb{R}^n$ by

$$u(t) = x(t),\ t \in [t_0,\beta), \quad u(t) = v(t),\ t \in [\beta,\beta+\delta].$$

Now assume that $s_1 \in [t_0,\beta)$ and $s_2 \in [\beta,\beta+\delta]$. Then for $k \in \mathbb{N}$ sufficiently large we have $t_k \in (s_1,\beta)$ and

$$\int_{s_1}^{s_2} DF(u(\tau),t) = \int_{s_1}^{\beta} DF(u(\tau),t) + \int_{\beta}^{s_2} DF(u(\tau),t) =$$

$$= \int_{s_1}^{t_k} DF(u(\tau),t) + \int_{t_k}^{\beta} DF(u(\tau),t) + u(s_2) - u(\beta) =$$

$$= u(t_k) - u(s_1) + \int_{t_k}^{\beta} DF(u(\tau),t) + u(s_2) - u(\beta) =$$

$$= x(t_k) - u(s_1) + u(s_2) - y + \int_{t_k}^{\beta} DF(u(\tau),t). \tag{4.9}$$

By Lemma 3.9 we have

$$\| \int_{t_k}^{\beta} DF(u(\tau), t) \| \leq h(\beta) - h(t_k).$$

Since h is continuous from the left and $\lim_{k\to\infty} t_k = \beta$, $t_k \leq \beta$ we have

$$\lim_{k\to\infty} \int_{t_k}^{\beta} DF(u(\tau), t) = 0.$$

Using this and $\lim_{k\to\infty} x(t_k) = y$ we take $k \to \infty$ in (4.9) and obtain

$$\int_{s_1}^{s_2} DF(u(\tau), t) = u(s_2) - u(s_1).$$

For all other possible positions of $s_1, s_2 \in [t_0, \beta + \delta]$ we obtain the same relation directly from the definition of u. In this way we obtain that $u : [t_0, \beta + \delta] \to \mathbb{R}^n$ is a solution of (4.1) on $[t_0, \beta + \delta]$ which is evidently a proper prolongation of the solution x which is assumed to be maximal. This contradiction proves the result. □

CHAPTER V

GENERALIZED DIFFERENTIAL EQUATIONS AND OTHER CONCEPTS OF DIFFERENTIAL SYSTEMS

Assume that

$$\overline{B} = \{x \in \mathbb{R}^n;\ \|x\| \le c\}$$

for some $c > 0$ and that $[a, b] \subset \mathbb{R}$ is a bounded closed interval.

Let $\omega : [0, +\infty) \to \mathbb{R}$ be a continuous increasing function with $\omega(0) = 0$ (the function ω has the character of a modulus of continuity).

Let μ be a finite positive regular measure on $[a, b]$ (see [21]).

5.1 Definition. A function $g : \overline{B} \times [a, b] \to \mathbb{R}^n$ belongs to the class $\mathcal{C}(\overline{B} \times [a, b], \mu, \omega)$ if

$$g(x, \cdot) \text{ is measurable with respect to the measure } \mu, \tag{5.1}$$

there exists a μ–measurable function $m : [a, b] \to \mathbb{R}$ such that $\int_a^b m(s)\, d\mu < +\infty$ and

$$\|g(s, x\| \le m(s) \tag{5.2}$$

for $(x, s) \in \overline{B} \times [a, b]$,
there exists a μ–measurable function $l : [a, b] \to \mathbb{R}$ such that $\int_a^b l(s)\, d\mu < +\infty$ and

$$\|g(s, x) - g(s, y)\| \le l(s)\omega(\|x - y\|) \tag{5.3}$$

for $(x, s), (y, s) \in \overline{B} \times [a, b]$.

5.2 *Remark.* Integrability here, and also in the subsequent parts of this chapter, has to be understood as the Lebesgue–Stieltjes integrability with respect to the finite positive regular measure μ.

Define

$$G(x,t) = \int_{t_0}^{t} g(x,s)\,d\mu \tag{5.4}$$

for $(x,t) \in \overline{B} \times [a,b]$ where $t_0 \in [a,b]$ and $g \in \mathcal{C}(\overline{B} \times [a,b], \mu, \omega)$.

By (5.1) and (5.2) it is clear that the function $G : \overline{B} \times [a,b] \to \mathbb{R}^n$ is well defined by (5.4).

5.3 Lemma. *If a function $g : \overline{B} \times [a,b] \to \mathbb{R}^n$ satisfies (5.1) and (5.2) then for the function G given by (5.4) we have*

$$\|G(x,t_2) - G(x,t_1)\| \le \int_{t_1}^{t_2} m(s)\,d\mu \tag{5.5}$$

for every $x \in \overline{B}$ and $t_1, t_2 \in [a,b]$.

Proof. Since (5.2) holds, we have

$$\|G(x,t_2) - G(x,t_1)\| = \|\int_{t_1}^{t_2} g(x,s)\,d\mu\| \le$$

$$\le \int_{t_1}^{t_2} \|g(x,s)\|\,d\mu \le \int_{t_1}^{t_2} m(s)\,d\mu$$

for every $x \in \overline{B}$ and $t_1, t_2 \in [a,b]$. □

5.4 Lemma. *If $g \in \mathcal{C}(\overline{B} \times [a,b], \mu, \omega)$ then for the function G given by (5.4) we have*

$$\|G(x,t_2) - G(x,t_1) - G(y,t_2) + G(y,t_1)\| \le \omega(\|x-y\|) \int_{t_1}^{t_2} l(s)\,d\mu \tag{5.6}$$

for every $x, y \in \overline{B}$ *and* $t_1, t_2 \in [a, b]$.

Proof. By the definition of the function G and by (5.3) we have

$$\|G(x,t_2) - G(x,t_1) - G(y,t_2) + G(y,t_1)\| =$$

$$= \| \int_{t_1}^{t_2} [g(x,s) - g(y,s)]\, d\mu \| \le \int_{t_1}^{t_2} \|g(x,s) - g(y,s)\|\, d\mu \le$$

$$\le \omega(\|x - y\|) \int_{t_1}^{t_2} l(s)\, d\mu$$

for every $x, y \in \overline{B}$ and $t_1, t_2 \in [a, b]$. □

Our next statement shows how the class $\mathcal{C}(\overline{B} \times [a,b], \mu, \omega)$ is related to the class $\mathcal{F}(\overline{B} \times [a,b], h, \omega)$ which was introduced in Definition 3.8.

5.5 Proposition. *If* $g \in \mathcal{C}(\overline{B} \times [a,b], \mu, \omega)$ *then the function* G *given by (5.4) belongs to the class* $\mathcal{F}(\overline{B} \times [a,b], h, \omega)$ *where*

$$h(t) = \int_{t_0}^{t} m(s)\, d\mu + \int_{t_0}^{t} l(s)\, d\mu$$

with $t_0 \in [a, b]$ *is a nondecreasing function.*

Proof. By Lemma 5.3 we have

$$\|G(x,t_2) - G(x,t_1)\| \le \int_{t_1}^{t_2} m(s)\, d\mu \le |h(t_2) - h(t_1)|$$

for every $x \in \overline{B}$ and $t_1, t_2 \in [a, b]$, and therefore (3.7) from Definition 3.8 is satisfied. From Lemma 5.4 we get

$$\|G(x,t_2) - G(x,t_1) - G(y,t_2) + G(y,t_1)\| \le$$

$$\le \omega(\|x - y\|) \int_{t_1}^{t_2} l(s)\, d\mu \le \omega(\|x - y\|)|h(t_2) - h(t_1)|$$

for every $x, y \in \overline{B}$ and $t_1, t_2 \in [a, b]$, and this shows that (3.8) from Definition 3.8 holds true. □

5.6 Definition. A function $g : \overline{B} \times [a,b] \to \mathbb{R}^n$ belongs to the class $\mathcal{C}ar(\overline{B} \times [a,b], \mu)$ if

$$g(x,\cdot) \text{ is measurable with respect to the measure } \mu, \tag{5.1}$$

there exists a μ–measurable function $m : [a,b] \to \mathbb{R}$ such that $\int_a^b m(s)\, d\mu < +\infty$ and

$$\|g(s,x\| \leq m(s) \tag{5.2}$$

for $(x,s) \in \overline{B} \times [a,b]$,

$$g(\cdot,s) \text{ is continuous for every } s \in [a,b]. \tag{5.7}$$

5.7 *Remark.* This definition of the class $\mathcal{C}ar(\overline{B} \times [a,b], \mu)$ of functions $g : \overline{B} \times [a,b] \to \mathbb{R}^n$ is closely related to the class of functions satisfying the Carathéodory conditions as they are described in Chap. II. Indeed, if μ is the Lebesgue measure on $[a,b]$ then they are the same except that (5.1) and (5.2) here are required to hold everywhere instead of μ–almost everywhere. Clearly this difference is not essential from the point of view of Carathéodory differential equations.

In the definition of $\mathcal{C}ar(\overline{B} \times [a,b], \mu)$, (5.3) from Definition 5.1 of $\mathcal{C}(\overline{B} \times [a,b], \mu, \omega)$ is replaced by (5.7). The condition expressed by (5.3) explicitly requires that the continuity from (5.7) has a given modulus ω. It is evident that

$$\mathcal{C}(\overline{B} \times [a,b], \mu, \omega) \subset \mathcal{C}ar(\overline{B} \times [a,b], \mu).$$

The assumption (5.7) was replaced here by (5.3) only for the sake of simplicity. In fact the following can be shown.

5.8 Proposition. *If $g \in \mathcal{C}ar(\overline{B} \times [a,b], \mu)$ then there exist an increasing continuous function $\omega : [0,2c] \to \mathbb{R}$, $\omega(0) = 0$ and a nonnegative μ–integrable function $p : [a,b] \to \mathbb{R}$ such that for the function G given by (5.4) we have*

$$\|G(x,t_2)-G(x,t_1)-G(y,t_2)+G(y,t_1)\| \le \omega(\|x-y\|)\int_{t_1}^{t_2} p(s)\,d\mu$$

for every $x, y \in \overline{B}$ and $t_1, t_2 \in [a,b]$.

The *proof* of this result is rather technical, it uses the so called Scorza–Dragoni property of functions which belong to $\mathcal{C}ar(\overline{B} \times [a,b], \mu)$. The detailed proof of this result is given in [150]

Therefore we actually can use the proof of Proposition 5.5 to show the following.

5.9 Proposition. *If $g \in \mathcal{C}ar(\overline{B} \times [a,b], \mu)$ then the function G given by (5.4) belongs to $\mathcal{F}(\overline{B} \times [a,b], h, \omega)$ with a nondecreasing function $h : [a,b] \to \mathbb{R}$ and a modulus of continuity ω.*

5.10 *Remark.* Let us mention that the previous results hold for a *positive* regular measure μ on $[a,b]$.

If μ is a general "signed" measure on $[a,b]$ then it can be represented in the form of the difference of two positive regular measures μ^+ and μ^-, i.e. $\mu = \mu^+ - \mu^-$. In this case for an arbitrary μ–measurable $M \subset [a,b]$ and $f : [a,b] \to \mathbb{R}^n$ we have

$$\int_M f\,d\mu = \int_M f\,d\mu^+ - \int_M f\,d\mu^-$$

provided the right hand side in this equality is defined. The integral $\int_M f\,d\mu$ converges if and only if the integrals $\int_M f\,d\mu^+$, $\int_M f\,d\mu^-$ converge and this is equivalent to the convergence of $\int_M f\,d\mu^*$ where $\mu^* = \mu^+ + \mu^-$ is the total variation of the measure μ. Clearly also the μ–measurability of a function is equivalent to its μ^+– and μ^-–measurability.

If the measure μ for the case introduced in Definition 5.1 is not assumed to be positive then we have

$$G(x,t) = \int_{t_0}^{t} g(x,s)\,d\mu =$$

$$= \int_{t_0}^{t} g(x,s)\,d\mu^+ - \int_{t_0}^{t} g(x,s)\,d\mu^- \tag{5.8}$$

where μ^+ and μ^- are positive measures and the results of Lemmas 5.3 and 5.4 apply to each term on the right hand side of (5.8).

Let us now assume that $u : [a,b] \to \mathbb{R}$ is of bounded variation on $[a,b]$. Let μ be the Legesgue–Stieltjes measure on $[a,b]$ which corresponds to the function $u : [a,b] \to \mathbb{R}$. The function u can be written in the form $u = u^+ - u^-$ where $u^+, u^- : [a,b] \to \mathbb{R}$ are bounded increasing functions, and if for a function $g : \overline{B} \times [a,b] \to \mathbb{R}^n$ the integral $\int_{t_0}^{t} g(x,s)\,d\mu$ exists then we write as usual $\int_{t_0}^{t} g(x,s)\,du(s)$ for this integral.

5.11 Proposition. *If $g : \overline{B} \times [a,b] \to \mathbb{R}^n$ is such that $g \in \mathcal{C}(\overline{B} \times [a,b], \mu, \omega)$ where μ is the Lebesgue–Stieltjes measure given by the function $u : [a,b] \to \mathbb{R}$ which is of bounded variation, then for the function*

$$G(x,t) = \int_{t_0}^{t} g(x,s)\,du(s), \quad x \in \overline{B},\ t, t_0 \in [a,b] \tag{5.9}$$

there is a nondecreasing function $h : [a,b] \to \mathbb{R}$ such that

$$\|G(x,t_2) - G(x,t_1)\| \le |h(t_2) - h(t_1)|$$

and

$$\|G(x,t_2) - G(x,t_1) - G(y,t_2) + G(y,t_1)\| \le$$

$$\leq \omega(\|x-y\|)|h(t_2)-h(t_1)|$$

for $t_1, t_2 \in [a,b]$, $x, y \in \overline{B}$, i.e. $G \in \mathcal{F}(\overline{B} \times [a,b], h, \omega)$.

Proof. Let $u = u^+ - u^-$ be the Jordan decomposition of the function u on $[a,b]$, the functions u^+, u^- being bounded and increasing on $[a,b]$.

Let us consider the function

$$G^+(x,t) = \int_{t_0}^{t} g(x,s)\,du^+(s), \quad x \in \overline{B},\ t, t_0 \in [a,b]. \tag{5.10}$$

By Lemma 5.3 we have

$$\|G^+(x,t_2) - G^+(x,t_1)\| \leq \int_{t_1}^{t_2} m(s)\,du^+(s)$$

for every $x \in \overline{B}$, $t_1, t_2 \in [a,b]$.

Similarly also for the function

$$G^-(x,t) = \int_{t_0}^{t} g(x,s)\,du^-(s), \quad x \in \overline{B},\ t, t_0 \in [a,b] \tag{5.11}$$

we have

$$\|G^-(x,t_2) - G^-(x,t_1)\| \leq \int_{t_1}^{t_2} m(s)\,du^+(s)$$

for every $x \in \overline{B}$, $t_1, t_2 \in [a,b]$.

Hence

$$\|G(x,t_2) - G(x,t_1)\| =$$

$$= \|G^+(x,t_2) - G^+(x,t_1) - G^-(x,t_2) + G^-(x,t_1)\| \leq$$

$$\leq \int_{t_1}^{t_2} m(s)\,du^+(s) + \int_{t_1}^{t_2} m(s)\,du^-(s) =$$

$$= \int_{t_1}^{t_2} m(s)\, d(\operatorname{var}_a^s u).$$

If we set

$$h_1(t) = \int_a^t m(s)\, d(\operatorname{var}_a^s u), \quad t \in [a,b]$$

then $h_1 : [a,b] \to \mathbb{R}$ is nondecreasing since m is nonnegative on $[a,b]$ and the function $s \in [a,b] \mapsto \operatorname{var}_a^s u$ is nondecreasing. Hence we have

$$\|G(x,t_2) - G(x,t_1)\| \le |h_1(t_2) - h_1(t_1)| \tag{5.12}$$

for $x \in \overline{B}$, $t_1, t_2 \in [a,b]$.

Similarly, Lemma 5.4 implies

$$\|G^+(x,t_2) - G^+(x,t_1) - G^+(y,t_2) + G^+(y,t_1)\| \le$$
$$\le \omega(\|x-y\|) \int_{t_1}^{t_2} l(s)\, du^+(s)$$

if $x, y \in \overline{B}$ and $t_1, t_2 \in [a,b]$, and a similar inequality holds for the function G^- which is given in (5.11). Hence the function G from (5.9) satisfies

$$\|G(x,t_2) - G(x,t_1) - G(y,t_2) + G(y,t_1)\| \le$$
$$\le \omega(\|x-y\|) \int_{t_1}^{t_2} l(s)\, du^+(s) + \omega(\|x-y\|) \int_{t_1}^{t_2} l(s)\, du^-(s) \le$$
$$\le \omega(\|x-y\|)|h_2(t_2) - h_2(t_1)| \tag{5.13}$$

for $x, y \in \overline{B}$ and $t_1, t_2 \in [a,b]$ where

$$h_2(t) = \int_a^t l(s)\, du^-(s) + \int_a^t l(s)\, du^+(s)$$

for $t \in [a,b]$.

The function h_2 is evidently nondecreasing on $[a,b]$. If we take $h(t) = h_1(t) + h_2(t)$ for $t \in [a,b]$ then (5.12) and (5.13) imply the statement. □

5.12 Proposition. *Assume that $g : \overline{B} \times [a,b] \to \mathbb{R}^n$ belongs to $\mathcal{C}(\overline{B} \times [a,b], \mu, \omega)$ where μ is the Lebesgue–Stieltjes measure given by the function $u : [a,b] \to \mathbb{R}$ which is of bounded variation on $[a,b]$, let $G : \overline{B} \times [a,b] \to \mathbb{R}^n$ be defined by (5.9).*

If $x : [\alpha, \beta] \to \overline{B}$, $[\alpha,\beta] \subset [a,b]$ is the pointwise limit of finite step functions then both the generalized Perron integral $\int_\alpha^\beta DG(x(\tau),t)$ and the Lebesgue–Stieltjes integral $\int_\alpha^\beta g(x(s),s)\,du(s)$ exist and have the same value.

Proof. By Proposition 5.12 we have $G \in \mathcal{F}(\overline{B} \times [a,b], h, \omega)$ where h is nondecreasing and ω has the character of a modulus of continuity. Therefore the existence of the integral $\int_\alpha^\beta DG(x(\tau),t)$ is guaranteed by Corollary 3.15. It is a matter of routine to show that for every finite step function $\varphi : [\alpha,\beta] \to \mathbb{R}^n$ the integral $\int_\alpha^\beta g(\varphi(s),s)\,du(s)$ exists and its value is the same as the value of $\int_\alpha^\beta DG(\varphi(\tau),t)$ (the technique of the proof of Corollary 3.15 can be used).

Assume now that $\varphi_k : [\alpha,\beta] \to \overline{B}$, $k \in \mathbb{N}$ is a sequence of finite step functions such that

$$\lim_{k\to\infty} \varphi_k(s) = x(s), \quad s \in [\alpha,\beta].$$

Then (5.3) yields

$$\lim_{k\to\infty} g(\varphi_k(s),s) = g(x(s),s) \quad s \in [\alpha,\beta]$$

and (5.2) enables us to use the Lebesgue dominated convergence theorem for showing that the integral $\int_\alpha^\beta g(x(s),s)\,du(s)$ exists and

$$\int_\alpha^\beta DG(x(\tau),t) = \lim_{k\to\infty} \int_\alpha^\beta DG(\varphi_k(\tau),t) =$$

$$\lim_{k\to\infty} \int_\alpha^\beta g(\varphi_k(s),s)\,du(s) = \int_\alpha^\beta g(x(s),s)\,du(s)$$

where Theorem 3.14 is used to establish the first equality. □

5.13 *Remark.* The results given above will be used in the sequel for the representation of some more conventional concepts of ordinary differential equations within the framework of the generalized ordinary differential equations. The way how to do it is based on the construction of the function G given in (5.4) for a function g.

Carathéodory equations

Assume that a function $f : \overline{B} \times [a, b] \to \mathbb{R}^n$ satisfies the following conditions:

$$f(x, \cdot) \text{ is Lebesgue measurable on } [a, b], \tag{5.14}$$

there exists a Lebesgue measurable function $m : [a, b] \to \mathbb{R}$ such that $\int_a^b m(s)\, ds < +\infty$ and

$$\|g(s, x\| \leq m(s) \tag{5.15}$$

for $(x, s) \in \overline{B} \times [a, b]$,
there exists a Lebesgue measurable function $l : [a, b] \to \mathbb{R}$ such that $\int_a^b l(s)\, ds < +\infty$ and

$$\|g(s, x) - g(s, y)\| \leq l(s)\omega(\|x - y\|) \tag{5.16}$$

for $(x, s), (y, s) \in \overline{B} \times [a, b]$.

Let us associate f with the function

$$F(x, t) = \int_{t_0}^{t} f(x, s)\, ds \quad x \in \overline{B},\ t, t_0 \in [a, b]. \tag{5.17}$$

Looking at (5.14) and (5.15) it is easy to see that the function $F : \overline{B} \times [a, b] \to \mathbb{R}^n$ is well defined and because all assumptions of Proposition 5.11 are satisfied with $u(t) = t$, $t \in [a, b]$ we know from Proposition 5.11 that $F \in \mathcal{F}(\overline{B} \times [a, b], h, \omega)$ where h and ω have all the properties required in Definition 3.8 of the class $\mathcal{F}(\overline{B} \times [a, b], h, \omega)$.

Let us recall the Carathéodory concept of a solution of the differential equation

$$\dot{x} = f(x, t). \tag{5.18}$$

A function $x : [\alpha, \beta] \to \mathbb{R}^n$ is called a solution of (5.18) on $[\alpha, \beta] \subset [a, b]$ if x is absolutely continuous on $[\alpha, \beta]$, $x(s) \in \overline{B}$ for almost all (in the sense of Lebesgue measure) $s \in [\alpha, \beta]$ and if for almost all $t \in [\alpha, \beta]$ the equality

$$\dot{x} = f(x(t), t) \tag{5.19}$$

is satisfied. These properties of the solution are equivalent to the requirement that for every $s_1, s_2 \in [\alpha, \beta]$ we have

$$x(s_2) - x(s_1) = \int_{s_1}^{s_2} f(x(s), s)\, ds. \tag{5.20}$$

The following result connects the Carathéodory theory with the theory of generalized differential equations.

5.14 Theorem. *A function $x : [\alpha, \beta] \to \mathbb{R}^n$, $[\alpha, \beta] \subset [a, b]$ is a solution of (5.18) (in the sense of Carathéodory) on $[\alpha, \beta]$ if and only if x is a solution of the generalized differential equation*

$$\frac{dx}{d\tau} = DF(x, t) \tag{5.21}$$

on $[\alpha, \beta]$ in the sense of Definition 3.1, where F is given by (5.17).

Proof. Assume that $x : [\alpha, \beta] \to \mathbb{R}^n$ is a solution of (5.18). By Proposition 5.12 the integral $\int_{s_1}^{s_2} DF(x(\tau), t)$ exists and

$$x(s_2) - x(s_1) = \int_{s_1}^{s_2} f(x(s), s)\, ds = \int_{s_1}^{s_2} DF(x(\tau), t)$$

for all $s_1, s_2 \in [\alpha, \beta]$. Hence x is a solution of (5.21). If, conversely, x is a solution of (5.21) then again Proposition 5.12 shows that $x : [\alpha, \beta] \to \mathbb{R}^n$ satisfies the equality (5.20) and x is absolutely continuous because by Lemma 3.10 we have

$$\|x(s_2) - x(s_1)\| \leq |h(s_2) - h(s_1)|$$

for every $s_1, s_2 \in [\alpha, \beta]$ and h can be chosen as an absolutely continuous function on $[a, b]$. □

5.15 *Remark.* Theorem 5.14 justifies the term *generalized* differential equation in the sense that for an ordinary differential equation we can find a generalized differential equation such that the two equations have the same set of solutions.

It has to be mentioned here that the condition (5.16) is stronger than the condition used for defining the class of Carathéodory functions.

If we replace (5.16) by

$$f(\cdot, s) \text{ is continuous for every } s \in [a, b], \tag{5.22}$$

i.e. if we assume that $f \in \mathcal{C}ar(\overline{B} \times [a, b]) = \mathcal{C}ar(\overline{B} \times [a, b], \mu)$ where μ is the Lebesgue measure on $[a, b]$, then Proposition 5.9 yields that the function F given by (5.17) belongs to a certain class $\mathcal{F}(\overline{B} \times [a, b], h, \omega)$ and all the above constructions can be repeated without any changes.

5.16 *Historical comments.* The generalized Perron integral described in Chapter I was introduced for the first time in 1957 in the paper [68] by J. Kurzweil and its introduction and motivation comes from the theory of ordinary differential equations. The problem goes back to the early fifties when the averaging principle for differential equations was substantiated by I.I. Gichman in his paper [31] I.I. Gichman showed that the averaging principle can be proved correctly by using some new results on continuous dependence of solutions of differential equations on a parameter. In 1955 the paper [66] of M.A. Krasnoselskij and S.G. Krejn on this problem appeared, being followed in a short time by the paper [88] of J. Kurzweil and Z. Vorel.

From these results it became clear that for continuous dependence of solutions on a parameter of a nonautonomous ordinary differential equation of the form

$$\dot{x} = f(x, t) \tag{5.23}$$

the indefinite integral

$$F(x, t) = \int_{t_0}^{t} f(x, s)\, ds \tag{5.24}$$

to the right hand side f of (5.23) is significant, i.e. if these independent integrals for two ordinary differential equations are close to each other, then the solutions starting e.g. from the same point are close to each other.

One of the problems motivating the introduction of generalized ordinary differential equations can be stated as follows:

Describe the solution of (5.23) in terms of F from (5.24).

The idea how to deal with this problem is the following. The initial value problem

$$\dot{x} = f(x, t), \quad x(t_0) = x^* \tag{5.25}$$

is (e.g. in the Carathéodory setting) equivalent to the integral equation

$$x(s) = x^* + \int_{t_0}^{s} f(x(\sigma), \sigma)\, d\sigma \tag{5.26}$$

with the Lebesgue integral on the right hand side.

If x is a solution of (5.25) on $[t_0, t]$ then x is absolutely continuous on $[t_0, t]$ and it can be approximated by a finite step function x_l which is constant on intervals of the form (α_{j-1}, α_j) where

$$t_0 = \alpha_0 < \alpha_1 < \cdots < \alpha_{k_l} = t$$

and which on (α_{j-1}, α_j) assumes the value $x(\tau_j)$ where $\tau_j \in [\alpha_{j-1}, \alpha_j]$, $j = 1, \ldots, k_l$. Moreover, this can be done in such a way that

$$\lim_{l \to \infty} x_l(s) = x(s) \tag{5.27}$$

uniformly on $[t_0, t]$. If the function $f(x, t)$ satisfies the current assumption of continuity in the variable x then

$$\lim_{l \to \infty} f(x_l(s), s) = f(x(s), s)$$

on $[t_0, t]$. If now e.g. the assumptions of the Lebesgue dominated convergence theorem are satisfied for the sequence of functions $f(x_l(s), s)$, $l = 1, 2, \ldots$ (of this kind is (5.15)) then it can be concluded that

$$\lim_{l \to \infty} \int_{t_0}^{t} f(x_l(s), s)\, ds = \int_{t_0}^{t} f(x(s), s)\, ds. \tag{5.28}$$

However, for a fixed $l \in \mathbb{N}$ we have

$$\int_{t_0}^{t} f(x_l(s), s)\, ds = \sum_{j=1}^{k_l} \int_{\alpha_{j-1}}^{\alpha_j} f(x(\tau_j), s)\, ds =$$

$$= \sum_{j=1}^{k_l} [F(x(\tau_j), \alpha_j) - F(x(\tau_j), \alpha_{j-1})]$$

and this together with (5.28) shows that the integral $\int_{t_0}^{t} f(x(s), s)\, ds$ can be approximated by sums of the form

$$\sum_{j=1}^{k_l} [F(x(\tau_j), \alpha_j) - F(x(\tau_j), \alpha_{j-1})],$$

i.e. using (5.26) the value $x(t)$ can be approximated by

$$x^* + \sum_{j=1}^{k_l} [F(x(\tau_j), \alpha_j) - F(x(\tau_j), \alpha_{j-1})]. \tag{5.29}$$

The only thing we have to do now is to fix the process of constructing a sufficiently fine division $t_0 = \alpha_0 < \alpha_1 < \cdots < \alpha_k = t$ and the choice of $\tau_j \in [\alpha_{j-1}, \alpha_j]$, $j = 1, \ldots, k$ in order to obtain the uniform convergence (5.27).

Looking at the sum in (5.29) it is easy to recognize that it is formally related to the integral sums treated in Chapter I. Moreover, if the problem is to describe the solutions of (5.25) with f fastly oscillating in the variable t then it is clear that we have to take the intervals $[\alpha_{j-1}, \alpha_j]$ of the division in dependence on the choice of the point τ_j because otherwise we would ignore the instantaneous influence of the fastly changing vector field of the right hand sides of the equation. This leads to the concept of a δ–fine partition for a given gauge as it was described in the introduction to Chapter I.

These few vague remarks are given here to describe the essence of the intuitive idea which was (probably) behind the theory of generalized ordinary differential equations created by J. Kurzweil in 1957 and, of course, also of the Kurzweil branch of the Henstock–Kurzweil approach to integration.

For the sake of historical truth it is necessary to intimate that the sum approach to the generalized Perron integral in Kurzweil's work [68] is secondary. The primary one is based on the concept of major and minor functions and is very close to the concept used in the theory of the Denjoy–Perron integral. Since it is shown in [68] that these two concepts are equivalent, the equivalence of the new sum integral and the Denjoy–Perron integral was clear from the very beginning of Kurzweil's work even if this fact was not sufficiently emphasized as concerns the theory of integration.

Measure differential equations

Measure differential equations have been investigated by many authors, e.g. W.W. Schmaedeke [124], P.C. Das and R.R. Sharma [19], M. Rama Mohana Rao and Sree Hari Rao [114], S.G. Pandit [P] and others. A survey of this approach to differential systems is given in the monograph [107] written by S.G. Pandit and S.G. Deo.

The main purpose of the concept of measure differential equations is the description of systems exhibiting discontinuous solutions caused by the impulsive behaviour of the differential system. The solutions of a measure differential equation are discontinuous functions of bounded variation, i.e. they have the same properties as generalized ordinary differential equations described in Chapter III.

Assume as before that

$$\overline{B} = \{x \in \mathbb{R}^n;\ \|x\| \leq c\}$$

and $G = \overline{B} \times [a,b]$, $-\infty < a < b + \infty$. Let $f : G \to \mathbb{R}^n$ be a function which satisfies (5.14) - (5.16) with some ω_1 in the role of ω. Further assume that $u : [a,b] \to \mathbb{R}$ is of bounded variation on $[a,b]$ and continuous from the left. Let $g : G \to \mathbb{R}^n$ be such that $g \in \mathcal{C}(G, \mu, \omega_2)$ where μ is the Lebesgue–Stieltjes measure generated by the function u (see Proposition 5.11).

The measure differential equation is formally written in the form

$$Dx = f(x,t) + g(x,t)Du \tag{5.30}$$

where Dx and Du stand for the distributional derivatives of the functions x and u in the sense of distributions of L. Schwartz. In the paper [19] Das and Sharma have shown that the concept of a solution of (5.30) satisfying the initial condition $x(t_0) = \widetilde{x}$, $t_0 \in [a,b]$, $\widetilde{x} \in \overline{B}$, is equivalent to the concept of a solution of the integral equation

$$x(t) = \widetilde{x} + \int_{t_0}^{t} f(x(s),s)\,ds + \int_{t_0}^{t} g(x(s),s)\,du(s) \tag{5.31}$$

for $t \in [a,b]$. In other words, a function $x : [\alpha,\beta] \to \mathbb{R}^n$ is a solution of the measure differential equation (5.30) if and only if $(x(s),s) \in G$ for $s \in [\alpha,\beta] and$

$$x(s_2) - x(s_1) = \int_{s_1}^{s_2} f(x(s),s)\,ds + \int_{s_1}^{s_2} g(x(s),s)\,du(s) \tag{5.32}$$

for every $s_1, s_2 \in [\alpha,\beta]$.

Since we have got some experience with the Carathéodory equation in the previous part it is now evident that for $x \in \overline{B}$, $t, t_0 \in [a,b]$ we have to define

$$F_1(x,t) = \int_{t_0}^{t} f(x(s),s)\,ds$$

and

$$F_2(x,t) = \int_{t_0}^{t} g(x(s),s)\,du(s).$$

The function $F_1 : G \to \mathbb{R}^n$ is the same as in the previous section on Carathéodory equations because it corresponds to f satisfying

(5.14) - (5.16) and we have $F_1 \in \mathcal{F}(G, h_1, \omega_1)$ where h_1 is absolutely continuous on $[a, b]$. To the function F_2 Proposition 5.11 can be applied for concluding that $F_2 \in \mathcal{F}(G, h_2, \omega_2)$ where h_2 is nondecreasing and continuous from the left because u is continuous from the left. (Using Remark 5.7 and Propositions 5.8 and 5.9 it can be assumed that f is a Carathéodory function and $g \in \mathcal{C}ar(G, \mu)$ where μ is the Lebesgue–Stieltjes measure generated by u on $[a, b]$ and the results are the same.)

Let us set

$$F(x,t) = F_1(x,t) + F_2(x,t) \tag{5.33}$$

for $(x, t) \in G$. It is a matter of routine to show by Definition 3.8) that $F \in \mathcal{F}(G, h, \omega)$ where $h = h_1 + h_2$ and $\omega = \omega_1 + \omega_2$. The functions h and ω have the properties required in Chapters III and IV for the generalized differential equation

$$\frac{dx}{d\tau} = DF(x,t). \tag{5.34}$$

Proposition 5.12 can now be used to show that for every function $x : [\alpha, \beta] \to \mathbb{R}^n$, $[\alpha, \beta] \subset [a, b]$ which is the pointwise limit of a sequence of finite step functions the integrals

$$\int_\alpha^\beta f(x(s), s)\, ds, \quad \int_\alpha^\beta g(x(s), s)\, du(s),$$

$$\int_\alpha^\beta DF_1(x(\tau), t), \quad \int_\alpha^\beta DF_2(x(\tau), t)$$

exist and

$$\int_\alpha^\beta g(x(s), s)\, du(s) = \int_\alpha^\beta DF_2(x(\tau), t),$$

$$\int_\alpha^\beta f(x(s), s)\, ds = \int_\alpha^\beta DF_1(x(\tau), t) \tag{5.35}$$

Looking at the integral form (5.32) of the measure differential equation (5.30) it is easy to observe that every solution is a function of bounded variation. Hence by (5.35) the relation (5.31) can be writen in the form

$$x(s_2) - x(s_1) = \int_{s_1}^{s_2} f(x(s), ds + \int_{s_1}^{s_2} g(x(s), du(s) =$$

$$= \int_{s_1}^{s_2} DF_1(x(\tau), t) + \int_{s_1}^{s_2} DF_2(x(\tau), t) = \int_{s_1}^{s_2} DF(x(\tau), t)$$

for every solution $x : [\alpha, \beta] \to \mathbb{R}^n$ of (5.30) and every $s_1, s_2 \in [\alpha, \beta]$. Hence $x : [\alpha, \beta] \to \mathbb{R}^n$ is a solution of (5.43). The converse statement, namely that every solution $x : [\alpha, \beta] \to \mathbb{R}^n$ of (5.34) with F given by (5.33) is also a solution of (5.30), can be shown quite analogously. In this way we obtain the following result.

5.17 Theorem. *A function $x : [\alpha, \beta] \to \mathbb{R}^n$, $[\alpha, \beta] \subset [a, b]$ is a solution of the measure differential equation (5.30) on $[\alpha, \beta]$ if and only if x is a solution of the generalized differential equation (5.34) on $[\alpha, \beta]$ with the function F given by (5.33).*

5.18 *Remark.* In [107] the function u involved in the measure differential equation (5.30) is assumed to be continuous from the right. There is no essential difference between our setting of left continuous u. One case can be easily transformed into the other.

5.19 *Example.* Let us consider the measure differential equation

$$Dx = 2(t+1)^{-1} x\, Du, \quad t \in [0, 2]$$

where $u(t) = t$ for $t \in [0, 1]$, $u(t) = t - 1$ for $t \in (1, 2]$ (see Example 2.1 in [107]). Let us define

$$F(x, t) = \int_0^t 2(s+1)^{-1} x\, du(s).$$

Then

$$F(x,t) = A(t)x,$$

where

$$A(t) = \int_0^t 2(s+1)^{-1}\, du(s) = \int_0^t 2(s+1)^{-1}\, ds = 2\ln(t+1)$$

for $t \in [0,1]$ and

$$A(t) = \int_0^1 2(s+1)^{-1}\, ds + \int_1^t 2(s+1)^{-1}\, du(s) =$$

$$= 2\ln 2 + 2\frac{1}{2}(u(1+) - u(1)) + \int_1^t 2(s+1)^{-1}\, ds = -1 + 2\ln(t+1)$$

for $t \in (1,2]$.

The generalized differential equation which is associated to the given measure differential equation has the form

$$\frac{dx}{d\tau} = D[A(t)x].$$

If we consider the initial value problem $x(0) = 0$ for this equation then

$$x(t) = 0 \text{ for } t \in [0,1],$$

$$x(1+) - x(1) = x(1)[A(1+) - A(1)] = -x(1), \text{ i.e. } x(1+) = 0$$

and for $t \in (1,2]$ we get

$$x(t) = x(1) + \int_1^t x(\tau)\, dA(\tau) = x(1) + x(1) + \lim_{\delta \to 0+} \int_\delta^t x(\tau)\, dA(\tau) =$$

$$= 2\int_1^t x(\tau)\, d\ln(1+\tau) = 2\int_1^t x(\tau)(1+\tau)^{-1}\, d\tau.$$

Therefore for $t \in (1,2]$ the function x is the solution of the initial value problem

$$\dot{y} = 2(1+t)^{-1}y, \quad y(1) = 0$$

and $x(t) = y(t) = 0$ for $t > 1$. The situation is different when we consider the initial value problem with $x(2) = 0$ for this equation. Then we have as in the previous case

$$x(t) = 0 \text{ for } t \in (1,2]$$

because for $t \in (1,2]$ we have

$$x(t) = \int_2^t x(\tau)\,dA(\tau) = -2\int_t^2 x(\tau)(1+\tau)^{-1}\,d\tau.$$

Further

$$x(1+) = 0$$

and

$$0 = x(1+) = x(1) + x(1)[A(1+) - A(1)] = x(1) - x(1).$$

This means that $x(1+) = 0$ for all possible values of $x(1) \in \mathbb{R}$, and therefore every function $x : [0,1] \to \mathbb{R}$ for which

$$x(t) = x(1) + 2\int_1^t x(\tau)\,dA(\tau) = x(1) - 2\int_1^t x(\tau)(1+\tau)^{-1}\,d\tau$$

holds is a solution of the equation. Hence

$$\dot{x}(t) = 2(1+t)^{-1}x(t), \quad x(1) = c,\ c \in \mathbb{R}$$

and therefore

$$x(t) = \frac{(t+1)^2}{4}c, \quad t \in [0,1]$$

where $c \in \mathbb{R}$ is arbitrary.

Differential equations with impulses

In this part we describe shortly another approach to systems exhibiting impulsive behaviour, namely ordinary differential equations with impulses. The theory of these equations was extensively studied e.g. by A.D. Myshkis, A.M. Samojlenko, N.A. Perestjuk and others (see [99],[103], [116], [117], [118], [119], [108] etc.). A survey of the state of art is given in the comprehensive monograph [120].

Assume again that $\overline{B} = \{x \in \mathbb{R}^n;\ \|x\| \leq c\}$ and $G = \overline{B} \times [a, b]$, $-\infty < a < b + \infty$. Let $f : G \to \mathbb{R}^n$ be a function which satisfies (5.14) – (5.16).

Further let a finite set of points $t_i \in [a, b]$, $i = 1, \dots, k$ be given with $t_i < t_{i+1}$ for $i = 1, 2, \dots, k-1$ and a system of k continuous maps

$$I_i : \overline{B} \to \mathbb{R}^n, \quad i = 1, 2, \dots, k.$$

The system with impulsive action at the fixed instants $t_1, t_2, \dots, t_k$ is usually written in the form

$$\dot{x} = f(x, t), \quad t \neq t_i, \tag{5.36}$$

$$\Delta x|_{t_i} = x(t_i+) - x(t_i) = I_i(x(t_i)), \quad i = 1, \dots, k. \tag{5.37}$$

The equation (5.36) describes the behaviour of the state at instants different from t_i, $i = 1, \dots, k$ and (5.37) represents the discontinuity from the right of the solution for $t = t_i$.

We describe the *differential system with impulses* (5.36), (5.37) more exactly by describing its solutions. A function $x : [\alpha, \beta] \to \mathbb{R}^n$, $[\alpha, \beta] \subset [a, b]$ is called a solution of the differential system with impulses (5.36), (5.37) if $(x(t), t) \in G$ for $t \in [\alpha, \beta]$, the function x is absolutely continuous on every interval $[a, t_1] \cap [\alpha, \beta]$, $(t_i, t_{i+1}] \cap [\alpha, \beta]$, $i = 1, \dots, k-1$, $(t_k, b] \cap [\alpha, \beta]$

$$\dot{x}(t) = f(x(t), t) \text{ for almost all } t \in [\alpha, \beta],$$

and

$$x(t_i+) = \lim_{t \to t_i+} x(t) = x(t_i) + I_i(x(t_i)) \text{ if } t_i \in [\alpha, \beta].$$

For a given $d \in [a, b)$ define

$$H_d(t) = 0 \text{ for } t \leq d, \; H_d(t) = 1 \text{ for } t > d$$

and define

$$F(x,t) = \int_a^t f(x,s)\,ds + \sum_{i=1}^k I_i(x) H_{t_i}(t). \tag{5.38}$$

Since $\overline{B}$ is compact and the mappings I_i are continuous, there exists a constant $K \geq 0$ such that $\|I_i(x)\| \leq K$ for all $x \in \overline{B}$ and $i = 1, \ldots, k$. Therefore if $x \in \overline{B}$ and $s_1, s_2 \in [a, b]$ we obtain

$$\|F(x, s_2) - F(x, s_1)\| \leq \| \int_{s_1}^{s_2} f(x,s)\,ds\| +$$

$$+K| \sum_{i=1}^k (H_{t_i}(s_2) - H_{t_i}(s_1))| \leq$$

$$\leq |h_1(s_2) - h_1(s_1)| + K|h_2(s_2) - h_2(s_1)|$$

where $h_1 : [a, b] \to \mathbb{R}$ is the nondecreasing, absolutely continuous function corresponding to f in the same way as in the part on Carathéodory equations (i.e. the function $F_1(x,t) = \int_a^t f(x,s)\,ds$ belongs to the class $\mathcal{F}(G, h_1, \omega_1)$) and

$$h_2(t) = \sum_{i=1}^k (H_{t_i}(t), \quad t \in [a, b].$$

Clearly, h_2 is nondecreasing and continuous from the left on $[a, b]$. If ω_2 is the common modulus of continuity of the finite system of mappings I_i, $i = 1, \dots, k$ then

$$\|I_i(x) - I_i(y)\| \leq \omega_2(\|x - y\|)$$

for $x, y \in \overline{B}$. Using the information from the part on Carathéodory equations we obtain easily

$$\|F(x, s_2) - F(x, s_1) - F(y, s_2) + F(y, s_1)\| \leq$$
$$\leq \omega_1(\|x - y\|)|h_1(s_2) - h_1(s_1)| + \omega_2(\|x - y\|)|h_2(s_2) - h_2(s_1)|$$

for $x, y \in \overline{B}$ and $s_1, s_2 \in [a, b]$. The first term corresponds to f and for the second term in (5.38) we have the following estimate:

$$\|\sum_{i=1}^{k}(I_i(x) - I_i(y))(H_{t_i}(s_2) - H_{t_i}(s_1))\| \leq$$

$$\leq \omega_2(\|x - y\|)|\sum_{i=1}^{k}(H_{t_i}(s_2) - H_{t_i}(s_1))| \leq$$
$$\leq \omega_2(\|x - y\|)|h_2(s_2) - h_2(s_1)|.$$

If we take $h(t) = h_1(t) + h_2(t)$ for $t \in [a, b]$ and $\omega(r) = \omega_1(r) + \omega_2(r)$ then we obtain that the function F defined by (5.38) belongs to the class $\mathcal{F}(g, h, \omega)$ and the result of Lemma 3.12 states

$$x(t_i+) - x(t_i) = F(x(t_i), t_i+) - F(x(t_i), t_i) =$$

$$= \lim_{t \to t_i+} \sum_{i=1}^{k} I_i(x(t_i))H_{t_i}(t) - \sum_{i=1}^{k} I_i(x(t_i))H_{t_i}(t_i) =$$
$$= I_i(x(t_i))(H_{t_i}(t_i+) - H_{t_i}(t_i)) = I_i(x(t_i))$$

for a solution of the generalized differential equation

$$\frac{dx}{d\tau} = DF(x, t). \tag{5.39}$$

After this we summarize the results as follows.

5.20 Theorem. *A function $x : [\alpha, \beta] \to \mathbb{R}^n$, $[\alpha, \beta] \subset [a, b]$ is a solution of the differential equation (5.36) with impulses (5.37) on $[\alpha, \beta]$ if and only if x is a solution of the generalized differential equation (5.39) on $[\alpha, \beta]$ with the function F given by (5.38).*

By the results given in the above sections in this chapter the Carathéodory equations, measure differential equations and differential equations with impulses are embedded in a quite natural way into the class of generalized ordinary differential equations with an appropriately chosen right hand side F which belongs to the class $\mathcal{F}(G, h, \omega)$.

5.21 *Remark.* From Chap. III we know that a solution of a generalized ordinary differential equation with $F \in \mathcal{F}(G, h, \omega)$ is a function of bounded variation and therefore it cannot be expected that this equation would be equivalent in general with a classical differential equation. Nevertheless the following problem can be posed.

Given a generalized ordinary differential equation

$$\frac{dx}{d\tau} = DF(x, t). \tag{5.40}$$

under what condition is this equation equivalent to a classical differential equation?

Let us consider the case when $F \in \mathcal{F}(G, h, \omega)$ where $\overline{B} = \{x \in \mathbb{R}^n; \|x\| \leq c\}$ and $G = \overline{B} \times [a, b]$, $-\infty < a < b + \infty$, $h : [a, b] \to \mathbb{R}^n$ is nondecreasing and continuous from the left and ω has the properties of a modulus of continuity as given in Chap. III.

Assume that $n = 1$, i.e. $F : G \to \mathbb{R}$. By Definition 3.8 of $\mathcal{F}(G, h, \omega)$ we have

$$|F(x, t_2) - F(x, t_1)| \leq |h(t_2) - h(t_1)| \tag{5.41}$$

for every $x \in \overline{B}$ and $t_1, t_2 \in [a, b]$.

In the usual way the function h defines a Lebesgue–Stieltjes measure on $[a,b]$ via the additive interval function given by $\mu([c,d)) = h(d) - h(c)$ for an interval $[c,d)$ with $a \le c \le d \le b$.

For a fixed $x \in \overline{B}$ we define

$$\mu_x([c,d)) = F(x,d) - F(x,c)$$

for $a \le c \le d \le b$. Clearly μ_x induces a measure on $[a,b]$ and by the inequality (5.41) we can see immediately that the measure μ_x is absolutely continuous with respect to the measure μ. Therefore by the Radon–Nikodym Theorem (see e.g. [21, III.10.7. Theorem] or [25, (8.19) Theorem]) there exists a μ–measurable function g_x on $[a,b]$ such that for each μ–measurable set $E \subset [a,b]$ we have

$$\mu_x(E) = \int_E g_c \, d\mu. \tag{5.42}$$

This means that

$$F(x,t_2) - F(x,t_1) = \int_{t_1}^{t_2} g_x \, d\mu,$$

i.e.

$$F(x,t) - F(x,a) = \int_a^t g_x \, d\mu = \int_a^t g_x(s) \, dh(s).$$

Let us set $f(x,t) = g_x(t)$. Then $f(x,\cdot) : [a,b] \to \mathbb{R}$ is the Radon–Nikodym derivative of μ_x with respect to the measure μ it is μ–integrable and has a finite integral ($= \mu_x(E)$) over every measurable set $E \subset [a,b]$. Moreover,

$$F(x,t) = F(x,a) + \int_a^t g_x \, d\mu = F(x,a) + \int_a^t f(x,s) \, dh(s).$$

Denote by $D\mu_x(t)$ the general derivative of the set function μ_x given by (5.42). Then we have

$$D\mu_x(t) = g_x(t) = f(x,t)$$

almost everywhere with respect to the measure μ. (See e.g. [25, (8.29) Theorem].)

Hence for μ–almost all $t \in [a,b]$ we have by (5.41)

$$|f(x,t)| = |g_x(t)| = |D\mu_x(t)| \leq \lim_{s\to t} \left|\frac{F(x,s)-F(x,t)}{h(s)-h(t)}\right| \leq$$

$$\leq \lim_{s\to t} \frac{|h(s)-h(t)|}{|h(s)-h(t)|} = 1. \tag{5.43}$$

Since $F \in \mathcal{F}(G,h,\omega)$ we have also

$$|F(x,t_2)-F(x,t_1)-F(y,t_2)+F(y,t_1)| \leq \omega(|x-y|)|h(t_2)-h(t_1)| \tag{5.44}$$

for $x,y \in \overline{B}$ and $t_1,t_2 \in [a,b]$. Essentially in the same way as above we can deduce from (5.44) that

$$|f(x,t)-f(y,t)| = |g_x(t)-g_y(t)| \leq \omega(|x-y|) \tag{5.45}$$

holds for $x,y \in \overline{B}$ and μ–almost all $t \in [a,b]$. From (5.43) and (5.45) we can see that (5.2) and (5.3) from the definition of the class $\mathcal{C}(G,\mu,\omega)$ is satisfied with $m(s) = l(s) = 1$ μ–almost everywhere in $[a,b]$.

Assume now that $x : [\alpha,\beta] \to \mathbb{R}$ is a solution of (5.40). Using the method of the proof of Proposition 5.12 it is easy to show that the integral $\int_\alpha^s f(x(\sigma),\sigma)\,d\mu = = \int_\alpha^s f(x(\sigma),\sigma)\,dh(\sigma)$ exists and equals to $\int_\alpha^s DF(x(\tau),t)$. Therefore

$$x(s) = x(\alpha) + \int_\alpha^s DF(x(\tau),t) = x(\alpha) + \int_\alpha^s f(x(\sigma),\sigma)\,dh(\sigma)$$

and $x : [\alpha,\beta] \to \mathbb{R}$ is a solution of the measure differential equation

$$Dx = f(x,t)Dh \tag{5.46}$$

(see (5.30) and (5.31)). Taking into account the result stated in Theorem 5.17 we obtain the following.

Every generalized ordinary differential equation (5.40) with $F \in \mathcal{F}(G,h,\omega)$ *is equivalent to a certain measure differential equation of the form (5.46).*

Assume in addition that the function h is absolutely continuous (with respect to the Lebesgue measure) on $[a,b]$. Then we have $h(t) = \int_a^t m(s)\,ds$ where m is Lebesgue integrable over $[a,b]$. Using the construction of the function f given in the more general case above we have

$$x(s) = x(\alpha) + \int_\alpha^s DF(x(\tau),t) = x(\alpha) + \int_\alpha^s f(x(\sigma),\sigma)\,d\sigma$$

for an arbitrary solution $x : [\alpha,\beta] \to \mathbb{R}$ of the generalized differential equation (5.40) and all $s \in [\alpha,\beta]$. This means that x is a solution of the Carathéodory differential equation

$$\dot{x} = f(x,t)$$

and it can be shown that the right hand side of this equation satisfies (5.14) - (5.16). In this way we arrived at the following result.

Every generalized ordinary differential equation (5.40) with $F \in \mathcal{F}(G,h,\omega)$ *where* h *is absolutely continuous on* $[a,b]$ *is equivalent to a certain classical Carathéodory differential equation.*

We have restricted our consideration of (5.40) to the case $n = 1$. It is clear that this restriction is not essential. All the reasoning can be used for the components of a general n–dimensional function F for obtaining the same results for an arbitrary $n \in \mathbb{N}$.

Finally it is worth to mention here that in a more general setting the question given at the beginning of this remark was treated by T.S. Chew in the paper [15]. Chew gives a necessary and sufficient condition for the right hand side of the equation (5.40) for being equivalent to a classical ordinary differential equation with the right hand side $f \in \mathcal{C}ar(G, \mu)$ where μ is the Lebesgue measure on $[a, b]$.

Remark. It should be mentioned that D. Fraňková [26], [30] developed another approach to generalized ordinary differential equations, the so called generalized ordinary differential equations with a substitution. Generalized ordinary differential equations with a substitution represent a powerful technical tool. The method of D. Fraňková was used e.g. in [27] and [145] for treating second order linear equations with impulses (see also the paper [67] of K. Kreith).

CHAPTER VI

GENERALIZED LINEAR DIFFERENTIAL EQUATIONS

Assume that $J \subset \mathbb{R}$ is an interval. Denote by $L(\mathbb{R}^n)$ the set of all $n \times n$–matrices with real components and assume that $A : J \to L(\mathbb{R}^n)$ is an $n \times n$–matrix valued function defined on the interval J. We assume further that A is locally of bounded variation, i.e. $\operatorname{var}_a^b A < \infty$ for every compact interval $[a, b] \in J$ where the variation is defined using the norm of the matrix which is the norm of the corresponding linear operator on $\mathbb{R}^n$, i.e. the operator norm in $L(\mathbb{R}^n)$.

Let us note that if $A(t) = (a_{ij}(t))_{i,j=1,\dots,n}$ then every component $a_{ij} : J \to \mathbb{R}, i, j = 1, \dots, n$ is locally of bounded variation provided A has this property.

For $x \in \mathbb{R}^n$, $t \in J$ we set

$$F(x,t) = A(t)x.$$

If $(\widetilde{x}, t_0) \in \mathbb{R}^n \times J$ is an arbitrary point then for every closed ball $\overline{B}_c(\widetilde{x}) = \{x \in \mathbb{R}^n; \|x - \widetilde{x}\| \leq c\}$ in $\mathbb{R}^n$ with radius $c \geq 1$ centered at $\widetilde{x}$ and $[a, b] \subset J$, $t_0 \in [a, b]$ we have

$$\|F(x,t_2) - F(x,t_1)\| \leq \|A(t_2) - A(t_1)\| \|x\| \leq |h(t_2) - h(t_1)|$$

for $x \in \overline{B}_c(\widetilde{x})$, $t_1, t_2 \in [a, b]$ where $h : [a, b] \to \mathbb{R}$ is given by $h(t) = (c + \|\widetilde{x}\|) \operatorname{var}_a^t A$ for $t \in [a, b]$, and for $x, y \in \overline{B}_c(\widetilde{x})$, $t_1, t_2 \in$

$[a,b]$ we have

$$\|F(x,t_2)-F(x,t_1)-F(y,t_2)+F(y,t_1)\|\leq$$

$$\leq \|A(t_2)-A(t_1)\|\|x-y\|\leq \|x-y\||h(t_2)-h(t_1)|.$$

The function h defined on $[a,b]$ is evidently nondecreasing because A is assumed to be of locally bounded variation.

The above inequalities show that the function $F:\mathbb{R}^n\times J\to\mathbb{R}^n$ locally satisfies the requirements of Definition 3.8, i.e. $F\in\mathcal{F}(G,h,\omega)$ where $G=\overline{B}_c(\widetilde{x})\times[a,b]$, $\omega(r)=r,\ r\geq 0$.

If $g:J\to\mathbb{R}^n$ is a function of locally bounded variation on J then for $x\in\mathbb{R}^n,\ t\in J$ we can set

$$F(x,t)=A(t)x+g(t)$$

and show in the same way as above that $F\in\mathcal{F}(G,h,\omega)$ for $G=\overline{B}_c(\widetilde{x})\times[a,b]$ where $h(t)=(c+\|\widetilde{x}\|)\operatorname{var}_a^t A+\operatorname{var}_a^t g$ for $t\in[a,b]$ and $\omega(r)=r,\ r\geq 0$.

In this chapter we will study the generalized linear differential equation

$$\frac{dx}{d\tau}=D[A(t)x+g(t)] \tag{6.1}$$

where $A,\ g$ have the properties given above.

Using Definition 3.1 of a solution we can see that a function $x:[\alpha,\beta]\to\mathbb{R}^n$ is a solution of (6.1) on $[\alpha,\beta]$ if for any $s_1,s_2\in[\alpha,\beta]$ we have

$$x(s_2)-x(s_1)=\int_{s_1}^{s_2}D[A(t)x(\tau)+g(t)]=$$

$$=\int_{s_1}^{s_2}D[A(t)x(\tau)]+g(s_2)-g(s_1). \tag{6.2}$$

Using the more conventional notation $\int_{s_1}^{s_2} d[A(\tau)]x(\tau)$ for $\int_{s_1}^{s_2} D[A(t)x(\tau)]$ we can rewrite (6.2) in the form

$$x(s_2) - x(s_1) = \int_{s_1}^{s_2} d[A(\tau)]x(\tau) + g(s_2) - g(s_1)$$

where the integral here is the Perron–Stieltjes integral (cf. Remark 1.5) with the usual convention on integration of $\mathbb{R}^n$–valued functions.

Remark. Sometimes instead of the notation given in (6.1) the notation

$$dx = d[A]x + dg$$

is used.

6.1 Lemma. *If $x : [\alpha, \beta] \to \mathbb{R}^n$ is a solution of (6.1) on $[\alpha, \beta] \subset J$ then x is of bounded variation on $[\alpha, \beta]$.*

Proof. From the existence of the solution x of (6.1) we have

$$x(t) = x(t_0) + \int_{t_0}^{t} d[A(\tau)]x(\tau) + g(t) - g(t_0)$$

for every $t, t_0 \in [\alpha, \beta]$ and the integral $\int_{t_0}^{t} d[A(\tau)]x(\tau)$ exists for $t, t_0 \in [\alpha, \beta]$. Hence by Theorem 1.16 the limits

$$\lim_{t \to t_0-} \int_{t_0}^{t} d[A(\tau)]x(\tau) \text{ and } \lim_{t \to t_0+} \int_{t_0}^{t} d[A(\tau)]x(\tau)$$

exist for $t_0 \in (\alpha, \beta]$, $t_0 \in [\alpha, \beta)$, respectively. Hence the solution x has onesided limits at every point in $[\alpha, \beta]$ because also the function g evidently has this property. Therefore for every $t_0 \in [\alpha, \beta]$ there exists a $\delta > 0$ and a constant M such that $\|x(t)\| \leq M$ for $t \in (t_0 - \delta, t_0 + \delta) \cap [\alpha, \beta]$. By the Heine - Borel Covering

Theorem there is a finite set of intervals of the type $(t_0 - \delta, t_0 + \delta)$ covering the interval $[\alpha, \beta]$. Therefore there is a constant $K > 0$ such that $\|x(t)\| \leq K$ for every $t \in [\alpha, \beta]$. If now $\alpha = s_0 < s_1 < \cdots < s_k = b$ is an arbitrary division of $[\alpha, \beta]$ then by Corollary 1.6 (see also Remark 1.37) we obtain

$$\|x(s_i) - x(s_{i-1})\| \leq K \operatorname{var}_{s_{i-1}}^{s_i} A + \|g(s_i) - g(s_{i-1})\|$$

for every $i = 1, \cdots, k$ and

$$\sum_{i=1}^{k} \|x(s_i) - x(s_{i-1})\| \leq K \operatorname{var}_{\alpha}^{\beta} A + \operatorname{var}_{\alpha}^{\beta} g.$$

Hence $\operatorname{var}_{\alpha}^{\beta} x \leq K \operatorname{var}_{\alpha}^{\beta} A + \operatorname{var}_{\alpha}^{\beta} g$ by the definition of the variation because $\alpha = s_0 < s_1 < \cdots < s_k = b$ was an arbitrary division of $[\alpha, \beta]$. □

Assume that $[a, b] \subset J$ is a compact interval. Let us denote by $BV([a, b]) = BV$ the set of all functions $x : [a, b] \to \mathbb{R}^n$ of bounded variation on $[a, b]$ and set

$$\|x\|_{BV} = \|x(a)\| + \operatorname{var}_{a}^{b} x. \tag{6.3}$$

It is known that $\|\cdot\|_{BV}$ is a norm on BV and that BV with the norm $\|\cdot\|_{BV}$ is a Banach space.

Assume further that $t_0 \in [a, b]$ is fixed.

Given $x \in BV([a, b])$ we define

$$Tx(t) = \int_{t_0}^{t} d[A(s)]x(s), \; t \in [a, b]. \tag{6.4}$$

By Corollary 1.34 the integral on the right hand side of (6.4) exists for every $t \in [a, b]$ because A is of bounded variation on $[a, b]$ and x is evidently regular because we assume $x \in BV$.

6.2 Proposition. *If $x \in BV$ then $Tx : [a,b] \to \mathbb{R}^n$ given by (6.4) is a function of bounded variation on $[a,b]$, i.e. $Tx \in BV([a,b])$.*

Moreover, the mapping $T : BV \to BV$ given by (6.4) is a bounded compact linear operator on BV.

Proof. Assume that $s_1, s_2 \subset [a,b]$, $s_1 < s_2$. Then by Corollary 1.36 we have

$$\|Tx(s_2)-Tx(s_1)\| = \|\int_{s_1}^{s_2} d[A(s)]x(s)\| \le \int_{s_1}^{s_2} \|x(s)\|\, d(\operatorname{var}_{t_0}^{s} A).$$

Therefore for an arbitrary division $a = s_0 < s_1 < \cdots < s_k = b$ of $[a,b]$ we have

$$\sum_{i=1}^{k} \|Tx(s_j) - Tx(s_{j-1})\| \le \sum_{i=1}^{k} \int_{s_{j-1}}^{s_j} \|x(s)\|\, d(\operatorname{var}_{t_0}^{s} A) =$$

$$= \int_a^b \|x(s)\|\, d(\operatorname{var}_{t_0}^{s} A) \le \|x\|_{BV} \int_a^b d(\operatorname{var}_{t_0}^{s} A) = \|x\|_{BV} \operatorname{var}_a^b A$$

because for every $s \in [a,b]$ we evidently have

$$\|x(s)\| \le \|x(s) - x(a)\| + \|x(a)\| \le \|x(a)\| + \operatorname{var}_a^b x = \|x\|_{BV}.$$

Hence passing to the supremum over all divisions $a = s_0 < s_1 < \cdots < s_k = b$ of $[a,b]$ we obtain

$$\operatorname{var}_a^b Tx \le \operatorname{var}_a^b A \|x\|_{BV} \tag{6.5}$$

and $Tx : [a,b] \to \mathbb{R}^n$ is of bounded variation on $[a,b]$.

Clearly, the mapping $T : BV \to BV$ is linear. Using the same argument as above we have

$$\|Tx(a)\| = \|\int_{t_0}^{a} d[A(s)]x(s)\| \le \int_a^{t_0} \|x(s)\|\, d(\operatorname{var}_{t_0}^{s} A) \le$$

$$\leq \|x\|_{BV} \operatorname{var}_a^{t_0} A \leq \|x\|_{BV} \operatorname{var}_a^b A,$$

and this together with (6.5) yields

$$\|Tx\|_{BV} = \|Tx(a)\| + \operatorname{var}_a^b Tx \leq 2 \operatorname{var}_a^b A \|x\|_{BV}. \tag{6.6}$$

Therefore the operator $T : BV \to BV$ is bounded.

Finally, we have to show that $T : BV \to BV$ is compact. Assume that $x_k \in BV$, $k = 1, 2, \dots$ is a bounded sequence in BV, i.e. $\|x_k\| \leq C$, $k = 1, 2, \dots$ for some constant $C \geq 0$. By Helly's Choice Theorem the sequence (x_k) contains a subsequence (x_{k_l}) which converges pointwise to a function $\widetilde{x} \in BV$, i.e

$$\lim_{l \to \infty} x_{k_l}(s) = \widetilde{x}(s), \ s \in [a, b]. \tag{6.7}$$

Define

$$y(t) = \int_{t_0}^{t} d[A(s)]\widetilde{x}(s), \ t \in [a, b];$$

then evidently $y \in BV$. Let us set further

$$z_l(s) = x_{k_l}(s) - \widetilde{x}(s), \ s \in [a, b].$$

Evidently $z_l \in BV$ and by (6.7) we get

$$\lim_{l \to \infty} z_l(s) = 0, \ s \in [a, b]. \tag{6.8}$$

Moreover,

$$\|z_l(s)\| \leq \|z_l(a)\| + \|z_l(s) - z_l(a)\| \leq \|z_l\|_{BV} =$$

$$= \|x_{k_l} - \widetilde{x}\|_{BV} \leq \|x_{k_l}\|_{BV} + \|\widetilde{x}\|_{BV} \leq C + \|\widetilde{x}\|_{BV} \tag{6.9}$$

for every $s \in [a, b]$. We have also

$$\|Tx_{k_l}(a) - y(a)\| = \| \int_{t_0}^{a} d[A(s)] z_l(s) \| \leq \int_{a}^{t_0} \|z_l(s)\| \, d(\operatorname{var}_{t_0}^{s} A).$$

From (6.8) and (6.9) we obtain by Corollary 1.32 the relation

$$\lim_{l\to\infty} \|Tx_{k_l}(a) - y(a)\| = 0. \tag{6.10}$$

Further, for $s_1, s_2 \in [a, b]$, $s_1 < s_2$ we obtain

$$\|Tx_{k_l}(s_2) - y(s_2) - (Tx_{k_l}(s_1) - y(s_1))\| = \|Tz_l(s_2) - Tz_l(s_1)\| =$$

$$= \| \int_{s_1}^{s_2} d[A(s)]z_l(s)\| \leq \int_{s_1}^{s_2} \|z_l(s)\| \, d(\operatorname{var}_{t_0}^{s} A)$$

and for an arbitrary division $a = s_0 < s_1 < \cdots < s_k = b$ of $[a, b]$ also

$$\sum_{i=1}^{k} \|Tx_{k_l}(s_i) - y(s_i) - (Tx_{k_l}(s_{i-1}) - y(s_{i-1}))\| \leq$$

$$\leq \sum_{i=1}^{k} \int_{s_{i-1}}^{s_i} \|z_l(s)\| \, d(\operatorname{var}_{t_0}^{s} A) = \int_a^b \|z_l(s)\| \, d(\operatorname{var}_{t_0}^{s} A).$$

This yields as usual

$$\operatorname{var}_a^b(Tx_{k_l} - y) \leq \int_a^b \|z_l(s)\| \, d(\operatorname{var}_{t_0}^{s} A)$$

and again (6.8), (6.9) and Corollary 1.32 imply

$$\lim_{l\to\infty} \int_a^b \|z_l(s)\| \, d(\operatorname{var}_{t_0}^{s} A) = 0$$

and consequently

$$\lim_{l\to\infty} \operatorname{var}_a^b(Tx_{k_l} - y) = 0.$$

This relation together with (6.10) yields

$$\lim_{l\to\infty} \|Tx_{k_l} - y\|_{BV} = 0,$$

i.e. the sequence (Tx_k) contains a subsequence which converges in BV (to $y \in BV$) and the operator T is compact. □

6.3 Proposition. *If $[a,b] \subset J$ is a compact interval, $t_0 \in [a,b]$ then either*

(A) *the equation*

$$x(t) = \int_{t_0}^{t} d[A(s)]x(s) + f(t), \; t \in [a,b] \tag{6.11}$$

admits a unique solution in $BV([a,b])$ for any $f \in BV$

or

(B) *the homogeneous equation*

$$x(t) = \int_{t_0}^{t} d[A(s)]x(s), \; t \in [a,b] \tag{6.12}$$

admits at least one nontrivial solution in $BV([a,b])$.

Proof. The equations (6.11), (6.12) can be written in the form

$$x - Tx = f, \quad x - Tx = 0,$$

respectively, where $T : BV \to BV$ is the operator defined by (6.4). This operator is compact by Proposition 6.2 and therefore the alternative given in the statement follows immediately from the well known Fredholm alternative for equations of this form (see e.g. [21], [121] or any other standard textbook on linear operator equations in Banach spaces). □

Let us now consider the case when in Proposition 6.3 (A) occurs. This is equivalent to the situation when the homogeneous equation (6.12) has only the trivial solution $x = 0$ in BV.

6.4 Proposition. *If $[a,b] \subset J$ is a compact interval, $t_0 \in [a,b]$ then (6.12) has only the trivial solution $x = 0$ in BV if and only if*

(C)

$$I - [A(t) - A(t-)] = I - \Delta^{-} A(t)$$

is regular for any $t \in (t_0, b]$

and

$$I + [A(t+) - A(t)] = I - \Delta^+ A(t)$$

is regular for any $t \in [a, t_0)$.

Here I denotes the $n \times n$ identity matrix.

In other words the condition (C) is equivalent to (A) from Proposition 6.3.

Proof. Define

$$\xi(t) = \operatorname{var}_{t_0}^t A, \ t \in [a, b].$$

$\xi : [a, b] \to \mathbb{R}$ is a nondecreasing function.

Assume that the condition (C) is satisfied and that $x : [a, b] \to \mathbb{R}^n$, $x \in BV$ is a solution of (6.12). Then evidently $x(t_0) = 0$. Suppose that $t_0 < b$; then there is a $c \in (t_0, b]$ such that $\xi(c) - \xi(t_0) < \frac{1}{2}$.

For arbitrary $s_1, s_2 \in [t_0, c]$, $s_1 < s_2$ we have

$$\|x(s_2) - x(s_1)\| = \|\int_{s_1}^{s_2} d[A(s)]x(s)\| \le \int_{s_1}^{s_2} \|x(s)\| d\xi(s)$$

and, as usual, this yields

$$\operatorname{var}_{t_0}^c x \le \int_{t_0}^c \|x(s)\| d\xi(s) =$$

$$= \|x(t_0)\| [\xi(t_0+) - \xi(t_0)] + \lim_{\delta \to 0+} \int_{t_0+\delta}^c \|x(s)\| d\xi(s) \le$$

$$\le \|x\|_{BV([t_0,c])} \lim_{\delta \to 0+} (\xi(c) - \xi(t_0 + \delta)) =$$

$$= \|x\|_{BV([t_0,c])} (\xi(c) - \xi(t_0+)).$$

Hence by the choice of c we obtain

$$\|x\|_{BV([t_0,c])} \le \frac{1}{2}\|x\|_{BV([t_0,c])},$$

i.e. $\|x\|_{BV([t_0,c])} = 0$ and $x(t) = 0$ for every $t \in [t_0, c]$.

Let us assume that $t^\star \in (t_0, b]$ is the supremum of all $c \in (t_0, b]$ such that $x(t) = 0$ on $[t_0, c]$. Then evidently $x(t) = 0$ for all $t \in [t_0, t^\star)$. Using Theorem 1.16 we get

$$x(t^\star) - x(t^\star-) = \lim_{\delta\to 0+} \int_{t^*-\delta}^{t^*} d[A(s)]x(s) =$$

$$= \lim_{\delta\to 0+} [A(t^\star) - A(t^\star - \delta)]x(t^\star) = \Delta^- A(t^\star)x(t^\star),$$

i.e.

$$0 = x(t^\star-) = [I - \Delta^- A(t^\star)]x(t^\star),$$

and the regularity of the matrix $I - \Delta^- A(t^\star)$ yields $x(t^\star) = 0$.

If $< b$ then in the same way as above we can show that there is a $c > t^\star$ in $[a, b]$ such that $x(t) = 0$ on $[t_0, c]$, but this contradicts the assumption that $t^\star$ is the supremum of all $c \in [t_0, b]$ for which $x(t) = 0$ on $[t_0, c]$. Hence necessarily $t^\star = b$ and $x(t) = 0$ on $[t_0, b]$. In a completely analogous way it can be shown that $x(t) = 0$ on $[a, t_0]$ and therefore $x(t) = 0$, $t \in [a, b]$ is the only solution of (6.12) on $[a, b]$.

To show the other implication let us assume that the condition (C) is not satisfied.

Since A is of bounded variation on $[a, b]$, there can be only a finite set of points $t^\star > t_0$ at which the matrix $I - \Delta^- A(t^\star)$ is not regular because $\|\Delta^- A(t)\| \ge \frac{1}{2}$ only for a finite set of points in $(t_0, b]$. At all other points in this interval we have $\|\Delta^- A(t)\| < \frac{1}{2}$ and consequently the matrix $I - \Delta^- A(t)$ has an inverse $[I - \Delta^- A(t)]^{-1}$ at these points and $I - \Delta^- A(t)$ is regular there.

Assume e.g. that there is a point $t^\star \in (t_0, b]$ such that the matrix $I - \Delta^- A(t^\star)$ is not regular and $I - \Delta^- A(t)$ is regular for every $t \in (t_0, t^\star)$. Then there exists $y \in \mathbb{R}^n$ such that the algebraic system

$$[I - \Delta^- A(t^\star)]z = y$$

has no solution in $\mathbb{R}^n$.

Define $f(t) = 0$ for $t \in [a, b]$, $t \neq t^\star$ and $f(t^\star) = y$. Then evidently $f \in BV$. Suppose that x is a solution of the nonhomogeneous equation (6.11) with this f on the right hand side. Then we can show as above that $x(t) = 0$ for $t \in [t_0, t^\star)$ and we have

$$x(t^\star) = x(t^\star) - x(t^\star -) =$$

$$= \Delta^- A(t^\star)x(t^\star) + f(t^\star) - f(t^\star -) = \Delta^- A(t^\star)x(t^\star) + y,$$

and this means that for the value $x(t^\star)$ we obtain

$$[I - \Delta^- A(t^\star)]x(t^\star) = y.$$

By the assumptions given above such a value cannot exist in $\mathbb{R}^n$ and therefore also the equation (6.11) cannot have a solution in BV for the given choice of f. Consequently, by Proposition 6.3 there is at least one nontrivial solution of the homogeneous equation (6.12) and this proves the desired implication as well as the statement of our proposition. $\square$

After this short excursion into the functional analysis of the operator $T : BV \to BV$ we turn our attention again to the generalized linear differential equation (6.1).

The result given in Proposition 6.4 motivates the following additional assumption on the matrix valued function $A : J \to L(\mathbb{R}^n)$:

$$I - [A(t) - A(t-)] = I - \Delta^- A(t)$$

is regular for any $t \in J$

and

$$I + [A(t+) - A(t)] = I - \Delta^+ A(t) \tag{6.13}$$

is regular for any $t \in J$.

By Propositions 6.4 and 6.3 the conditions (6.13) (or equivalently (C)) assure that the equation (6.11) has a unique solution of bounded variation on $[a, b]$ for an arbitrary choice of $f \in BV([a, b])$, i.e. that the case (A) from Proposition 6.3 occurs.

6.5 Theorem. *If $J \subset \mathbb{R}$ is an interval (finite or infinite), $A : J \to L(\mathbb{R}^n)$, $g : J \to \mathbb{R}^n$ are functions of locally bounded variation in J and (6.13) holds, then for every $(\widetilde{x}, t_0) \in \mathbb{R}^n \times J$ there exists a unique solution $x(t)$ of*

$$\frac{dx}{d\tau} = D[A(t)x + g(t)] \tag{6.1}$$

satisfying the initial condition $x(t_0) = \widetilde{x}$. This solution exists on the whole interval J and is locally of bounded variation on J.

Proof. By definition we have

$$x(t) = \widetilde{x} + \int_{t_0}^{t} d[A(s)]x(s) + g(t) - g(t_0) \tag{6.14}$$

for the solution of the equation (6.1). Put $f(t) = \widetilde{x} + g(t) - g(t_0)$. Then $f : J \to \mathbb{R}^n$ is locally of bounded variation on J and on every compact interval $[a, b] \subset J$ with $t_0 \in [a, b]$ the equation (6.14) is equivalent to the equation (6.11). Using Proposition 6.3 and 6.4 we obtain that there is a unique solution $x \in BV([a, b])$ of (6.11) with this function f and therefore we have also the existence of a solution of (6.14) (or equivalently of (6.1)) on $[a, b]$. This applies for any compact interval $[a, b] \subset J$ and consequently our statement holds because evidently the whole interval J can be exhausted by intervals of the form $[a, b] \subset J$ and the resulting prolongation of the solution is unique by Propositions 6.3 and 6.4. □

6.6 Corollary. *If $J \subset \mathbb{R}$ is an interval (finite or infinite), $A : J \to L(\mathbb{R}^n)$ a function of locally bounded variation in J and (6.13) holds, then the homogeneous generalized linear differential equation*

$$\frac{dx}{d\tau} = D[A(t)x] \tag{6.15}$$

with the initial condition $x(t_0) = 0$, $t_0 \in J$ has only the trivial solution $x(t) = 0$ for $t \in J$.

Proof. The result easily follows from Theorem 6.5. □

6.7 Theorem. *Assume that $J \subset \mathbb{R}$ is an interval, $A : J \to L(\mathbb{R}^n)$, $g : J \to \mathbb{R}^n$ are of locally bounded variation in J and (6.13) holds.*

If $x, y : J \to \mathbb{R}^n$ are solutions of (6.15) and $\alpha, \beta \in \mathbb{R}$, then $\alpha x + \beta y$ is also a solution of (6.15).

If $x : J \to \mathbb{R}^n$ is a solution of (6.15) and $z : J \to \mathbb{R}^n$ is a solution of (6.1) then $x + z : J \to \mathbb{R}^n$ is a solution of (6.1).

Proof. For every $s_1, s_2 \in J$ we have by definition

$$x(s_2) - x(s_1) = \int_{s_1}^{s_2} d[A(s)]x(s)$$

and similarly for $y : J \to \mathbb{R}^n$. Hence by the linearity of the integral we get

$$\alpha x(s_2) + \beta y(s_2) - \alpha x(s_1) - \beta y(s_1) =$$

$$= \alpha \int_{s_1}^{s_2} d[A(s)]x(s) + \beta \int_{s_1}^{s_2} d[A(s)]x(s) =$$

$$= \int_{s_1}^{s_2} d[A(s)](\alpha x(s) + \beta y(s)),$$

and $\alpha x + \beta y$ is a solution of (6.15) by the definition. For $z : J \to \mathbb{R}^n$ we have

$$z(s_2) - z(s_1) = \int_{s_1}^{s_2} d[A(s)]z(s) + g(s_2) - g(s_1)$$

for every $s_1, s_2 \in J$. Hence

$$z(s_2) + x(s_2) - z(s_1) - x(s_1) =$$

$$= \int_{s_1}^{s_2} d[A(s)](z(s) + x(s)) + g(s_2) - g(s_1)$$

for every $s_1, s_2 \in J$, and $x + z$ is a solution of (6.1). □

6.8 Theorem. *If $J \subset \mathbb{R}$ is an interval (finite or infinite), $A : J \to L(\mathbb{R}^n)$ a function of locally bounded variation in J and (6.13) holds, then the set of all solutions of the homogeneous generalized linear differential equation (6.15) is an n–dimensional subspace in the space $BV_{loc}(J)$ of all $\mathbb{R}^n$– valued functions on J which are locally of bounded variation in J.*

Proof. The linearity of the set of all solutions of the homogeneous generalized linear differential equation (6.15) is established by Theorem 6.7. Every solution of (6.15) belongs to the space $BV_{loc}(J)$ by Theorem 6.5.

If $t_0 \in J$ then to a solution x of (6.15) we assign its value $x(t_0) \in \mathbb{R}^n$. It is easy to see (by the uniqueness of the solutions of (6.15)) that in this way a one–to–one map between $\mathbb{R}^n$ and the set of solutions of (6.15) is defined and this implies that the solutions of (6.15) form an n–dimensional subspace in the space $BV_{loc}(J)$. □

6.9 Theorem. *Assume that $J \subset \mathbb{R}$ is an interval, $A : J \to L(\mathbb{R}^n)$, $g : J \to \mathbb{R}^n$ are of locally bounded variation in J and (6.13) holds. Let $z : J \to \mathbb{R}^n$ be a fixed solution of (6.1).*

Then every solution of (6.1) can be written in the form $x + z$ where $x : J \to \mathbb{R}^n$ is a solution of (6.15).

Proof. The difference of any two solutions of (6.1) is a solution of the homogeneous equation (6.15). The result is a consequence of Theorem 6.7. □

6.10 *Remark.* The results of Theorems 6.7 - 6.9 show that the set of all solutions of the generalized linear differential equation (6.1) has the usual linear structure which is known for the case of classical linear differential equations.

The fundamental matrix

6.11 Theorem. *Assume that $A : J \to L(\mathbb{R}^n)$ is of locally bounded variation in J and satisfies (6.13).If $t_0 \in J$ then for every $n \times n$–matrix $\widetilde{X} \in L(\mathbb{R}^n)$ there exists a uniquely determined $n \times n$–matrix valued function $X : J \to L(\mathbb{R}^n)$ such that*

$$X(t) = \widetilde{X} + \int_{t_0}^{t} d[A(s)]X(s) \tag{6.16}$$

for $t \in J$.

Proof. It is easy to see that $X : J \to L(\mathbb{R}^n)$ satisfies (6.16) if and only if any column of X satisfies (6.15), i.e. if for every $k = 1, \ldots, n$ the k–th column X_k of X satisfies

$$X_k(t) = \widetilde{X}_k + \int_{t_0}^{t} d[A(s)]X_k(s).$$

The result now easily follows from Theorem 6.5. □

Let us introduce some notions analogous to the case of classical linear ordinary differential equations.

A matrix valued function $X : J \to L(\mathbb{R}^n)$ is called a *solution of the matrix equation*

$$\frac{dX}{d\tau} = D[A(t)X] \tag{6.17}$$

if for every $s_1, s_2 \in J$ the identity

$$X(s_2) - X(s_1) = \int_{s_1}^{s_2} d[A(s)]X(s)$$

holds.

A matrix valued function $X : J \to L(\mathbb{R}^n)$ is called a *fundamental matrix* of the equation (6.15) if X is a solution of the matrix equation (6.17) and if the matrix $X(t)$ is regular for at least one value $t \in J$.

6.12 Theorem. *Assume that $A : J \to L(\mathbb{R}^n)$ is of locally bounded variation in J and satisfies (6.13). Then every fundamental matrix $X : J \to L(\mathbb{R}^n)$ of the equation (6.15) is regular for all $t \in J$.*

Proof. By definition the fundamental matrix X is a solution of the matrix equation (6.17) and there is a $t_0 \in J$ duch that the matrix $\widetilde{X} = X(t_0)$ is regular. Assume that for some $t_1 \in J$ the matrix $X(t_1)$ is not regular. Then there exist constants $c_1, \ldots, c_n \in \mathbb{R}$, $c_k \neq 0$ for at least one $k = 1, \ldots, n$ such that

$$\sum_{k=1}^{n} c_k X_k(t_1) = 0$$

where X_k denotes the k–th column of the matrix X. Since every column of X is a solution of the homogeneous equation (6.15), by

Theorem 6.7 the linear combination $x(t) = \sum_{k=1}^{n} c_k X_k(t)$ is also a solution of (6.15) where $x(t_1) = \sum_{k=1}^{n} c_k X_k(t_1) = 0$. Hence by the uniqueness of solutions we have $x(t) = 0$ for $t \in J$ and therefore $x(t_0) = \sum_{k=1}^{n} c_k X_k(t_0) = 0$. Consequently, we get $c_k = 0$ for all $k = 1, \ldots, n$ because the columns $X_k(t_0)$ of the regular matrix $X(t_0)$ are linearly independent. This contradiction proves that the matrix $X(t)$ is regular for all $t \in J$. □

Now we reformulate the result given in Theorem 6.11.

6.13 Theorem. *Assume that $A : J \to L(\mathbb{R}^n)$ is of locally bounded variation in J and satisfies (6.13). Then there exists a uniquely determined $n \times n$–matrix valued function $U : J \times J \to L(\mathbb{R}^n)$ such that*

$$U(t,s) = I + \int_s^t d[A(r)]U(r,s) \tag{6.18}$$

for $t, s \in J$. For every fixed $s \in J$ the $n \times n$–matrix valued function $U(\cdot, s)$ is locally of bounded variation in J.

Proof. For a given $s \in J$ the matrix U is a solution of the equation

$$X(t) = I + \int_s^t d[A(r)]X(r).$$

By Theorem 6.11 this solution exists for all $t \in J$ and is uniquely determined for every fixed $s \in J$. The local boundedness of the variation of this solution on J is guaranteed by Theorem 6.5. □

6.14 Theorem. *Suppose that $A : J \to L(\mathbb{R}^n)$ is of locally bounded variation in J and satisfies (6.13). Then the unique solution $x : J \to \mathbb{R}^n$ of the initial value problem for the generalized linear differential equation*

$$\frac{dx}{d\tau} = D[A(t)x] \tag{6.15}$$

with the initial condition $x(s) = \widetilde{x} \in \mathbb{R}^n$, $s \in J$ is given by the relation

$$x(t) = U(t,s)\widetilde{x}, \quad t \in J \tag{6.19}$$

where $U : J \times J \to L(\mathbb{R}^n)$ is given by Theorem 6.13.

Proof. The function $x : J \to \mathbb{R}^n$ given by (6.19) is evidently of locally bounded variation on J by Theorem 6.5. Hence for every $t \in J$ the integral $\int_s^t d[A(r)]x(r)$ exists and we have

$$\int_s^t d[A(r)]x(r) = \int_s^t d[A(r)]U(r,s)\widetilde{x} =$$

$$= (U(t,s) - I)\widetilde{x} = x(t) - \widetilde{x}$$

for $t \in J$ and $x(s) = U(s,s)\widetilde{x} = \widetilde{x}$. This means that x is a solution of the initial value problem from the statement, and it is of course unique by Theorem 6.5. □

Now we give a survey of basic properties of the $n \times n$–matrix valued function $U : J \times J \to L(\mathbb{R}^n)$ introduced in Theorem 6.13.

6.15 Theorem. *Suppose that $A : J \to L(\mathbb{R}^n)$ is of locally bounded variation in J and satisfies (6.13). Then the $n \times n$–matrix valued function $U : J \times J \to L(\mathbb{R}^n)$, which is uniquely determined by (6.18) has the following properties.*

(a) *$U(t,t) = I$ for $t \in J$,*

(b) *for every compact interval $[a,b] \subset J$ there is a constant $M \geq 0$ such that*

$$\|U(t,s)\| \leq M \text{ for all } t,s \in [a,b],$$

$$\operatorname{var}_a^b U(t,\cdot) \leq M \text{ for } t \in [a,b],$$

$$\operatorname{var}_a^b U(\cdot,s) \leq M \text{ for } s \in [a,b],$$

(c) *for* $r, s, t \in J$ *the relation*

$$U(t,s) = U(t,r)U(r,s)$$

holds,

(d) $U(t,s) \in L(\mathbb{R}^n)$ *is regular for every* $t, s \in J$,

(e)

$$U(t+,s) = [I + \Delta^+ A(t)]U(t,s),$$

$$U(t-,s) = [I - \Delta^- A(t)]U(t,s),$$

$$U(t,s+) = U(t,s)[I + \Delta^+ A(s)]^{-1},$$

$$U(t,s-) = U(t,s)[I - \Delta^- A(s)]^{-1}$$

for $t, s \in J$ *whenever the limits involved make sense,*

(f) *for* $t, s \in J$ *the relation*

$$[U(t,s)]^{-1} = U(s,t)$$

holds.

Proof. The property (a) is an immediate consequence of the definition given by (6.18).

Denoting by U_k the k–th column of the matrix U we can write (6.18) in the form

$$U_k(t,s) = e_k + \int_s^t d[A(r)]U_k(r,s)$$

for $k = 1, \ldots, n$ where e_k is the k–th column of the identity matrix $I \in L(\mathbb{R}^n)$.

Assume that $s \in [a,b)$ is fixed and $s < t \leq b$. For $r \in [s,b]$ define $\widehat{A}(r)$ by the relations $\widehat{A}(s) = A(s)$ and $\widehat{A}(r) = A(r-)$ when

$r \in (s, b]$. It is not difficult to check that for any $y \in BV([s, b])$ and $t \in (s, b]$ we have

$$\int_s^t d[A(r)]y(r) = \int_s^t d[\widehat{A}(r)]y(r) + [A(t) - A(t-)]y(t) =$$

$$= \int_s^t d[\widehat{A}(r)]y(r) + \Delta^- A(t)y(t).$$

Therefore

$$U_k(t, s) = e_k + \int_s^t d[\widehat{A}(r)]U_k(r, s) + \Delta^- A(t)U_k(t, s)$$

and also

$$[I - \Delta^- A(t)]U_k(t, s) = e_k + \int_s^t d[\widehat{A}(r)]U_k(r, s).$$

Hence

$$U_k(t, s) = [I - \Delta^- A(t)]^{-1}(e_k + \int_s^t d[\widehat{A}(r)]U_k(r, s))$$

and also

$$\|U_k(t, s)\| \le \|[I - \Delta^- A(t)]^{-1}\|(\|e_k\| + \int_s^t \|U_k(r, s))\| d \operatorname{var}_s^r \widehat{A}$$

for every $t \in (s, b]$. Since $\widehat{A}$ is continuous from the left in $(s, b]$ the function $\operatorname{var}_s^r \widehat{A}$, $r \in (s, b]$ is also continuous from the left in its domain of definition. Since the matrix valued function $[I - \Delta^- A(t)]$ is assumed to be regular on J and A is of bounded variation on $[a,b]$ there exists a constant $C > 0$ such that $\|[I - \Delta^- A(t)]^{-1}\| \leq C$ for all $t \in [s,b]$. Therefore we obtain

$$\|U_k(t,s)\| \leq C\|e_k\| + \int_s^t \|U_k(r,s))\| d\operatorname{var}_s^r \widehat{A}$$

and Corollary 1.43 yields the estimate

$$\|U_k(t,s)\| \leq C\|e_k\| e^{C \operatorname{var}_s^t \widehat{A}} \leq$$

$$\leq C\|e_k\| e^{C \operatorname{var}_a^b A} = L_k$$

if $s \leq t \leq b$. If $s \in (a,b]$ is fixed and $a \leq t < s$ then for $r \in [a,s]$ we can define $\widehat{A}(r)$ by the relations $\widehat{A}(s) = A(s)$ and $\widehat{A}(r) = A(r+)$ if $r \in [a,s]$. Essentially the same reasoning as above makes it possible to use Corollary 1.44 for showing that

$$\|U_k(t,s)\| \leq C\|e_k\| e^{C \operatorname{var}_a^b A} = L_k$$

for $a \leq t \leq s$ with some constant C. Since this can be done for every column of U we have the estimate

$$\|U(t,s)\| \leq M_1, \quad t,s \in [a,b].$$

Using this estimate we obtain for arbitrary t_1, t_2 with $a \leq t_1 < t_2 \leq b$ the inequality

$$\|U(t_2,s) - U(t_1,s)\| = \| \int_{t_1}^{t_2} d[A(r)]U(r,s)\| \leq$$

$$\leq \int_{t_1}^{t_2} \|U(r,s)\| d\operatorname{var}_s^r A \leq M_1 \operatorname{var}_{t_1}^{t_2} A$$

and therefore

$$\operatorname{var}_a^b U(\cdot, s) \le M_1 \operatorname{var}_a^b A = M_2$$

for $s \in [a, b]$.

If $a \le s_1 \le s_2 \le b$ then

$$U(t, s_2) - U(t, s_1) = \int_{s_2}^{t} d[A(r)]U(r, s_2) - \int_{s_2}^{t} d[A(r)]U(r, s_1) -$$

$$- \int_{s_1}^{s_2} d[A(r)]U(r, s_1) = - \int_{s_1}^{s_2} d[A(r)]U(r, s_1) +$$

$$+ \int_{s_2}^{t} d[A(r)](U(r, s_2) - U(r, s_1)),$$

and this means that the difference $X(t) = U(t, s_2) - U(t, s_1)$, $t \in [a, b]$ satisfies on $[a, b]$ the matrix equation (6.17) with the initial condition

$$X(s_2) = - \int_{s_1}^{s_2} d[A(r)]U(r, s_1),$$

and by Theorem 6.14 (when applied to the columns of X) we get

$$X(t) = U(t, s_2) - U(t, s_1) = -U(t, s_2) \int_{s_1}^{s_2} d[A(r)]U(r, s_1)$$

for $t \in [a, b]$. Hence

$$\|U(t, s_2) - U(t, s_1)\| \le \|U(t, s_2)\| \int_{s_1}^{s_2} \|U(r, s_1)\| d \operatorname{var}_a^r A \le$$

$$\le M_1^2 \operatorname{var}_{s_1}^{s_2} A$$

and therefore

$$\operatorname{var}_a^b U(t,\cdot) \leq M_1^2 \operatorname{var}_a^b A = M_3$$

for every $t \in [a,b]$. Putting $M = \max(M_1, M_2, M_3)$ we obtain (b).

By (6.18) which defines U we have

$$U(t,s) = I + \int_s^t d[A(\tau)]U(\tau,s) =$$

$$= I + \int_s^r d[A(\tau)]U(\tau,s) + \int_r^t d[A(\tau)]U(\tau,s) =$$

$$= U(r,s) + \int_r^t d[A(\tau)]U(\tau,s).$$

By this relation the k–th column ($k = 1,\dots,n$) $U_k(t,s)$ of $U(t,s)$ is a solution od the initial value problem

$$\frac{dx}{d\tau} = D[A(t)x], \quad x(r) = U_k(r,s)$$

and therefore by Theorem 6.14 we have

$$U_k(t,s) = U(t,r)U_k(r,s), \quad k = 1,\dots,n$$

and (c) evidently holds.

The property (d) is the content of Theorem 6.12.

The property (e) is a consequence of Theorem 1.16. Let us prove for example the second identity in (e). By definition we have

$$U(t,s) - U(t-\delta,s) = \int_{t-\delta}^t d[A(r)]U(r,s)$$

whenever $\delta > 0,\ t, t-\delta \in J$. Theorem 1.16 now gives

$$U(t,s) - U(t-,s) = \lim_{\delta\to 0+}[U(t,s) - U(t-\delta,s)] =$$

$$= \lim_{\delta\to 0+}\int_{t-\delta}^{t} d[A(r)]U(r,s) =$$

$$= \lim_{\delta\to 0+}[A(t) - A(t-\delta)]U(t,s) = \Delta^- A(t)U(t,s),$$

and therefore

$$U(t-,s) = [I - \Delta^- A(t)]U(t,s).$$

The other identities can be proved in the same way.

The identity in (f) is an easy consequence of the identity $U(t,s)U(s,t) = U(t,t) = I$ which holds by (c) and (a).

The variation of constants formula

□

6.16 Lemma. *Assume that $A : J \to L(\mathbb{R}^n)$ is of locally bounded variation in J and that (6.13) is satisfied. Let $\varphi : J \to \mathbb{R}^n$ be of locally bounded variation in J and let $K \in L(\mathbb{R}^n, \mathbb{R}^m)$ be an $m\times n$–matrix valued function which is of bounded variation on $[a,b]$.*

Then for $U : J\times J \to L(\mathbb{R}^n)$ given by (6.18) in Theorem 6.13 the equality

$$\int_{t_0}^{t} d[K(r)](\int_{t_0}^{r} d_s[U(r,s)]\varphi(s)) =$$

$$= int_{t_0}^{t} d_s[K(s)]\varphi(s) + \int_{t_0}^{t} d_s[\int_{s}^{t} d_r[K(r)]U(r,s)]\varphi(s) \qquad (6.20)$$

holds for $t_0, t \in J$.

Proof. Assume that $t \geq t_0$ and define

$$V(r,s) = U(r,s) \text{ for } t_0 \leq s \leq r,$$

$$V(r,s) = U(r,r) = I \text{ for } t_0 \leq r \leq s.$$

If $t_0 \leq r \leq t$ then

$$\int_r^t d_s[V(r,s)]\varphi(s) = \int_r^t d_s[U(r,r)]\varphi(s) = 0$$

and

$$\int_{t_0}^r d_s[V(r,s)]\varphi(s) = \int_{t_0}^r d_s[U(r,s)]\varphi(s)$$

and therefore for $t_0 \leq r \leq t$ we get

$$\int_{t_0}^r d_s[U(r,s)]\varphi(s) = \int_{t_0}^t d_s[V(r,s)]\varphi(s). \tag{6.21}$$

For $\rho, s, \sigma \in [t_0, t]$ now define

$$W(\rho,\sigma,s) = V(\rho,s)\varphi(\sigma).$$

By (6.21) we have

$$\int_{t_0}^r d_s[U(\rho,s)]\varphi(s) = \int_{t_0}^t D_s[W(\rho,\sigma,s)]$$

and also

$$\int_{t_0}^t d[K(r)](\int_{t_0}^r d_s[U(r,s)]\varphi(s)) =$$

$$= \int_{t_0}^t D_r[\int_{t_0}^t D_s[K(r)V(\rho,s)]\varphi(\sigma)] =$$

$$= \int_{t_0}^{t} D_r[\int_{t_0}^{t} D_s[K(r)W(\rho,\sigma,s)]].$$

Our goal is to interchange the order of integration in this double integral. We will use the Tonelli–type Theorem 1.45 to this end. Theorem 1.45 is here used in fact for the components of a vector valued function.

By Theorem 6.15 the function $U(\cdot, s)$ is of bounded variation on $[t_0, t]$.By Corollary 1.34 this yields the existence of the integral $\int_{t_0}^{t} D[K(r)V(\rho, s)]$ for any $s \in [t_0, t]$.

Therefore the integral

$$\int_{t_0}^{t} D_r[K(r)V(\rho,s)\varphi(\sigma)] = \int_{t_0}^{t} D_r[K(r)W(\rho,\sigma,s)]$$

exists for every $(\sigma, s) \in [t_0, t] \times [t_0, t]$.

The next thing we have to do is to prove that the system of functions $W(\rho, \cdot, \cdot) = V(\rho, \cdot)]\varphi(\cdot)$ is equiintegrable over $[t_0, t]$ for $\rho \in [t_0, t]$. Using (c) from Theorem 6.15 we obtain by (6.21) the equality

$$\int_{t_0}^{t} D_s W(\rho,\sigma,s) = \int_{t_0}^{\rho} d_s[U(\rho,s)]\varphi(s) =$$

$$= U(\rho,t_0)\int_{t_0}^{\rho} d_s[U(t_0,s)]\varphi(s) \tag{6.22}$$

for $\rho \in [t_0, t]$. Let $\eta > 0$ be arbitrary.

The function $U(t_0, \cdot)$ is of bounded variation in the interval $[t_0, t]$ (see (b) in Theorem 6.15), the integral $\int_{t_0}^{t} d_s[U(t_0, s)]\varphi(s)$ exists and therefore by definition there is a gauge δ on $[t_0, t]$ such that

$$\|\sum_{j=1}^{m}(U(t_0,\beta_j) - U(t_0,\beta_{j-1}))\varphi(\sigma_j) - \int_{t_0}^{t} d_s[U(t_0,s)]\varphi(s)\| < \eta \tag{6.23}$$

for every δ–fine partition

$$D = \{\beta_0, \sigma_1, \beta_1, \dots, \beta_{m-1}, \sigma_m, \beta_m\}$$

of $[t_0, t]$ with tags σ_j in $[\beta_{j-1}, \beta_j]$. Let $\rho \in [t_0, t]$ be given and let D be a δ–fine partition of $[t_0, t]$ of the form described above. Then there is an index $k \in \{1, \dots, m\}$ such that $\rho \in [\beta_{k-1}, \beta_k]$ and

$$\|\sum_{j=1}^{m}(W(\rho, \sigma_j, \beta_j) - W(\rho, \sigma_j, \beta_{j-1})) - \int_{t_0}^{t} D_s W(\rho, \sigma, s)\| =$$

$$= \|\sum_{j=1}^{m}(V(\rho, \beta_j) - V(\rho, \beta_{j-1}))\varphi(\sigma_j) -$$

$$-U(\rho, t_0)\int_{t_0}^{\rho} d_s[U(t_0, s)]\varphi(s)\| =$$

$$= \|\sum_{j=1}^{k-1}(U(\rho, \beta_j) - U(\rho, \beta_{j-1}))\varphi(\sigma_j) +$$

$$+(U(\rho, \rho) - U(\rho, \beta_{k-1}))\varphi(\sigma_k) - U(\rho, t_0)\int_{t_0}^{\rho} d_s[U(t_0, s)]\varphi(s)\| \leq$$

$$\leq \|U(\rho, t_0)\| . \|\sum_{j=1}^{k-1}(U(t_0, \beta_j) - U(t_0, \beta_{j-1}))\varphi(\sigma_j) -$$

$$-\int_{t_0}^{\beta_{j-1}} d_s[U(t_0, s)]\varphi(s)\| +$$

$$+\|U(\rho, t_0)\| . \|(U(t_0, \rho) - U(t_0, \beta_{k-1}))\varphi(\sigma_k) -$$

$$-\int_{\beta_{k-1}}^{\rho} d_s[U(t_0, s)]\varphi(s)\|. \tag{6.24}$$

The Saks–Henstock Lemma 1.13 yields by (6.23) the estimate

$$\|\sum_{j=1}^{k-1}(U(t_0,\beta_j)-U(t_0,\beta_{j-1}))\varphi(\sigma_j)-\int_{t_0}^{\beta_{j-1}} d_s[U(t_0,s)]\varphi(s)\| \le \eta. \tag{6.25}$$

If $\sigma_k \le \rho$ then $[\beta_{k-1},\rho] \subset [\beta_{k-1},\beta_k] \subset (\sigma_k - \delta(\sigma_k), \sigma_k + \delta(\sigma_k))$ and again by the Saks–Henstock Lemma we get

$$\|(U(t_0,\rho) - U(t_0,\beta_{k-1}))\varphi(\sigma_k) - \int_{\beta_{k-1}}^{\rho} d_s[U(t_0,s)]\varphi(s)\| \le \eta. \tag{6.26}$$

If $\rho < \sigma_k$ then the Saks–Henstock Lemma similarly yields

$$\|(U(t_0,\beta_k) - U(t_0,\rho))\varphi(\sigma_k) - \int_{\rho}^{\beta_k} d_s[U(t_0,s)]\varphi(s)\| \le \eta$$

and

$$\|(U(t_0,\rho) - U(t_0,\beta_{k-1}))\varphi(\sigma_k) - \int_{\beta_{k-1}}^{\rho} d_s[U(t_0,s)]\varphi(s)\| \le$$

$$\le \|(U(t_0,\beta_k) - U(t_0,\beta_{k-1}))\varphi(\sigma_k) - \int_{\beta_{k-1}}^{\beta_k} d_s[U(t_0,s)]\varphi(s)\| +$$

$$+ \|(U(t_0,\beta_k) - U(t_0,\rho))\varphi(\sigma_k) - \int_{\rho}^{\beta_k} d_s[U(t_0,s)]\varphi(s)\| \le 2\eta. \tag{6.27}$$

Taking into account the inequalities (6.24) – (6.27) we arrive at

$$\|\sum_{j=1}^{m}(W(\rho,\sigma_j,\beta_j) - W(\rho,\sigma_j,\beta_{j-1})) - \int_{t_0}^{t} D_s W(\rho,\sigma,s)\| \le$$

$$\le 3\eta\|U(\rho,t_0)\| \le 3\eta M$$

where M is the bound of U on $[t_0, t] \times [t_0, t]$ given by (b) from Theorem 6.15. Since the partition D was independent of ρ and $\eta > 0$ was arbitrary, this last inequality shows the required equi-integrability of the system of functions $W(\rho, \cdot, \cdot) = V(\rho, \cdot)]\varphi(\cdot)$. This enables us to utilize Theorem 1.45 to state that

$$\int_{t_0}^{t} d_r[K(r)](\int_{t_0}^{r} d_s[U(r,s)]\varphi(s)) =$$

$$= \int_{t_0}^{t} D_r[\int_{t_0}^{t} D_s[K(r)V(\rho,s)]\varphi(\sigma)] =$$

$$= \int_{t_0}^{t} D_r[\int_{t_0}^{t} D_s[K(r)W(\rho,\sigma,s)]] =$$

$$= \int_{t_0}^{t} D_s[\int_{t_0}^{t} D_r[K(r)W(\rho,\sigma,s)]]$$

By the definition of the function W we have

$$\int_{t_0}^{t} D_r[K(r)W(\rho,\sigma,s)] = \int_{t_0}^{t} D_r[K(r)V(\rho,s)\varphi(\sigma)] =$$

$$= \int_{t_0}^{s} D_r[K(r)V(\rho,s)\varphi(\sigma)] + \int_{s}^{t} D_r[K(r)V(\rho,s)\varphi(\sigma)] =$$

$$= \int_{t_0}^{s} D_r[K(r)]\varphi(\sigma) + \int_{s}^{t} D_r[K(r)U(\rho,s)\varphi(\sigma)] =$$

$$= \int_{t_0}^{s} d_r[K(r)]\varphi(\sigma) + \int_{s}^{t} d_r[K(r)]U(r,s)\varphi(\sigma) =$$

$$= [K(s) - K(t_2)]\varphi(\sigma)] + \int_{s}^{t} d_r[K(r)]U(r,s)\varphi(\sigma).$$

This yields

$$\int_{t_0}^{t} D_s[\int_{t_0}^{t} D_r[K(r)W(\rho,\sigma,s)]] =$$

$$= \int_{t_0}^{t} d_s[K(s)]\varphi(s) + \int_{t_0}^{t} d_s[\int_{s}^{t} d_r[K(r)]U(r,s)]\varphi(s),$$

i.e.

$$\int_{t_0}^{t} d_r[K(r)](\int_{t_0}^{r} d_s[U(r,s)]\varphi(s)) =$$

$$= \int_{t_0}^{t} d_s[K(s)]\varphi(s) + \int_{t_0}^{t} d_s[\int_{s}^{t} d_r[K(r)]U(r,s)]\varphi(s)$$

and (6.20) is proved for the case $t \geq t_0$. The case $t < t_0$ can be treated similarly. □

6.16a Corollary. *Assume that $A : J \to L(\mathbb{R}^n)$ is of locally bounded variation in J and that is satisfied (6.13) is satisfied. Let $\varphi : J \to \mathbb{R}^n$ be of locally bounded variation in J.*

Then for $U : J \times J \to L(\mathbb{R}^n)$ given by (6.18) in Theorem 6.13 the equality

$$\int_{t_0}^{t} d[A(r)](\int_{t_0}^{r} d_s[U(r,s)]\varphi(s)) =$$

$$= \int_{t_0}^{t} d[A(s)]\varphi(s) + \int_{t_0}^{t} d[U(t,s)]\varphi(s))$$

holds for $t_0, t \in J$.

Proof. If we set $K = A$ in Lemma 6.16 then by (6.18) we have

$$\int_{s}^{t} d_r[K(r)]U(r,s) = \int_{s}^{t} d_r[A(r)]U(r,s) = U(t,s) - I$$

and

$$\int_{t_0}^{t} d_s[\int_{s}^{t} d_r[K(r)]U(r,s)]\varphi(s) =$$

$$= \int_{t_0}^{t} d_s[U(t,s) - I]\varphi(s) = \int_{t_0}^{t} d_s[U(t,s)]\varphi(s).$$

This proves the formula given in the corollary. □

6.17 Theorem (variation of constants formula). *Assume that $A : J \to L(\mathbb{R}^n)$ is of locally bounded variation in J and that (6.13) is satisfied. Then for every $t_0 \in J$, $\widetilde{x} \in \mathbb{R}^n$ and $g : J \to \mathbb{R}^n$ of locally bounded variation in J the unique solution of the initial value problem*

$$\frac{dx}{d\tau} = D[A(t)x + g(t)], \quad x(t_0) = \widetilde{x} \tag{6.28}$$

can be written in the form

$$x(t) = U(t,t_0)\widetilde{x} + g(t) - g(t_0) - \int_{t_0}^{t} d_s[U(t,s)](g(s) - g(t_0)) \tag{6.29}$$

for $t \in J$ where $U : J \times J \to L(\mathbb{R}^n)$ is given by (6.18) in Theorem 6.13.

Proof. The function $x(t)$ given in (6.29) is well defined by the properties of the function U described in Theorem 6.15. Assume that $t \geq t_0$, $t \in J$. Using (6.18) and (6.29) we obtain

$$\int_{t_0}^{t} d[A(r)]x(r) = \int_{t_0}^{t} d[A(r)]U(r,t_0)\widetilde{x} + \int_{t_0}^{t} d[A(r)](g(r) - g(t_0)) -$$

$$- \int_{t_0}^{t} d[A(r)] \int_{t_0}^{r} d_s[U(r,s)](g(s) - g(t_0)) =$$

$$= (U(t,t_0) - I)\widetilde{x} + \int_{t_0}^{t} d[A(r)]\varphi(r) -$$

$$- \int_{t_0}^{t} d[A(r)] \int_{t_0}^{r} d_s[U(r,s)]\varphi(s)$$

with $\varphi(s) = g(s) - g(t_0)$ for $s \in J$. The function φ evidently preserves the property of g being of locally bounded variation on J. Therefore Corollary 6.16a can be used for the last integral in this relation to obtain

$$\int_{t_0}^{t} d[A(r)]x(r) = U(t,t_0)\widetilde{x} - \widetilde{x} + \int_{t_0}^{t} d[A(r)]\varphi(r) -$$

$$- \int_{t_0}^{t} d[A(r)]\varphi(r) - \int_{t_0}^{t} d[U(t,s)]\varphi(s)) =$$

$$= U(t,t_0)\widetilde{x} - \widetilde{x} - \int_{t_0}^{t} d[U(t,s)]\varphi(s)) =$$

$$= U(t,t_0)\widetilde{x} - \widetilde{x} - \int_{t_0}^{t} d[U(t,s)](g(s) - g(t_0)) =$$

$$= -\widetilde{x} - (g(t) - g(t_0)) + x(t),$$

and this means that

$$x(t) = \widetilde{x} + \int_{t_0}^{t} d[A(r)]x(r) + g(t) - g(t_0)$$

holds for $t \in J$, $t \geq t_0$ and the function x given by (6.29) is a solution of (6.28) for these values of t. For the remaining case $t \in J$, $t < t_0$ the result can be proved in a completely analogous way. □

6.18 Lemma. *If $A : J \to L(\mathbb{R}^n)$ is of locally bounded variation in J , (6.13) is satisfied and $X : J \to L(\mathbb{R}^n)$ is an arbitrary fundamental matrix of (6.15) then*

$$U(t,s) = X(t)X^{-1}(s) \tag{6.30}$$

for every $t, s \in J$ where $U : J \times J \to L(\mathbb{R}^n)$ is given by (6.18) in Theorem 6.13.

Proof. By Theorem 6.13 the matrix $X(s)$ is regular for every $s \in J$ and therefore the product $X(t)X^{-1}(s)$ is well defined and regular for $t, s \in J$. Since X is a solution of the matrix equation (6.17) we have

$$\int_s^t d[A(r)]X(r) = X(t) - X(s)$$

and this means that

$$\int_s^t d[A(r)]X(r)X^{-1}(s) = X(t)X^{-1}(s) - X(s)X^{-1}(s) =$$

$$= X(t)X^{-1}(s) - I.$$

Hence the product $X(t)X^{-1}(s)$ satisfies the equation (6.18) and the uniqueness of U stated in Theorem 6.13 yields the result. □

6.19 Corollary. *If $A : J \to L(\mathbb{R}^n)$ is of locally bounded variation in J, (6.13) is satisfied, $t_0 \in J$, $\widetilde{x} \in \mathbb{R}^n$, $g : J \to \mathbb{R}^n$ is of locally bounded variation in J and $X : J \to L(\mathbb{R}^n)$ is an arbitrary fundamental matrix of the equation (6.15) then the unique solution of the initial value problem (6.28) can be represented in the form*

$$x(t) = g(t) - g(t_0) +$$

$$+X(t)\left(X^{-1}(t_0)\widetilde{x} - \int_{t_0}^{t} d_s[X^{-1}(s)](g(s) - g(t_0))\right). \qquad (6.31)$$

Proof. The result follows immediately from the variation of constants formula (6.29) if the equality (6.30) from Lemma 6.18 is taken into account. □

6.20 *Example.* Consider the linear differential system with impulses

$$\dot{x} = F(t)x, \qquad (6.32)$$

$$\Delta x|_{t_i} = x(t_i+) - x(t_i) = B_i x(t_i) \qquad (6.33)$$

on an interval $J \subset \mathbb{R}$ where $F : J \to L(\mathbb{R}^n)$ is an $n \times n$ matrix–valued function which has locally integrable (in the sense of Lebesgue) components on J. We assume that $t_i \in J$, $t_i < t_{i+1}$ for $i = 1, 2, \ldots$ and that $B_i \in L(\mathbb{R}^n)$ are $n \times n$ matrices for $i = 1, 2, \ldots$ such that $I + B_i$ are regular matrices for $i = 1, 2, \ldots$. For the concept of a differential system with impulses see Chapter V. Linear differential systems with impulses of the form (6.32), (6.33) have been considered by A.M. Samojlenko and others (see e.g. [116], [117], [118], [120] etc.).

Given a fixed $a \in J$ define

$$A(t) = \int_a^t F(s)\,ds + \sum_{i=1}^{\infty} B_i H_{t_i}(t) \qquad (6.34)$$

for $t \in J$ where $H_{t_i}(t) = 0$ for $t \le t_i$ and $H_{t_i}(t) = 1$ for $t > t_i$. The integral in (6.34) is the integral in the Lebesgue sense.

If it is supposed that in every compact interval $[a, b] \subset J$ the set of points t_i such that $t_i \in [a, b]$ is finite, then the matrix valued function A given by (6.34) is locally of bounded variation on J, continuous from the left (in spite of the definition of the function H_{t_i}), and we have

$$I + \Delta^+ A(t) = I \text{ if } t \ne t_i,$$

$$I + \Delta^+ A(t) = I + B_i \text{ if } t = t_i.$$

Hence the matrix $I + \Delta^+ A(t)$ is regular for every $t \in J$ and of course also $I - \Delta^- A(t) = I$ is regular for every $t \in J$ because A is continuous from the left. Therefore A satisfies (6.13).

By the results stated in Chapter V we know that the linear differential system with impulses of the form (6.32), (6.33) is equivalent to the homogeneous generalized linear differential equation

$$\frac{dx}{d\tau} = D[A(t)x]. \tag{6.35}$$

Let us consider the fundamental matrix $U : J \times J \to L(\mathbb{R}^n)$ of (6.35) which satisfies

$$U(t,s) = I + \int_s^t d[A(r)]U(r,s)$$

(see Theorem 6.13)).

If $t, \tau \in (t_i, t_{i+1}]$ then

$$U(t,s) = U(\tau,s) + \int_\tau^t d[A(r)]U(r,s) =$$

$$= U(\tau,s) + \int_\tau^t F(r)U(r,s)\,dr,$$

i.e.

$$U(t,s) = \Phi(t,\tau)U(\tau,s), \tag{6.36}$$

where $\Phi : J \times J \to L(\mathbb{R}^n)$ is the classical fundamental matrix of the linear system of ordinary differential equations

$$\dot{x} = F(t)x$$

which is defined by the relation

$$\frac{d}{dt}(\Phi(t,\tau)) = F(t)\Phi(t,\tau), \quad \Phi(t,t) = I.$$

(6.36) and the property (e) from Theorem 6.15 imply for $t, \tau \in (t_i, t_{i+1}]$ the relation

$$U(t,s) = \lim \tau \to t_i + \Phi(t,\tau)U(\tau,s) = \Phi(t,t_i)U(t_i+,s) =$$

$$= \Phi(t,t_i)[I + \Delta^+ A(t_i)]U(t_i,s) = \Phi(t,t_i)[I + B_i]U(t_i,s).$$

This procedure can be repeated for $U(t_i, s)$ to obtain

$$U(t,s) = \Phi(t,t_i)[I + B_i]U(t_i,s) =$$

$$= \Phi(t,t_i)[I + B_i]\Phi(t_i,t_{i-1})[I + B_{i-1}]U(t_{i-1},s).$$

A continuation of this procedure leads to the relation

$$U(t,s) = \Phi(t,t_i)[I + B_i]\Phi(t_i,t_{i-1})[I + B_{i-1}]\cdots[I + B_j]\Phi(t_j,s) =$$

$$= \Phi(t,t_i)\prod_{k=j+1}^{i}[I + B_k]\Phi(t_k,t_{k-1})[I + B_j]\Phi(t_j,s)$$

whenever $s \in J$, $s \le t$, $s \in (t_{j-1}, t_j]$ for some j or $j = 1$ if $s \le t_1$.

If $t < s$, $s \in [t_i, t_{i+1}]$, $t \in (t_{j-1}, t_j]$ or $t \le t_1$ then by (f) from Theorem 6.15 we obtain

$$U(t,s) = [U(s,t)]^{-1} =$$

$$= [\Phi(s,t_i)[I + B_i]\Phi(t_i,t_{i-1})[I + B_{i-1}]\cdots[I + B_j]\Phi(t_j,t)]^{-1} =$$

$$= [\Phi(t_j,t)]^{-1}[I + B_j]^{-1}\cdots[I + B_i]^{-1}[\Phi(s,t_i)]^{-1} =$$

$$= \Phi(t,t_j)[I + B_j]^{-1}\cdots[I + B_i]^{-1}\Phi(t_i,s).$$

Now by Theorem 6.14 the solution of the initial value problem (6.35) with the initial condition $x(s) = \widetilde{x}$ can be expressed in the form

$$x(t) = U(t,s)\widetilde{x} =$$

$$= \Phi(t,t_i) \prod_{k=j+1}^{i} [I + B_k]\Phi(t_k, t_{k-1})[I + B_j]\Phi(t_j, s)\widetilde{x}$$

for $t \geq s$, $t \in (t_i, t_{i+1}]$, $s \in (t_{j-1}, t_j]$ or $j = 1$ and $s \leq t_1$.

In this way we obtain the formula for solutions of the initial value problem for the linear differential systems with impulses (6.32), (6.33) which is known from the work of A.M. Samojlenko and N.A. Perestjuk [120] and can be also found in a slightly different setting in Chap. 4 of [107]. It should be noted that if $B_i = 0$ for all i then we have an ordinary linear differential equation and the fundamental matrix in this "impulse" setting coincides with the classical one.

It is clear that the results presented in this chapter apply also to the case of a nonhomogeneous differential system with impulses of the form

$$\dot{x} = F(t)x + h(t),$$

$$\Delta x|_{t_i} = x(t_i+) - x(t_i) = B_i x(t_i) + a_i$$

where $h : J \to \mathbb{R}^n$ is locally integrable in J and $a_i \in \mathbb{R}^n$, $i = 1, 2, \ldots$.

Boundary value problems

Let us consider the following problem:

Given a generalized linear ordinary differential equation

$$\frac{dx}{d\tau} = D[A(t)x + g(t)] \tag{6.1}$$

with $A : [a,b] \to L(\mathbb{R}^n)$ satisfying the condition (6.13) find a solution of (6.1) which satisfies the side condition

$$\int_a^b d[K(s)]x(s) = r \tag{6.37}$$

where $K : [a,b] \to L(\mathbb{R}^n, \mathbb{R}^m)$ is an $m \times n$–matrix valued function of bounded variation on the interval $[a,b]$ and $r \in \mathbb{R}^m$.

This is a general form of a boundary value problem for generalized linear ordinary differential equations.

It should be mentioned that a side condition of the form

$$Mx(a) + Nx(b) + \int_a^b d[L(s)]x(s) = r \tag{6.38}$$

with $M, N \in L(\mathbb{R}^n, \mathbb{R}^m)$, $L : [a,b] \to L(\mathbb{R}^n, \mathbb{R}^m)$ of bounded variation on the interval $[a,b]$ and $r \in \mathbb{R}^m$ assumes the form (6.37) if we take

$$\begin{aligned} K(t) &= -M + L(a) && \text{for } t = a \\ K(t) &= L(t) && \text{for } a < t < b \\ K(t) &= N + L(b) && \text{for } t = b. \end{aligned}$$

This follows from an easy computation. Moreover, if L is a constant function then (6.38) has the conventional form

$$Mx(a) + Nx(b) = r.$$

Assume that $\Psi : [a,b] \to L(\mathbb{R}^n)$ is the fundamental matrix of the equation

$$\frac{dx}{d\tau} = DA(t)x$$

such that $\Psi(a) = I$.

6.21 Lemma. *The boundary value problem (6.1), (6.37) has a solution if and only if*

$$\gamma^T \left\{ \int_a^b d[K(t)]g(t) - \int_a^b d[K(t)]\Psi(t) \int_a^t d[(\Psi(s))^{-1}]g(s) \right\} =$$

$$= \gamma^T r \tag{6.39}$$

holds for every $\gamma \in \mathbb{R}^m$ such that

$$\gamma^T \int_a^b d[K(t)]\Psi(t) = 0 \tag{6.40}$$

where by γ^T the transpose to $\gamma \in \mathbb{R}^m$ is denoted.

Proof. By the variation of constants formula (6.31) $x : [a, b] \to \mathbb{R}^n$ is a solution of (6.1) if and only if

$$x(t) = \Psi(t)c + g(t) - g(a) - \Psi(t) \int_a^t d[(\Psi(s))^{-1}](g(s) - g(a)) \tag{6.41}$$

for some $c \in \mathbb{R}^n$ and all $t \in [a, b]$. Inserting (6.41) into the left hand side of the side condition (6.37) we obtain

$$\int_a^b d[K(t)]x(t) = (\int_a^b d[K(t)]\Psi(t))c + \int_a^b d[K(t)](g(t) - g(a)) -$$

$$- \int_a^b d[K(t)]\Psi(t) \int_a^t d[(\Psi(s))^{-1}](g(s) - g(a))$$

$$= (\int_a^b d[K(t)]\Psi(t))c + \int_a^b d[K(t)](g(t) - g(a)) -$$

$$- \int_a^b d[K(t)]\Psi(t) \int_a^t d[(\Psi(s))^{-1}]g(s) +$$

$$+\int_a^b d[K(t)]\Psi(t)[(\Psi(t))^{-1} - I]g(a) =$$

$$= (\int_a^b d[K(t)]\Psi(t))c + \int_a^b d[K(t)]g(t)-$$

$$-\int_a^b d[K(t)]\Psi(t)\int_a^t d[(\Psi(s))^{-1}]g(s) - \int_a^b d[K(t)]\Psi(t)g(a).$$

This implies that x is a solution to the boundary value problem (6.1), (6.37) if and only if x is given by (6.41) where $c \in \mathbb{R}^n$ is such that

$$\left(\int_a^b d[K(t)]\Psi(t)\right) c = r + \int_a^b d[K(t)]\Psi(t)g(a)-$$

$$-\int_a^b d[K(t)]g(t) + \int_a^b d[K(t)]\Psi(t)\int_a^t d[(\Psi(s))^{-1}]g(s). \quad (6.42)$$

Using the well known facts about linear systems of algebraic equations we conclude in particular that the boundary value problem (6.1), (6.37) possesses a solution if and only if the right hand side of (6.42) is orthogonal to the null space of the adjoint linear system (6.39), i.e. if and only if

$$\gamma^T r = \gamma^T \left\{ -\int_a^b d[K(t)]\Psi(t)g(a) + \int_a^b d[K(t)]g(t)- \right.$$

$$\left. -\int_a^b d[K(t)]\Psi(t)\int_a^t d[(\Psi(s))^{-1}]g(s) \right\}$$

holds for every $\gamma \in \mathbb{R}^n$ for which (6.40) is satisfied. Since

$$\gamma^T \left\{ -\int_a^b d[K(t)]\Psi(t)g(a) \right\} = 0$$

by (6.40), the assertion of the lemma follows readily. □

Now we modify slightly the condition of solvability of the boundary value problem (6.1), (6.37) given in Lemma 6.21.

6.22 Lemma. *The boundary value problem (6.1), (6.37) has a solution if and only if*

$$\int_a^b d_s[\int_s^b d_t[\gamma^T K(t)\Psi(t)(\Psi(s))^{-1}]g(s) + \gamma^T r = 0$$

for every solution $\gamma \in \mathbb{R}^n$ of the equation (6.40)

Proof. Let us turn our attention to the relation (6.39). By the identity (6.20) from Lemma 6.16 we have

$$\int_a^b d[K(t)]\Psi(t)\int_a^t d[(\Psi(s))^{-1}]g(s) =$$

$$= \int_a^b d[K(s)]g(s) + \int_a^b d_s[\int_s^b d_t[\gamma^T K(t)\Psi(t)(\Psi(s))^{-1}]g(s).$$

Inserting this into the left hand side of (6.39) we obtain

$$\gamma^T\left\{\int_a^b d[K(t)]g(t) - \int_a^b d[K(t)]\Psi(t)\int_a^t d[(\Psi(s))^{-1}]g(s)\right\} =$$

$$= \gamma^T\left\{-\int_a^b d_s[\int_s^b d_t[\gamma^T K(t)\Psi(t)(\Psi(s))^{-1}]g(s)\right\}.$$

and this proves the statement. □

The results given in Lemmas 6.21 and 6.22 represent solvability conditions for the boundary value problem (6.1), (6.37). They form the basis for more detailed studies of boundary value problems of the type (6.1), (6.37) from the point of view of functional analysis. The reader interested in a more detailed theory of such boundary value problems (the adjoint problem, Green's function, etc.) should consult e.g. [155], [154]. It is worth to mention that the method of proving Lemma 6.21 can be used for the construction of a solution of the boundary value problem provided the condition of Lemma 6.21 or that of Lemma 6.22 is satisfied.

CHAPTER VII

PRODUCT INTEGRATION AND GENERALIZED LINEAR DIFFERENTIAL EQUATIONS

Multiplicative integration has a long history. Vito Volterra [169] invented product integration at the end of the last century. Product integration was studied by L. Schlesinger [122], [123] in connection with linear differential equations. The present state of art is well described in the book [20] of J.D. Dollard and C.N. Friedman and in a good survey [32] of R.D. Gill and S. Johansen which tends to applications in statistics and Markov processes. This was also one of the points in the book of V. Volterra and B. Hostinský [170] from 1938.

The original approach to product integrals is based on partial integral products which are similar to Riemann integral sums. This way leads also to a "Lebesgue type" product integration by the extension of the integral product of "step functions", see e.g. [125] or [20].

Here we present the theory when the classical Riemann approach is replaced by the Kurzweil – Henstock concept of δ–fine partitions which are fine with respect to gauges. This idea was used for the first time by J. Jarník and J. Kurzweil in [53] and we present here a certain refinement of their approach on the basis of [148].

We denote by $L(\mathbb{R}^n)$ the set of all linear operators from $\mathbb{R}^n$ to $\mathbb{R}^n$ (the set of all $n \times n$–matrices) and assume that $\|\cdot\|$ is the operator norm on $L(\mathbb{R}^n)$ which corresponds to the norm used in the space $\mathbb{R}^n$.

Let an interval $[a, b] \subset \mathbb{R}$, $-\infty < a < b < +\infty$ be given. Let $\mathfrak{J}$ be the set of all compact subintervals in $[a, b]$, i.e. of intervals of the form $[\alpha, \beta]$, where $a \leq \alpha \leq \beta \leq b$.

Assume that a function $V : [a, b] \times \mathfrak{J} \to L(\mathbb{R}^n)$ is given; V is an $n \times n$–matrix valued point–interval function. Let us assume that a partition

$$D = \{(\tau_j, J_j),\ j = 1, \dots, k\} = \{(\tau_j, [\alpha_{j-1}, \alpha_j]),\ j = 1, \dots, k\} =$$

$$= \{\alpha_0, \tau_1, \alpha_1, \dots, \alpha_{k-1}, \tau_k, \alpha_k\}$$

is given where

$$a = \alpha_0 < \alpha_1 < \dots < \alpha_k = b,$$

$$\alpha_{j-1} \leq \tau_j \leq \alpha_j,\ j = 1, 2, \dots, k$$

and

$$[\alpha_{j-1}, \alpha_j] \subset [\tau_j - \delta(\tau_j), \tau_j + \delta(\tau_j)],\ j = 1, 2, \dots, k.$$

For the function $V : [a, b] \times \mathfrak{J} \to L(\mathbb{R}^n)$ and a given partition D of the interval $[a, b]$ we denote

$$P(V, D) = V(\tau_k, J_k)V(\tau_{k-1}, J_{k-1}) \dots V(\tau_1, J_1) =$$

$$= V(\tau_k, [\alpha_{k-1}, \alpha_k])V(\tau_{k-1}, [\alpha_{k-2}, \alpha_{k-1}]) \dots V(\tau_1, [\alpha_0, \alpha_1]),$$

the ordered product of elements of $L(\mathbb{R}^n)$.

7.1 Definition. A function $V : [a,b] \times \mathfrak{J} \to L(\mathbb{R}^n)$ is called *Perron product integrable* if there exists $Q \in L(\mathbb{R}^n)$ such that for every $\varepsilon > 0$ there is a gauge $\delta : [a,b] \to (0,+\infty)$ on $[a,b]$ such that

$$\|P(V,D) - Q\| < \varepsilon \tag{7.1}$$

for every δ–fine partition D of $[a,b)$.

$Q \in L(\mathbb{R}^n)$ is called *the Perron product integral* of V over $[a,b]$ and we use the notation $Q = \prod_a^b V(t,dt)$.

The following statement is based on the fact that $L(\mathbb{R}^n)$ with the operator norm is complete because it is a Banach space. The proof of the result follows exactly the lines of the proof of Theorem 1.7.

7.2 Proposition. *Let $V : [a,b] \times \mathfrak{J} \to L(\mathbb{R}^n)$ be given. The following two conditions are equivalent.*

(i) There is a $Q \in L(\mathbb{R}^n)$ such that for every $\varepsilon > 0$ there is a gauge $\delta : [a,b] \to (0,+\infty)$ on $[a,b]$ such that $\|P(V,D) - Q\| < \varepsilon$ for every δ–fine partition D of $[a,b)$.

(ii) For every $\varepsilon > 0$ there is a gauge $\delta : [a,b] \to (0,+\infty)$ on $[a,b]$ such that $\|P(V,D_1) - P(V,D_2)\| < \varepsilon$ for every δ–fine partitions D_1, D_2 of $[a,b)$.

7.3 *Remark.* It is easy to see that the concept of the Perron product integral is defined in a manner very similar to Definition 1.2 (or 1.2n) where the generalized Perron integral is described. The role of summation is replaced by multiplication and because the product of two matrices is not commutative in general, the order of multiplication in $P(V,D)$ is fixed.

Now we turn our attention to some more specific requirements concerning the $n \times n$–matrix valued point–interval function $V : [a,b] \times \mathfrak{J} \to L(\mathbb{R}^n)$. These conditions play an essential role in our subsequent studies:

$$V(\tau, [\tau, \tau]) = I \text{ for every } \tau \in [a, b], \tag{7.2}$$

where $I \in L(\mathbb{R}^n)$ is the identity operator in $L(\mathbb{R}^n)$,

for every $\tau \in [a, b]$ and $\zeta > 0$ there exists $\sigma > 0$ such that

$$\|V(\tau, [\alpha, \beta]) - V(\tau, [\tau, \beta])V(\tau, [\alpha, \tau])\| < \zeta \tag{7.3}$$

for all $\alpha, \beta \in [a, b]$, $\tau - \sigma < \alpha \leq \tau \leq \beta < \tau + \sigma$;

for every $\tau \in [a, b)$ there is an invertible $V_+(\tau) \in L(\mathbb{R}^n)$ such that

$$\lim_{\beta \to \tau+} \|V(\tau, [\tau, \beta]) - V_+(\tau)\| = 0, \text{ i.e. } \lim_{\beta \to \tau+} V(\tau, [\tau, \beta]) = V_+(\tau), \tag{7.4+}$$

and

for every $\tau \in (a, b]$ there is an invertible $V_-(\tau) \in L(\mathbb{R}^n)$ such that

$$\lim_{\alpha \to \tau-} \|V(\tau, [\alpha, \tau]) - V_-(\tau)\| = 0, \text{ i.e. } \lim_{\alpha \to \tau-} V(\tau, [\alpha, \tau]) = V_-(\tau). \tag{7.4–}$$

7.4 Definition. If the function $V : [a, b] \times \mathfrak{J} \to L(\mathbb{R}^n)$ satisfies (7.2), (7.3),(7.4–) and (7.4+) then we say that V satisfies *condition C.*

7.5 Lemma. *Assume that the function $V : [a, b] \times \mathfrak{J} \to L(\mathbb{R}^n)$ satisfies the condition*

for every $\tau \in [a, b]$ and $\zeta > 0$ *there exists* $\sigma > 0$ such that

$$\|V(\tau, [\alpha, \beta]) - I\| < \zeta \tag{7.5}$$

for all $\alpha, \beta \in [a, b]$, $\tau - \sigma < \alpha \leq \tau \leq \beta < \tau + \sigma$.

Then the function V satisfies condition C.

Proof. Since we have

$$V(\tau,[\alpha,\beta]) - V(\tau,[\tau,\beta])V(\tau,[\alpha,\tau]) =$$

$$= V(\tau,[\alpha,\beta]) - V(\tau,[\tau,\beta]) - V(\tau,[\alpha,\tau]) + I -$$

$$-(V(\tau,[\tau,\beta]) - I)(V(\tau,[\alpha,\tau]) - I) =$$

$$= V(\tau,[\alpha,\beta]) - I + I - V(\tau,[\tau,\beta]) + I - V(\tau,[\alpha,\tau]) -$$

$$-(V(\tau,[\tau,\beta]) - I)(V(\tau,[\alpha,\tau]) - I)$$

we have also

$$\|V(\tau,[\alpha,\beta]) - V(\tau,[\tau,\beta])V(\tau,[\alpha,\tau])\| \leq$$

$$\leq \|V(\tau,[\alpha,\beta]) - I\| + \|V(\tau,[\tau,\beta]) - I\| + \|V(\tau,[\alpha,\tau]) - I\| +$$

$$+\|V(\tau,[\tau,\beta]) - I\|\|V(\tau,[\alpha,\tau]) - I\|.$$

This inequality implies that if the function $V : [a,b] \times \mathfrak{J} \to L(\mathbb{R}^n)$ satisfies the condition (7.5) then

$$\|V(\tau,[\alpha,\beta]) - V(\tau,[\tau,\beta])V(\tau,[\alpha,\tau])\| < 3\zeta + \zeta^2$$

for all $\alpha, \beta \in [a,b]$, $\tau - \sigma < \alpha \leq \tau \leq \beta < \tau + \sigma$ and this implies that (7.3) is fulfilled.

Moreover, (7.5) evidently yields $\lim_{\beta \to \tau+} V(\tau,[\tau,\beta]) = I$ for $\tau \in [a,b)$ and $\lim_{\alpha \to \tau-} V(\tau,[\alpha,\tau]) = I$ for $\tau \in (a,b]$ and therefore (7.4+) and (7.4–) hold with $V_+(\tau) = V_-(\tau) = I$ and it is easy to see that also (7.2) is satisfied. This means that the condition C is satisfied and the lemma is proved. □

7.6 Lemma. *Assume that for the function $V : [a,b] \times \mathfrak{J} \to L(\mathbb{R}^n)$ the condition C is satisfied. Then for every $\tau \in [a,b]$ there exists a $\sigma_1 = \sigma_1(\tau) > 0$ such that $V(\tau,[\alpha,\beta]) \in L(\mathbb{R}^n)$ is invertible (the inverse matrix $(V(\tau,[\alpha,\beta]))^{-1} \in L(\mathbb{R}^n)$ exists) provided $\alpha, \beta \in [a,b]$, $\tau - \sigma_1 < \alpha \le \tau \le \beta < \tau + \sigma_1$.*

Proof. Let $\tau \in [a,b]$ be given. For a given $\zeta > 0$ let $\sigma_1(\tau) > 0$ be such that we have

$$\|V(\tau,[\alpha,\beta]) - V(\tau,[\tau,\beta])V(\tau,[\alpha,\tau])\| < \zeta \tag{7.6}$$

and

$$\|V(\tau,[\tau,\beta]) - V_+(\tau)\| < \zeta, \ \|V(\tau,[\alpha,\tau]) - V_-(\tau)\| < \zeta \tag{7.7}$$

provided $\alpha, \beta \in [a,b]$, $\tau - \sigma_1 < \alpha \le \tau \le \beta < \tau + \sigma_1$. The possibility of choosing such a $\sigma_1 > 0$ is ensured by (7.3), (7.4–) and (7.4+).

Since $V_-(\tau)$ and $V_+(\tau)$ are invertible operators (we define $V_-(a) = I$ and $V_+(b) = I$) their product $V_+(\tau)V_-(\tau)$ is also invertible with $(V_+(\tau)V_-(\tau))^{-1} = (V_-(\tau))^{-1}(V_+(\tau))^{-1}$.

It is easy to verify that

$$V(\tau,[\alpha,\beta]) - V_+(\tau)V_-(\tau) =$$

$$= V(\tau,[\alpha,\beta]) - V(\tau,[\tau,\beta])V(\tau,[\alpha,\tau]) +$$

$$+(V(\tau,[\tau,\beta]) - V_+(\tau))(V(\tau,[\alpha,\tau]) - V_-(\tau)) +$$

$$+V_+(\tau)(V(\tau,[\alpha,\tau]) - V_-(\tau)) + (V(\tau,[\tau,\beta]) - V_+(\tau))V_-(\tau).$$

Hence

$$\|V(\tau,[\alpha,\beta]) - V_+(\tau)V_-(\tau)\| \le$$

$$\le \|V(\tau,[\alpha,\beta]) - V(\tau,[\tau,\beta])V(\tau,[\alpha,\tau])\| +$$

$$+\|V(\tau,[\tau,\beta])-V_+(\tau)\|\|V(\tau,[\alpha,\tau])-V_-(\tau)\|+$$

$$+\|V_+(\tau)\|\|V(\tau,[\alpha,\tau])-V_-(\tau)\|+\|V(\tau,[\tau,\beta])-V_+(\tau)\|\|V_-(\tau)\|,$$

and if $\alpha,\beta\in[a,b]$, $\tau-\sigma_1<\alpha\le\tau\le\beta<\tau+\sigma_1$ then by (7.6) and (7.7) we obtain

$$\|V(\tau,[\alpha,\beta])-V_+(\tau)V_-(\tau)\|\le$$

$$\le\zeta+\zeta^2+\zeta(\|V_+(\tau)\|+\|V_-(\tau)\|)=$$

$$=\zeta(1+\zeta+\|V_+(\tau)\|+\|V_-(\tau)\|).$$

Since $\zeta>0$ can be chosen arbitrarily small, the operator $V(\tau,[\alpha,\beta])$ is invertible. (It is sufficient to take $\zeta>0$ such that $\zeta(1+\zeta+\|V_+(\tau)\|+\|V_-(\tau)\|)>\|(V_+(\tau))^{-1}(V_-(\tau))^{-1}\|^{-1}$.) If e.g. $\alpha=\tau<\beta$ then the result comes immediately from the second inequality in (7.7) for a sufficiently small ζ. The case $\alpha<\tau=\beta$ is a consequence of the first relation in (7.7), and finally for $\alpha=\tau=\beta$ we have $V(\tau,[\alpha,\beta])=I$ and $V(\tau,[\alpha,\beta])$ is evidently invertible. □

7.7 Lemma. *Assume that for the function $V:[a,b]\times\mathfrak{J}\to L(\mathbb{R}^n)$ the condition C is satisfied. Then for every $\tau\in[a,b]$ there exists a $\sigma_2=\sigma_2(\tau)>0$ such that*

$$\|V(\tau,[\alpha,\tau])\|\le\|V_-(\tau)\|+\frac{1}{2}\|(V_-(\tau))^{-1}\|,$$

$$\|(V(\tau,[\alpha,\tau]))^{-1}\|\le 2\|(V_-(\tau))^{-1}\| \tag{7.8}$$

for all $\alpha\in[a,b]$ such that $\tau-\sigma_2<\alpha<\tau$ and

$$\|V(\tau,[\tau,\beta])\|\le\|V_+(\tau)\|+\frac{1}{2}\|(V_+(\tau))^{-1}\|,$$

$$\|(V(\tau,[\tau,\beta]))^{-1}\|\le 2\|(V_+(\tau))^{-1}\| \tag{7.9}$$

for all $\beta \in [a, b]$ such that $\tau < \beta < \tau + \sigma_2$.

Proof. We prove the result stated in (7.8), the proof of (7.9) is analogous. If $\tau = a$ then there is nothing to prove because there is no $\alpha \in [a, b]$ such that $\alpha < \tau$. Assume therefore that $\tau \in (a, b]$. The matrix $V_-(\tau) \in L(\mathbb{R}^n)$ is invertible by (7.4–). If now $B \in L(\mathbb{R}^n)$ and

$$\|B - V_-(\tau)\| < \frac{1}{2}\|(V_-(\tau))^{-1}\|$$

then $B^{-1} \in L(\mathbb{R}^n)$ exists (see the general result given in [D S ,VII.6.1] and

$$B^{-1} = (V_-(\tau))^{-1} \sum_{k=0}^{\infty} [(V_-(\tau) - B)(V_-(\tau))^{-1}]^k.$$

Therefore

$$\|B^{-1}\| = \|(V_-(\tau))^{-1}\| \sum_{k=0}^{\infty} [\|(V_-(\tau) - B)\| \|(V_-(\tau))^{-1}\|]^k =$$

$$= \frac{\|(V_-(\tau))^{-1}\|}{1 - \|(V_-(\tau) - B)\| \|(V_-(\tau))^{-1}\|}.$$

Since in this case $\|(V_-(\tau) - B)\| \|(V_-(\tau))^{-1}\| < \frac{1}{2}$ we have

$$1 - \|(V_-(\tau) - B)\| \|(V_-(\tau))^{-1}\| > \frac{1}{2}$$

and therefore

$$\|B^{-1}\| < 2\|(V_-(\tau))^{-1}\|. \tag{7.10}$$

By (7.4–) there is a $\sigma_2^-(\tau) > 0$ such that if $\alpha \in [a,b]$, $\tau - \sigma_2^- < \alpha < \tau$, then

$$\|V(\tau,[\alpha,\tau]) - V_-(\tau)\| < \frac{1}{2}\|(V_-(\tau))^{-1}\|. \tag{7.11}$$

Hence (7.10) yields

$$\|(V(\tau,[\alpha,\tau]))^{-1}\| < 2\|(V_-(\tau))^{-1}\|$$

and (7.11) implies also

$$\|V(\tau,[\alpha,\tau])\| \le \|V(\tau,[\alpha,\tau]) - V_-(\tau)\| + \|V_-(\tau)\| <$$

$$< \frac{1}{2}\|(V_-(\tau))^{-1}\| + \|V_-(\tau)\|$$

provided $\tau - \sigma_2^- < \alpha < \tau$, i.e. (7.8) holds for such $\alpha \in [a,b]$.

For the case $\tau \in [a,b)$ we can find a $\sigma_2^+(\tau) > 0$ such that (7.9) holds if $\beta \in [a,b]$, $\tau < \beta < \tau + \sigma_2^+$. Taking $\sigma_2 = \min(\sigma_2^+, \sigma_2^-)$ we obtain the statement of the lemma. □

7.8 Theorem. *Let $V : [a,b] \times \mathfrak{J} \to L(\mathbb{R}^n)$ be Perron product integrable over $[a,b]$ with $\prod_a^b V(t,dt) = Q$ where $Q \in L(\mathbb{R}^n)$ is invertible and assume that the condition C is satisfied.*

Then there exists a constant $K > 0$ such that for every $s \in [a,b]$ the Perron product integrals $\prod_a^s V(t,dt)$, $\prod_s^b V(t,dt)$ exist, the equality

$$\prod_s^b V(t,dt) \prod_a^s V(t,dt) = \prod_a^b V(t,dt)$$

holds and

$$\|\prod_a^s V(t,dt)\| \le K, \ \|(\prod_a^s V(t,dt))^{-1}\| \le K.$$

Proof. Let $\zeta > 0$ be arbitrary. Let $\delta_0 : [a,b] \to (0,\infty)$ be a gauge on $[a,b]$ such that $\delta_0(t) \le \min(\sigma_1(t), \sigma_2(t))$ for $t \in [a,b]$ where $\sigma_1(t), \sigma_2(t)$ are given in Lemma 7.6 and 7.7, respectively, and such that

$$\|P(V,D) - Q\| < \frac{1}{2}\|Q^{-1}\|^{-1} \tag{7.12}$$

holds for every δ_0–fine partition D of the interval $[a,b]$. Assume further (by (7.3)) that

$$\|V(\tau,[\alpha,\beta]) - V(\tau,[\tau,\beta])V(\tau,[\alpha,\tau])\| < \zeta \tag{7.13}$$

for all $\tau, \alpha, \beta \in [a,b]$, $\tau - \delta_0(\tau) < \alpha \le \tau \le \beta < \tau + \delta_0(\tau)$.

The proof of the theorem will be divided into several steps. First we prove the following assertion.

For every $\tau \in [a,b]$ there is a $K_1(\tau) > 0$ such that

if $s \in (\tau - \delta_0(\tau), \tau] \cap [a,b]$ and D_1 is a δ_0–fine partition of $[a,s]$ then

$$\max\{\|P(V,D_1)\|, \|(P(V,D_1))^{-1}\|\} \le K_1(\tau), \tag{7.14–}$$

and

if $s \in (\tau, \tau + \delta_0(\tau)] \cap [a,b]$ and D_2 is a δ_0–fine partition of $[s,b]$ then

$$\max\{\|P(V,D_2)\|, \|(P(V,D_2))^{-1}\|\} \le K_1(\tau). \tag{7.14+}$$

For proving this statement let us first mention that because we have $\delta_0(\tau) \le \sigma_1(\tau)$, Lemma 7.6 implies that $V(\tau,[\alpha,\beta]) \in L(\mathbb{R}^n)$ is invertible for every $\tau, \alpha, \beta \in [a,b]$ such that $\tau - \delta_0(\tau) < \alpha \le \tau \le \beta < \tau + \delta_0(\tau)$. In order to prove (7.14–) let D_3 be a δ_0–fine partition of the interval $[\tau, b]$. Let

$$D_1 = \{\alpha_0, \tau_1, \alpha_1, \dots, \alpha_{l-1}, \tau_l, \alpha_l\}$$

be a δ_0–fine partition of $[a,s]$ and let

$$D_3 = \{\alpha_{l+1}, \tau_{l+2}, \alpha_{l+2}, \dots, \alpha_{k-1}, \tau_k, \alpha_k\}$$

be a δ_0–fine partition of $[t,b]$. Set

$$D = \{\alpha_0, \tau_1, \alpha_1, \dots, \alpha_{l-1}, \tau_l, \alpha_l = s, \tau_{l-1} = \tau,$$

$$\alpha_{l+1} = \tau, \tau_{l+2}, \alpha_{l+2}, \dots, \alpha_{k-1}, \tau_k, \alpha_k\}.$$

In the sequel we will use the notation $D = D_1 \circ (\tau, [s,\tau]) \circ D_3$ for this construction of a partition of the interval $[a,b]$. This partition D is in fact the union of ordered finite sets in which the ordering preserves the ordering of the components $D_1, \{s,\tau,\tau\}, D_3$; $\circ$ denotes the union of ordered sets. It is evident that D is a δ_0–fine partition of $[a,b]$ and that $V(\tau_i, [\alpha_{i-1}, \alpha_i]) \in L(\mathbb{R}^n)$ is invertible for every $i = 1, \dots, k$. Therefore

$$P(V, D_1) = V(\tau_l, [\alpha_{l-1}, \alpha_l])V(\tau_{l-1}, [\alpha_{l-2}, \alpha_{l-1}])\dots$$

$$V(\tau_1, [\alpha_0, \alpha_1]) \in L(\mathbb{R}^n)$$

and

$$P(V, D_3) = V(\tau_k, [\alpha_{k-1}, \alpha_k])V(\tau_{k-1}, [\alpha_{k-2}, \alpha_{k-1}])\dots$$

$$V(\tau_{l+2}, [\alpha_{l+1}, \alpha_{l+2}]) \in L(\mathbb{R}^n)$$

are invertible and also the inequality (7.12) holds where by definition we have

$$P(V, D) = P(V, D_3)V(\tau_{l+1}, [\alpha_l, \alpha_{l+1}])P(V, D_1) =$$

$$= P(V, D_3)V(\tau, [s,\tau])P(V, D_1)$$

and

$$\|P(V,D_1)-(V(\tau,[s,\tau]))^{-1}(P(V,D_3))^{-1}Q\| =$$

$$= \|(V(\tau,[s,\tau]))^{-1}(P(V,D_3))^{-1}.$$

$$.[(P(V,D_3)V(\tau,[s,\tau])P(V,D_1)-Q]\| \leq$$

$$\leq \|(V(\tau,[s,\tau]))^{-1}\|\|(P(V,D_3))^{-1}\|.\frac{1}{2}\|Q^{-1}\|^{-1}.$$

Consequently, by Lemma 7.7 (see also (7.11)) we obtain

$$\|P(V,D_1)\| \leq \|P(V,D_1)-(V(\tau,[s,\tau]))^{-1}(P(V,D_3))^{-1}Q\|+$$

$$+\|(V(\tau,[s,\tau]))^{-1}\|\|(P(V,D_3))^{-1}\|\|Q\| \leq$$

$$\leq \|(V(\tau,[s,\tau]))^{-1}\|\|(P(V,D_3))^{-1}\|(\frac{1}{2}\|Q^{-1}\|^{-1}+\|Q\|) \leq$$

$$2\|(V_-(\tau))^{-1}\|\|(P(V,D_3))^{-1}\|(\frac{1}{2}\|Q^{-1}\|^{-1}+\|Q\|) = K_0(\tau) > 0. \tag{7.15}$$

On the other hand, we have

$$\|(P(V,D_1))^{-1}-Q^{-1}P(V,D_3)V(\tau,[s,\tau])\| =$$

$$= \|Q^{-1}(Q-P(V,D_3)V(\tau,[s,\tau])P(V,D_1))(P(V,D_1))^{-1}\| \leq$$

$$\leq \|Q^{-1}\|\|(Q-P(V,D)\|\|(P(V,D_1))^{-1}\| \leq$$

$$\leq \|Q^{-1}\|\frac{1}{2}\|Q^{-1}\|^{-1}\|(P(V,D_1))^{-1}\| = \frac{1}{2}\|(P(V,D_1))^{-1}\|$$

and by Lemma 7.7 we therefore obtain

$$\|(P(V,D_1))^{-1}\| \leq \|(P(V,D_1))^{-1}-Q^{-1}P(V,D_3)V(\tau,[s,\tau])\|$$

$$+\|Q^{-1}\|\|P(V,D_3)\|\|V(\tau,[s,\tau])\| \leq$$

$$\leq \frac{1}{2}\|(P(V,D_1))^{-1}\| + \|Q^{-1}\|\|P(V,D_3)\|(\|V_-(\tau)\|+$$

$$+\frac{1}{2}\|(V_-(\tau))^{-1}\|),$$

i.e. we obtain the inequality

$$\|(P(V,D_1))^{-1}\| \leq$$

$$\leq 2\|Q^{-1}\|\|P(V,D_3)\|(\|V_-(\tau)\| + \frac{1}{2}\|(V_-(\tau))^{-1}\|) = K^0(\tau) > 0. \tag{7.16}$$

Taking $K_-(\tau) = \max(K_0(\tau), K^0(\tau)) > 0$ we can conclude by (7.15) and (7.16) that

$$\max\{\|P(V,D_1)\|, \|(P(V,D_1))^{-1}\|\} \leq K_-(\tau)$$

holds. A fully analogous reasoning gives also that if $s \in [\tau, \tau + \delta_0(\tau)) \cap [a,b]$ and D_2 is a δ_0–fine partition of the interval $[s,b]$ then

$$\max\{\|P(V,D_2)\|, \|(P(V,D_2))^{-1}\|\} \leq K_+(\tau)$$

where $K_+(\tau) > 0$. Putting $K_1(\tau) = \max(K_-(\tau), K_+(\tau))$ we obtain (7.14–) and (7.14+).

Now we show that the following is holds.

For every $\tau \in [a,b]$ there is a $K_2(\tau) > 0$ such that

$$\max\{\|P(V,D_1)\|, \|(P(V,D_1))^{-1}\|,$$

$$\|P(V,D_2)\|, \|(P(V,D_2))^{-1}\|\} \leq K_2(\tau) \tag{7.17}$$

if $s \in (\tau - \delta_0(\tau), \tau + \delta_0(\tau)) \cap [a,b]$ and D_1, D_2 are arbitrary δ_0–fine partitions of $[a,s]$, $[s,b]$, respectively.

Let us take e.g. $s \in [\tau, \tau + \delta_0(\tau))$ and set $D = D_1 \circ D_2$. Then

$$P(V, D) = P(V, D_2)P(V, D_1)$$

and $P(V, D_1), P(V, D_2) \in L(\mathbb{R}^n)$ are invertible by Lemma 7.6. Since (7.12) is assumed to be satisfied we have

$$\|P(V, D_2)P(V, D_1) - Q\| < \frac{1}{2}\|Q^{-1}\|^{-1}$$

and

$$\|P(V, D_1) - (P(V, D_2))^{-1}Q\| =$$
$$= \|(P(V, D_2))^{-1}(P(V, D_2)P(V, D_1) - Q)\| \leq$$
$$\leq \|(P(V, D_2))^{-1}\|\frac{1}{2}\|Q^{-1}\|^{-1}.$$

Therefore

$$\|P(V, D_1)\| \leq$$
$$\leq \|P(V, D_1) - (P(V, D_2))^{-1}Q\| + \|(P(V, D_2))^{-1}\|\|Q\| \leq$$
$$\leq \|(P(V, D_2))^{-1}\|(\frac{1}{2}\|Q^{-1}\|^{-1} + \|Q\|). \tag{7.18}$$

On the other hand, we have

$$\|(P(V, D_1))^{-1} - Q^{-1}P(V, D_2)\| =$$
$$= \|Q^{-1}(Q - P(V, D_2)P(V, D_1))(P(V, D_1))^{-1}\| \leq$$
$$\leq \|Q^{-1}\|\|Q - P(V, D_2)P(V, D_1)\|\|(P(V, D_1))^{-1}\| <$$
$$< \frac{1}{2}\|(P(V, D_1))^{-1}\|,$$

henceforth

$$\|(P(V, D_1))^{-1}\| \leq$$

$$\le \|(P(V,D_1))^{-1} - Q^{-1}P(V,D_2)\| + \|Q^{-1}\|\|P(V,D_2)\| \le$$
$$\le \frac{1}{2}\|(P(V,D_1))^{-1}\| + \|Q^{-1}\|\|P(V,D_2)\|$$

and therefore we have

$$\|(P(V,D_1))^{-1}\| \le 2\|Q^{-1}\|\|P(V,D_2)\|. \tag{7.19}$$

Using (7.14+), (7.18) and (7.19) we get the estimate

$$\max\{\|P(V,D_1)\|, \|(P(V,D_1))^{-1}\|\} \le$$
$$\le K_1(\tau)[2\|Q^{-1}\| + \frac{1}{2}\|Q^{-1}\|^{-1} + \|Q\|] = K_L(\tau) > 0.$$

Similarly it can be also shown that

$$\max\{\|P(V,D_2)\|, \|(P(V,D_2))^{-1}\|\} \le K_R(\tau)$$

where $K_R(\tau) > 0$. Putting now $K_2(\tau) = \max\{K_L(\tau), K_R(\tau)\}$ we obtain (7.17) from the two inequalities given above.

The intervals of the form $(\tau - \delta_0(\tau), \tau + \delta_0(\tau))$ with $\tau \in [a,b]$ form an open covering of the compact interval $[a,b]$. Therefore there is a finite set $\{t_1, \dots, t_l\} \subset [a,b]$ such that

$$[a,b] \subset \cup_{j=1}^{l}(t_j - \delta_0(t_j), t_j + \delta_0(t_j)).$$

Define $K = \max\{1, K_2(t_1), K_2(t_2), \dots, K_2(t_l)\}$ where $K_2(\tau)$ is given by (7.17). Then (7.17) implies the following statement.

There exists a constant $K \ge 1$ such that

$$\max\{\|P(V,D_1)\|, \|(P(V,D_1))^{-1}\|\} \le K \tag{7.20-}$$

if $s \in (a,b]$ and D_1 is an arbitrary δ_0-fine partition of $[a,s]$

and

$$\max\{\|P(V,D_2)\|, \|(P(V,D_2))^{-1}\|\} \le K \tag{7.20+}$$

if $s \in [a,b)$ and D_2, is an arbitrary δ_0-fine partitions of $[s,b]$

Now let us prove the following statement.

Let $\varepsilon \in (0, \frac{1}{2}\|Q^{-1}\|^{-1})$ be given and let δ be a gauge on $[a,b]$ such that $\delta(\tau) \le \delta_0(\tau)$ for $\tau \in [a,b]$ and

$$\|P(V,D) - Q\| < \varepsilon$$

for every δ–fine partition D of $[a,b]$.

If $s \in (a,b]$ and D_1, D_3 are arbitrary δ–fine partitions of $[a,s]$, then

$$\|P(V,D_1) - P(V,D_3)\| \le 2K\varepsilon. \tag{7.21–}$$

If $s \in [a,b)$ and D_2, D_4 are arbitrary δ–fine partitions of $[s,b]$, then

$$\|P(V,D_2) - P(V,D_4)\| \le 2K\varepsilon. \tag{7.21+}$$

K is the constant given in (7.20).

Let us prove (7.21+) only; the proof of (7.21–) is similar. Let $s \in [a,b)$. Denote by D_1 an arbitrary δ–fine partition of $[a,s]$ and let us put $D_5 = D_1 \circ D_2$ and $D_6 = D_1 \circ D_4$. Evidently D_5 and D_6 are δ–fine partitions of the interval $[a,b]$. Hence

$$\|P(V,D_2)P(V,D_1) - P(V,D_4)P(V,D_1)\| \le$$

$$\le \|P(V,D_5) - Q\| + \|P(V,D_6) - Q\| \le 2\varepsilon$$

and

$$\|P(V,D_2) - P(V,D_4)\| =$$

$$= \|[P(V,D_2)P(V,D_1) - P(V,D_4)P(V,D_1)](P(V,D_1))^{-1}\| \le$$

$$\|P(V,D_2)P(V,D_1) - P(V,D_4)P(V,D_1)\|\|(P(V,D_1))^{-1}\| \le 2K\varepsilon$$

by (7.20–). This yields (7.21+), and (7.21–) can be shown similarly.

Using (7.21) and Proposition 7.2 we have the following result.

If $s \in (a,b)$ then there exist $Q^-, Q^+ \in L(\mathbb{R}^n)$ such that for every $\varepsilon \in (0, \frac{1}{2}\|Q^{-1}\|^{-1})$ there is a gauge $\delta_1 : [a,b] \to (0,+\infty)$ on the interval $[a,b]$ such that

$$\|P(V, D_1) - Q^-\| < \varepsilon \tag{7.22–}$$

for every δ_1–fine partition D_1 of $[a,s]$ and

$$\|P(V, D_2) - Q^+\| < \varepsilon \tag{7.22+}$$

for every δ_1–fine partition D_2 of $[s,b]$.

Now we are able to complete the proof of the theorem.

Assume that $s \in (a,b)$. Let us choose a gauge δ_2 on $[a,b]$ such that $\delta_2(\tau) \le \min(\delta(\tau), \delta_0(\tau), \delta_1(\tau), |\tau - s|)$ for $\tau \ne s$ and $\delta_2(s) \le \delta_1(s)$. By this choice every δ_2–fine partition

$$D = \{(\tau_j, J_j),\, j = 1, \dots, k\} = \{(\tau_j, [\alpha_{j-1}, \alpha_j]),\, j = 1, \dots, k\}$$

has the property that there is a $j \in \{1, 2, \dots, k\}$ such that $\tau_j = s$. For a δ_2–fine partition D of $[a,b]$ and δ_2–fine partitions D_1, D_2 of $[a,s]$, $[s,b]$, respectively, we have by (7.20) the inequality

$$\|P(V,D) - Q^+Q^-\| \le \|P(V,D) - P(V,D_2)P(V,D_1)\| +$$

$$+ \|P(V,D_2)P(V,D_1) - Q^+Q^-\| \le$$

$$\le \|P(V,D) - P(V,D_2)P(V,D_1)\| +$$

$$+ \|P(V,D_2)P(V,D_1) - Q^+P(V,D_1) + Q^+(P(V,D_1) - Q^-)\| \le$$

$$\le \|P(V,D) - P(V,D_2)P(V,D_1)\| +$$

$$+\|P(V,D_2)-Q^+\|\|P(V,D_1)\|+$$

$$+\|Q^+-P(V,D_2\|\|P(V,D_1)-Q^-\|+$$

$$+\|P(V,D_2\|\|P(V,D_1)-Q^-\|\le$$

$$\le\|P(V,D)-P(V,D_2)P(V,D_1)\|+\varepsilon(2K+\varepsilon). \qquad (7.23)$$

For a given δ_2–fine partition

$$D=\{\alpha_0,\tau_1,\alpha_1,\dots,\alpha_{j-1},\tau_j=s,\alpha_j,\tau_{j+1},\alpha_{j+1},$$

$$\dots,\alpha_{k-1},\tau_k,\alpha_k\}$$

we put

$$D_-=\{\alpha_0,\tau_1,\alpha_1,\dots,\tau_{j-1},\alpha_{j-1}\}$$

and

$$D_+=\{\alpha_j,\tau_{j+1},\alpha_{j+1},\dots,\alpha_{k-1},\tau_k,\alpha_k\}$$

and also

$$D_1=D_-\circ\{\alpha_{j-1},\widetilde{t}_j=s,\widetilde{\alpha}_j=s\},$$

$$D_2=\{\widetilde{\alpha}_{j-1}=s,\widetilde{t}_j=s,\alpha_j\}\circ D_+.$$

It is easy to see that D_1, D_2 are δ_2–fine partitions of $[a,s]$, $[s,b]$, respectively, and that

$$P(V,D)=P(V,D_+)V(s,[\alpha_{j-1},\alpha_j])P(V,D_-),$$

$$P(V,D_1)=V(s,[\alpha_{j-1},s])P(V,D_-),$$

$$P(V,D_2)=P(V,D_+)V(s,[s,\alpha_j]).$$

Moreover,

$$\|P(V,D)-P(V,D_2)P(V,D_1)\|=$$

$$=\|P(V,D_+)V(s,[\alpha_{j-1},\alpha_j])P(V,D_-)-$$

$$-P(V,D_+)V(s,[s,\alpha_j])V(s,[\alpha_{j-1},s])P(V,D_-)\| =$$
$$= \|P(V,D_+)[V(s,[\alpha_{j-1},\alpha_j])-$$
$$-V(s,[s,\alpha_j])V(s,[\alpha_{j-1},s])]P(V,D_-)\| \le K^2\zeta$$

by (7.20) and (7.13) because we have $\alpha_{j-1},\alpha_j \in [a,b]$ and

$$s-\delta_0(s) < s-\delta_2(s) < \alpha_{j-1} \le s \le \alpha_j < s+\delta_2(s) < s+\delta_0(s).$$

By the inequality (7.23) we therefore obtain

$$\|P(V,D) - Q^+Q^-\| < K^2\zeta + \varepsilon(2K+\varepsilon).$$

If we take for example $\zeta = \dfrac{\varepsilon}{K^2}$ and use the inequality

$$\|P(V,D) - Q\| < \varepsilon$$

then for every δ_2–fine partition D of $[a,b]$ we get (see (7.21))

$$\|Q - Q^+Q^-\| \le \|Q - P(V,D)\| + \|P(V,D) - Q^+Q^-\| <$$
$$< \varepsilon + \varepsilon + \varepsilon(2K+\varepsilon) = \varepsilon(2+2K+\varepsilon)$$

and because $\varepsilon > 0$ can be chosen arbitrarily small we finally obtain

$$Q = Q^+Q^-. \tag{7.24}$$

Since $Q \in L(\mathbb{R}^n)$ is invertible, we have by (7.24) the identity

$$Q^{-1}Q^+Q^- = I$$

and this means that $Q^{-1}Q^+ \in L(\mathbb{R}^n)$ is the inverse to Q^- (in fact this shows that $Q^{-1}Q^+$ is the left inverse to Q^- but we have also $Q^-Q^{-1}Q^+Q^- = Q^-$ and consequently $Q^-Q^{-1}Q^+ = I$; this means that $Q^{-1}Q^+$ is also the right inverse to Q^-). Similarly it can be also shown that $Q^+ \in L(\mathbb{R}^n)$ is also invertible with $(Q^+)^{-1} = Q^- Q^{-1}$.

This yields by (7.21) that the Perron product integrals $\prod_a^s V(t,dt) = Q^-$, $\prod_s^b V(t,dt) = Q^+$ exist and have invertible "values" Q^-, Q^+, respectively. Moreover, (7.24) is in fact the equality

$$\prod_s^b V(t,dt) \prod_a^s V(t,dt) = \prod_a^b V(t,dt) \tag{7.25}$$

given in the statement of the theorem.

The estimates

$$\|\prod_a^s V(t,dt)\| \le K, \ \|(\prod_a^s V(t,dt))^{-1}\| \le K$$

are simple consequences of (7.20) and of (7.25). □

7.9 Lemma. *Let $V : [a,b] \times \mathfrak{J} \to L(\mathbb{R}^n)$ be Perron product integrable over $[a,b]$ with $\prod_a^b V(t,dt) = Q$ where $Q \in L(\mathbb{R}^n)$ is invertible and assume that the condition C is satisfied.*

Let us define $\Phi : [a,b] \to L(\mathbb{R}^n)$ by the relations

$$\Phi(a) = I,$$

$$\Phi(s) = \prod_a^s V(t,dt) \text{ for } s \in (a,b]. \tag{7.26}$$

The function Φ is well defined and its values are invertible elements of $L(\mathbb{R}^n)$, $\Phi(b) = Q$.

For a given $\varepsilon > 0$ let $\delta : [a,b] \to (0,+\infty)$ be a gauge on $[a,b]$ such that

$$\|P(V,D) - \Phi(b)\| = \|P(V,D) - \prod_a^b V(t,dt)\| < \varepsilon \tag{7.27}$$

holds for every δ–fine partition D of the interval $[a,b]$. If

$$a \le \beta_1 \le \xi_1 \le \gamma_1 \le \beta_2 \le \xi_2 \le \gamma_2 \le \dots$$

$$\dots \le \beta_m \le \xi_m \le \gamma_m \le b$$

represents a δ–fine system $\{(\xi_j,[\beta_j,\gamma_j]),\ j=1,\dots,m\}$, i.e.

$$\xi_j \in [\beta_j,\gamma_j] \subset [\xi_j - \delta(\xi_j), \xi_j + \delta(\xi_j)], \quad j=1,\dots,m$$

then

$$\begin{aligned} &\|(\Phi(\gamma_m))^{-1}V(\xi_m,[\beta_m,\gamma_m])\Phi(\beta_m). \\ &(\Phi(\gamma_{m-1}))^{-1}V(\xi_{m-1},[\beta_{m-1},\gamma_{m-1}])\Phi(\beta_{m-1})\cdots \\ &\cdots(\Phi(\gamma_1))^{-1}V(\xi_1,[\beta_1,\gamma_1])\Phi(\beta_1) - I\| \le \|(\Phi(b))^{-1}\|\varepsilon. \end{aligned} \tag{7.28}$$

Proof. The function $\Phi : [a,b] \to L(\mathbb{R}^n)$ given by (7.26) is well defined by Theorem 7.8 and the same theorem yields also the invertibility of $\Phi(s)$ for every $s \in [a,b]$. By Theorem 7.8 also the Perron product integral $\prod_c^d V(t,dt)$ exists over every interval $[c,d] \subset [a,b]$.

Let us denote $\gamma_0 = a$ and $\beta_{m+1} = b$.

Since the product integral $\prod_{\gamma_j}^{\beta_{j+1}} V(t,dt)$ exists for every $j = 0,1,\dots,m$ we have by definition the following assertion: *For every $\eta > 0$ there is a gauge $\delta_j : [\gamma_j,\beta_{j+1}] \to (0,\infty)$ such that $\delta_j(t) < \delta(t)$ for $t \in [\gamma_j,\beta_{j+1}]$ and*

$$\|P(V,D_j) - \prod_{\gamma_j}^{\beta_{j+1}} V(t,dt)\| =$$

$$= \|P(V,D_j) - \Phi(\beta_{j+1})(\Phi(\gamma_j))^{-1}\| < \eta \tag{7.29}$$

for every δ_j–fine partition D_j of $[\gamma_j,\beta_{j+1}]$ and for every $j = 0,1,\dots,m$.

For δ_j–fine partitions D_j of $[\gamma_j, \beta_{j+1}]$, $j = 0, 1, \dots, m$ let us set

$$D = D_0 \circ (\xi_1, [\beta_1, \gamma_1]) \circ D_1 \circ (\xi_2, [\beta_2, \gamma_2]) \circ D_3 \circ \dots$$

$$\circ D_{m-1} \circ (\xi_m, [\beta_m, \gamma_m]) \circ D_m.$$

D given in this way evidently forms a δ–fine partition of $[a, b]$ and therefore (7.27) holds for this partition. Hence

$$\|(\Phi(b))^{-1} P(V, D) - I\| = \|(\Phi(b))^{-1}[P(V, D) - \Phi(b)]\| <$$

$$< \|(\Phi(b))^{-1}\|\varepsilon. \tag{7.30}$$

Further, we have

$$P(V, D) = P(V, D_m) V(\xi_m, [\beta_m, \gamma_m]) P(V, D_{m-1}) \dots$$

$$P(V, D_1) V(\xi_1, [\beta_1, \gamma_1]) P(V, D_0)$$

and

$$(\Phi(b))^{-1} P(V, D) =$$

$$= (\Phi(b))^{-1} P(V, D_m) V(\xi_m, [\beta_m, \gamma_m]) P(V, D_{m-1}) \dots$$

$$P(V, D_1) V(\xi_1, [\beta_1, \gamma_1]) P(V, D_0) =$$

$$= (\Phi(b))^{-1} P(V, D_m) V(\xi_m, [\beta_m, \gamma_m]) \Phi(\gamma_m) (\Phi(\gamma_m))^{-1}.$$

$$V(\xi_m, [\beta_m, \gamma_m]) \Phi(\beta_m) (\Phi(\beta_m))^{-1}.$$

$$P(V, D_{m-1}) \Phi(\gamma_{m-1}) (\Phi(\gamma_{m-1}))^{-1}. \cdots$$

$$\cdots \Phi(\beta_2)(\Phi(\beta_2))^{-1} P(V, D_1) \Phi(\gamma_1)(\Phi(\gamma_1))^{-1} V(\xi_1, [\beta_1, \gamma_1]) \Phi(\beta_1).$$

$$(\Phi(\beta_1))^{-1} P(V, D_0) \Phi(\gamma_0).$$

Denoting

$$(\Phi(\beta_{j+1}))^{-1} P(V, D_j) \Phi(\gamma_j) = A_j + I$$

for $j = 0, 1, \ldots, m$ and

$$(\Phi(\gamma_j))^{-1} V(\xi_j, [\beta_j, \gamma_j]) \Phi(\beta_j) = Z_j + I$$

for $j = 1, 2, \ldots, m$ we obtain

$$(\Phi(b))^{-1} P(V, D) =$$

$$= (I + A_m)(I + Z_m)(I + A_{m-1})(I + Z_{m-1}) \cdots$$

$$(I + A_1)(I + Z_1)(I + A_0)$$

and we can write (7.30) in the form

$$\|(I + A_m)(I + Z_m)(I + A_{m-1})(I + Z_{m-1}) \cdots$$

$$(I + A_1)(I + Z_1)(I + A_0) - I\| < \|(\Phi(b))^{-1}\| \varepsilon. \tag{7.31}$$

By (7.29) we have

$$\|A_j\| = \|(\Phi(\beta_{j+1}))^{-1} P(V, D_j) \Phi(\gamma_j) - I\| =$$

$$= \|(\Phi(\beta_{j+1}))^{-1} [P(V, D_j) - \Phi(\beta_{j+1})(\Phi(\gamma_j))^{-1}] \Phi(\gamma_j)\| \le K^2 \eta \tag{7.32}$$

where K is the constant given by Theorem 7.8 and $j = 0, 1, \ldots, m$.

Using the estimate (7.32) we obtain the following estimate: for every $\theta > 0$ there is an $\eta > 0$ such that

$$\|(I + A_m)(I + Z_m)(I + A_{m-1})(I + Z_{m-1}) \cdots (I + A_1)(I + Z_1).$$

$$(I + A_0) - (I + Z_m)(I + Z_{m-1}) \cdots (I + Z_1)\| < \eta.$$

Hence by (7.31) we have

$$\|(I + Z_m)(I + Z_{m-1}) \cdots (I + Z_1) - I\| \le$$

$$\leq \|(I+A_m)(I+Z_m)(I+A_{m-1})(I+Z_{m-1})\cdots$$
$$(I+A_1)(I+Z_1)(I+A_0)-(I+Z_m)(I+Z_{m-1})\cdots(I+Z_1)\|+$$
$$+\|(I+A_m)(I+Z_m)(I+A_{m-1})(I+Z_{m-1})\cdots$$
$$(I+A_1)(I+Z_1)(I+A_0)-I\|<$$
$$<\theta+\|(\Phi(b))^{-1}\|\varepsilon$$

where $\theta > 0$ is arbitrary, and consequently we arrive at

$$\|(I+Z_m)(I+Z_{m-1})\cdots(I+Z_1)-I\| \leq \|(\Phi(b))^{-1}\|\varepsilon$$

and by the definition of Z_j, $j = 1,2,\ldots,m$ we obtain (7.28). □

Remark. The result given in the second part of Lemma 7.9 plays a similar role for the Perron product integral as the Saks–Henstock Lemma 1.13 does for the generalized Perron integral.

7.10 Corollary. *Let $V : [a,b]\times\mathfrak{J} \to L(\mathbb{R}^n)$ be Perron product integrable over $[a,b]$ with $\prod_a^b V(t,dt) = Q$ where $Q \in L(\mathbb{R}^n)$ is invertible and assume that the condition C is satisfied.*

Then for every $\eta > 0$, $\tau \in [a,b]$ there exists a $\delta > 0$ such that

$$\|(\Phi(\gamma))^{-1}V(\tau_m,[\beta_m,\gamma_m])\Phi(\beta)-I\| < \eta \tag{7.33}$$

and

$$\|V(\tau_m,[\beta_m,\gamma_m])-\Phi(\gamma)(\Phi(\beta))^{-1}\| \leq K^2\eta \tag{7.34}$$

whenever $\beta,\gamma \in [a,b]$, $\tau-\delta < \beta \leq \tau \leq \gamma < \tau+\delta$, where $\Phi : [a,b] \to L(\mathbb{R}^n)$ is given by (7.26) and K is the constant from Theorem 7.8.

Proof. Taking $\varepsilon = \eta\|(\Phi(b))^{-1}\|^{-1} > 0$ we obtain (7.33) directly from (7.28) in Lemma 7.9 when $\delta : [a,b] \to (0,+\infty)$ is the gauge on $[a,b]$ corresponding to this choice of ε by the definition of the Perron product integral (see (7.27)).

Since we have

$$\|V(\tau_m, [\beta_m, \gamma_m]) - \Phi(\gamma)(\Phi(\beta))^{-1}\| =$$

$$= \|\Phi(\gamma)[(\Phi(\gamma))^{-1}V(\tau_m, [\beta_m, \gamma_m])\Phi(\beta) - I](\Phi(\beta))^{-1}\| \leq$$

$$\leq \|\Phi(\gamma)\|\|(\Phi(\beta))^{-1}\|\|(\Phi(\gamma))^{-1}V(\tau_m, [\beta_m, \gamma_m])\Phi(\beta) - I\|,$$

we obtain the inequality (7.34) from (7.33) and from the inequalities

$$\|\Phi(\tau)\| \leq K, \ \|(\Phi(\tau))^{-1}\| \leq K$$

which hold for every $\tau \in [a, b]$ by Theorem 7.8. □

7.11 Lemma. *Assume that* $A, A_k \in L(\mathbb{R}^n)$, $k = 1, 2, \ldots$ *are invertible and such that*

$$\lim_{k \to \infty} A_k = A. \tag{7.35}$$

Then

$$\lim_{k \to \infty} (A_k)^{-1} = A^{-1}. \tag{7.36}$$

Proof. By (7.35) there is a $k_0 \in \mathbb{N}$ such that for $k > k_0$ we have $\|A - A_k\| < \|A^{-1}\|^{-1}$ and therefore

$$\|I - A_k A^{-1}\| = \|(A - A_k)A^{-1}\| \leq \|(A - A_k)\|\|A^{-1}\| < 1.$$

Hence $A_k A^{-1}$ has an inverse given by

$$(A_k A^{-1})^{-1} = \sum_{l=0}^{\infty}(I - A_k A^{-1})^l = \sum_{l=0}^{\infty}((A - A_k)A^{-1})^l = A(A_k)^{-1}.$$

Consequently

$$(A_k)^{-1} = A^{-1}\sum_{l=0}^{\infty}((A - A_k)A^{-1})^l =$$

$$= A^{-1} + A^{-1} \sum_{l=1}^{\infty} ((A - A_k)A^{-1})^l,$$

i.e.

$$(A_k)^{-1} - A^{-1} = A^{-1} \sum_{l=1}^{\infty} ((A - A_k)A^{-1})^l$$

and

$$\|(A_k)^{-1} - A^{-1}\| \leq \|A^{-1}\| \sum_{l=1}^{\infty} (\|A - A_k\| \|A^{-1}\|^l \leq$$

$$\leq \|A^{-1}\| \frac{\|A - A_k\| \|A^{-1}\|}{1 - \|A - A_k\| \|A^{-1}\|}$$

for $k > k_0$.

Since $\|A - A_k\| \to 0$ for $k \to \infty$ we obtain from this estimate that $\|(A_k)^{-1} - A^{-1}\| \to 0$ for $k \to \infty$, i.e. (7.36) holds. □

7.12 Lemma. *Let $V : [a, b] \times \mathfrak{J} \to L(\mathbb{R}^n)$ be Perron product integrable over $[a, b]$ with $\prod_a^b V(t, dt) = Q$ where $Q \in L(\mathbb{R}^n)$ is invertible and assume that the condition C is satisfied.*

Then

$$\lim_{\beta \to \tau-} \Phi(\beta) = (V_-(\tau))^{-1} \Phi(\tau) \tag{7.37}$$

for $\tau \in (a, b]$ and

$$\lim_{\gamma \to \tau+} \Phi(\gamma) = (V_+(\tau))^{-1} \Phi(\tau) \tag{7.38}$$

for $\tau \in [a, b)$.

Proof. It follows immediately from Corollary 7.10 that

$$\lim_{\beta \to \tau-} \|(\Phi(\tau))^{-1} V(\tau, [\beta, \tau]) \Phi(\beta) - I\| = 0 \tag{7.39}$$

for $\tau \in (a, b]$ and

$$\lim_{\gamma \to \tau+} \|(\Phi(\gamma))^{-1} V(\tau, [\tau, \gamma])\Phi(\tau) - I\| = 0 \tag{7.40}$$

for $\tau \in [a, b)$.

By (7.4–) and (7.4+) from the condition C we also have

$$\lim_{\beta \to \tau-} \|V(\tau, [\beta, \tau]) - V_-(\tau)\| = 0 \tag{7.41}$$

for $\tau \in (a, b]$ and

$$\lim_{\gamma \to \tau+} \|V(\tau, [\tau, \gamma]) - V_+(\tau)\| = 0 \tag{7.42}$$

for $\tau \in [a, b)$ where $V_-(\tau), V_+(\tau) \in L(\mathbb{R}^n)$ are invertible. Since by Theorem 7.8 we have $\|\Phi(\tau)\| \leq K, \|(\Phi(\tau))^{-1}\| \leq K$, we get for $\tau \in (a, b]$, $\beta < \tau$ the inequality

$$\|(\Phi(\beta))^{-1} - (\Phi(\tau))^{-1} V_-(\tau)\| =$$

$$= \|(\Phi(\beta))^{-1} - (\Phi(\tau))^{-1} V(\tau, [\beta, \tau]) + (\Phi(\tau))^{-1} V(\tau, [\beta, \tau]) -$$
$$-(\Phi(\tau))^{-1} V_-(\tau)\| =$$
$$= \|[I - (\Phi(\tau))^{-1} V(\tau, [\beta, \tau])\Phi(\beta)](\Phi(\beta))^{-1} +$$
$$+(\Phi(\tau))^{-1} V(\tau, [\beta, \tau]) - (\Phi(\tau))^{-1} V_-(\tau)\| \leq$$
$$\leq K[\|I - (\Phi(\tau))^{-1} V(\tau, [\beta, \tau])\Phi(\beta)\| + \|V(\tau, [\beta, \tau]) - V_-(\tau)\|].$$

This inequality together with (7.39) and (7.41) implies

$$\lim_{\beta \to \tau-} (\Phi(\beta))^{-1} = (\Phi(\tau))^{-1} V_-(\tau)$$

and Lemma 7.11 immediately yields (7.37).

Similarly for $\tau \in [a, b)$, $\gamma > \tau$ we have

$$\|\Phi(\gamma) - V_+(\tau)\Phi(\tau)\| =$$

$$= \|\Phi(\gamma) - V(\tau, [\tau, \gamma])\Phi(\tau) + V(\tau, [\tau, \gamma])\Phi(\tau) - V_+(\tau)\Phi(\tau)\| \le$$

$$\le \|\Phi(\gamma)[I - (\Phi(\gamma))^{-1}V(\tau, [\tau, \gamma])\Phi(\tau)]\| +$$

$$+ \|[V(\tau, [\tau, \gamma]) - V_+(\tau)]\Phi(\tau)\| \le$$

$$\le K[\|I - (\Phi(\gamma))^{-1}V(\tau, [\tau, \gamma])\Phi(\tau)\| + \|V(\tau, [\tau, \gamma]) - V_+(\tau)\|]$$

and (7.40) together with (7.42) directly imply (7.38). □

7.13 Lemma. *Assume that $Y_1, Y_2, \dots, Y_k \in L(\mathbb{R}^n)$ where $\sum_{i=1}^k \|Y_i\| \le 1$. Let us set*

$$X = (I + Y_k)(I + Y_{k-1}) \dots (I + Y_1) - I$$

and

$$Z = X - \sum_{i=1}^k Y_i.$$

Then

$$\|X\| \le 2 \sum_{i=1}^k \|Y_i\|$$

and

$$\|Z\| \le (\sum_{i=1}^k \|Y_i\|)^2.$$

Proof. Put $\lambda_i = \|Y_i\|$ for $i = 1, 2, \dots, k$ and $\lambda = \sum_{i=1}^k \lambda_i \le 1$. We have

$$(I + \lambda_k)(I + \lambda_{k-1}) \dots (I + \lambda_1) =$$

$$= 1 + \sum_{i=1}^{k} \lambda_i + \sum_{j_2 > j_1} \lambda_{j_2}\lambda_{j_1} + \sum_{j_3 > j_2 > j_1} \lambda_{j_3}\lambda_{j_2}\lambda_{j_1} + \cdots +$$

$$+ \lambda_k \lambda_{k-1} \dots \lambda_1 \le e^{\lambda_k} e^{\lambda_{k-1}} \dots e^{\lambda_1} = e^{\lambda}.$$

Hence

$$\sum_{i=1}^{k} \lambda_i + \sum_{j_2 > j_1} \lambda_{j_2}\lambda_{j_1} + \cdots + \lambda_k \lambda_{k-1} \dots \lambda_1 \le e^{\lambda} - 1 < 2\lambda$$

and

$$\sum_{j_2 > j_1} \lambda_{j_2}\lambda_{j_1} + \sum_{j_3 > j_2 > j_1} \lambda_{j_3}\lambda_{j_2}\lambda_{j_1} + \cdots + \lambda_k \lambda_{k-1} \dots \lambda_1 \le$$

$$e^{\lambda} - 1 - \lambda \le \lambda^2$$

because $\lambda \le 1$. We evidently have

$$X = \sum_{i=1}^{k} Y_i + \sum_{j_2 > j_1} Y_{j_2} Y_{j_1} + \cdots + Y_k Y_{k-1} \dots Y_1$$

and

$$Z = \sum_{j_2 > j_1} Y_{j_2} Y_{j_1} + \sum_{j_3 > j_2 > j_1} Y_{j_3} Y_{j_2} Y_{j_1} + \cdots + Y_k Y_{k-1} \dots Y_1.$$

Hence

$$\|X\| \le \sum_{i=1}^{k} \|Y_i\| + \sum_{j_2 > j_1} \|Y_{j_2}\| \|Y_{j_1}\| + \cdots + \|Y_k\| \|Y_{k-1}\| \dots \|Y_1\| =$$

$$= \sum_{i=1}^{k} \lambda_i + \sum_{j_2 > j_1} \lambda_{j_2}\lambda_{j_1} + \cdots + \lambda_k \lambda_{k-1} \dots \lambda_1 \le$$

$$\leq e^{\lambda} - 1 < 2\lambda = 2\sum_{i=1}^{k} \|Y_i\|,$$

and similarly also

$$\|Z\| \leq \sum_{j_2 > j_1} \|Y_{j_2}\|\|Y_{j_1}\| + \sum_{j_3 > j_2 > j_1} \|Y_{j_3}\|\|Y_{j_2}\|\|Y_{j_1}\| + \cdots +$$

$$+\|Y_k\|\|Y_{k-1}\| \ldots \|Y_1\| =$$

$$= \sum_{j_2 > j_1} \lambda_{j_2}\lambda_{j_1} + \sum_{j_3 > j_2 > j_1} \lambda_{j_3}\lambda_{j_2}\lambda_{j_1} + \cdots + \lambda_k\lambda_{k-1} \ldots \lambda_1 \leq$$

$$e^{\lambda} - 1 - \lambda \leq \lambda^2 = (\sum_{i=1}^{k} \|Y_i\|)^2.$$

□

7.14 Theorem. *Assume that $V : [a,b] \times \mathfrak{J} \to L(\mathbb{R}^n)$ satisfies the condition C and that for every $c \in [a,b)$ the Perron product integral $\prod_a^c V(t,dt)$ exists and has invertible values.*

Let the limit

$$\lim_{c \to b-} V(b,[c,b]) \prod_a^c V(t,dt) = Q \tag{7.43}$$

exist, where $Q \in L(\mathbb{R}^n)$.

Then $V : [a,b] \times \mathfrak{J} \to L(\mathbb{R}^n)$ is Perron product integrable over $[a,b]$ and

$$\prod_a^b V(t,dt) = Q. \tag{7.44}$$

Proof. Let $\varepsilon \in (0,1)$ be given. Since the limit (7.43) exists, there is a $B \in [a,b)$ such that for every $c \in [B,b)$ we have

$$\|V(b,[c,b]) \prod_a^c V(t,dt) - Q\| < \varepsilon. \tag{7.45}$$

Let us have a sequence $a = c_0 < c_1 < \ldots$, $\lim_{p\to\infty} c_p = b$. Over every $[a, c_p]$, $p = 1, 2, \ldots$ the function V is Perron product integrable and therefore there exists a gauge $\delta_p : [a, c_p] \to (0, +\infty)$, $p = 1, 2, \ldots$ such that for every δ_p–fine partition D of $[a, c_p]$ we have

$$\|P(V, D) - \prod_a^{c_p} V(t, dt)\| \le \frac{\varepsilon}{2^{p+1}\|(\prod_a^{c_p} V(t, dt))^{-1}\|} \tag{7.46}$$

for $p = 1, 2, \ldots$.

For every $\tau \in [a, b)$ there is exactly one $p(\tau) \in \mathbb{N}$ such that $\tau \in [c_{p(\tau)-1}, c_{p(\tau)})$. For $\tau \in [a, b)$ let us choose $\delta^0(\tau) > 0$ such that $\delta^0(\tau) \le \delta_{p(\tau)}(\tau)$ and $[\tau - \delta^0(\tau), \tau + \delta^0(\tau)] \cap [a, b) \subset [a, c_{p(\tau)})$.

If $c \in [a, b)$ and

$$D^- = \{\alpha_0, \tau_1, \alpha_1, \ldots \alpha_{k-2}, \tau_{k-1}, \alpha_{k-1}\}$$

is a δ^0–fine partition of $[a, c]$, then if $p(\tau_j) = p$, we have

$$[\alpha_{j-1}, \alpha_j] \subset (\tau_j - \delta^0(\tau_j), \tau_j + \delta^0(\tau_j)) \subset [a, c_p]$$

and also

$$[\alpha_{j-1}, \alpha_j] \subset (\tau_j - \delta_p(\tau_j), \tau_j + \delta_p(\tau_j)). \tag{7.47}$$

For the partition D^- we have

$$P(V, D^-) = V(\tau_{k-1}, [\alpha_{k-2}, \alpha_{k-1}])V(\tau_{k-2}, [\alpha_{k-3}, \alpha_{k-2}]) \cdots \times$$

$$\times V(\tau_1, [\alpha_0, \alpha_1]) = A_m A_{m-1} \ldots A_1$$

where A_j, $j = 1, 2, \ldots m$ is the ordered product of all factors $V(\tau_l, [\alpha_{l-1}, \alpha_l])$, $1 \le l \le k-1$ with $\tau_l \in [c_{p_j - 1}, c_{p_j}]$, i.e.

$$A_j = V(\tau_{r_j + s_j}, [\alpha_{r_j + s_j - 1}, \alpha_{r_j + s_j}])\times$$

$$\times V(\tau_{r_j+s_j-1},[\alpha_{r_j+s_j-2},\alpha_{r_j+s_j-1}])\dots V(\tau_{r_j},[\alpha_{r_j-1},\alpha_{r_j}])$$

and $\tau_{r_j},\tau_{r_j+1},\dots,\tau_{r_j+s_j}\in[c_{p_j-1},c_{p_j}]$ with $1\le r_j\le r_j+s_j\le k-1$. Since the partition D^- satisfies (7.47) we also have

$$[\alpha_{i-1},\alpha_i]\subset(\tau_i-\delta_{p_j}(\tau_i),\tau_i+\delta_{p_j}(\tau_i)),$$

$i=r_j,r_j+1,\dots,r_j+s_j$. Using (7.46) and Lemma 7.9 we obtain

$$\|(\prod_a^{\tau_{r_j+s_j}} V(t,dt))^{-1}V(\tau_{r_j+s_j},[\alpha_{r_j+s_j-1},\alpha_{r_j+s_j}])\dots$$

$$\dots V(\tau_{r_j},[\alpha_{r_j-1},\alpha_{r_j}])\prod_a^{\tau_{r_j}}V(t,dt)-I\|=$$

$$=\|(\prod_a^{\tau_{r_j+s_j}} V(t,dt))^{-1}A_j\prod_a^{\tau_{r_j}}V(t,dt)-I\|\le$$

$$\le\frac{\varepsilon\|(\prod_a^{c_{p_j}} V(t,dt))^{-1}\|}{2^{p_j+1}\|(\prod_a^{c_{p_j}} V(t,dt))^{-1}\|}=\frac{\varepsilon}{2^{p_j+1}}$$

for every $j=1,2,\dots,m$. Hence

$$\sum_{j=1}^m\|(\prod_a^{\tau_{r_j+s_j}} V(t,dt))^{-1}A_j\prod_a^{\tau_{r_j}}V(t,dt)-I\|\le\sum_{j=1}^m\frac{\varepsilon}{2^{p_j+1}}<\varepsilon. \tag{7.48}$$

Denoting

$$Y_j=\|(\prod_a^{\tau_{r_j+s_j}} V(t,dt))^{-1}A_j\prod_a^{\tau_{r_j}}V(t,dt)-I\|$$

for $j = 1, 2, \ldots, m$ we have by (7.48) and by the choice of ε the inequality

$$\sum_{j=1}^{m} \|Y_j\| < \varepsilon < 1,$$

and for

$$X = (I + Y_m)(I + Y_{m-1}) \ldots (I + Y_1) - I =$$

$$= \Big(\prod_a^{\alpha_{k-1}} V(t, dt)\Big)^{-1} A_m A_{m-1} \ldots A_1 \prod_a^{a} V(t, dt) - I =$$

$$= \Big(\prod_a^{\alpha_{k-1}} V(t, dt)\Big)^{-1} A_m A_{m-1} \ldots A_1 - I =$$

$$= \Big(\prod_a^{c} V(t, dt)\Big)^{-1} P(V, D^-) - I$$

we obtain by Lemma 7.13 the estimate

$$\|X\| = \Big\|\Big(\prod_a^{c} V(t, dt)\Big)^{-1} P(V, D^-) - I\Big\| \le 2\sum_{j=1}^{m} \|Y_j\| < 2\varepsilon, \quad (7.49)$$

and this estimate does not depend on $c \in [a, b)$.

Define now a gauge δ on $[a, b]$ such that

$$0 < \delta(\tau) < \min(b - \tau, \delta^0(\tau))$$

for $\tau \in [a, b)$ and

$$0 < \delta(b) < b - B.$$

If

$$D = \{\alpha_0, \tau_1, \alpha_1, \ldots \alpha_{k-2}, \tau_{k-1}, \alpha_{k-1}, \tau_k, \alpha_k\}$$

is an arbitrary δ–fine partition of $[a,b]$ then by the choice of the gauge δ we necessarily have $\tau_k = \alpha_k = b$ and $\alpha_{k-1} \in (B,b)$. We have also $D = D^- \circ (b, [\alpha_{k-1}, b])$ where

$$D^- = \{\alpha_0, \tau_1, \alpha_1, \dots \alpha_{k-2}, \tau_{k-1}, \alpha_{k-1}\}$$

and

$$P(V,D) = V(b, [\alpha_{k-1}, b])P(V, D^-).$$

Hence

$$\|P(V,D) - Q\| = \|V(b,[\alpha_{k-1},b])P(V,D^-) - Q\| =$$

$$= \|V(b,[\alpha_{k-1},b]) \prod_a^{\alpha_{k-1}} V(t,dt)(\prod_a^{\alpha_{k-1}} V(t,dt))^{-1} P(V,D^-) - Q\| =$$

$$= \|V(b,[\alpha_{k-1},b]) \prod_a^{\alpha_{k-1}} V(t,dt)[(\prod_a^{\alpha_{k-1}} V(t,dt))^{-1} P(V,D^-) - I] +$$

$$+ V(b,[\alpha_{k-1},b]) \prod_a^{\alpha_{k-1}} V(t,dt) - Q\| \leq$$

$$\leq [\|V(b,[\alpha_{k-1},b]) \prod_a^{\alpha_{k-1}} V(t,dt) - Q\| + \|Q\|].$$

$$\|(\prod_a^{\alpha_{k-1}} [V(t,dt))^{-1} P(V,D^-) - I\| +$$

$$+ \|V(b,[\alpha_{k-1},b]) \prod_a^{\alpha_{k-1}} V(t,dt) - Q\|. \tag{7.50}$$

Since $B < \alpha_{k-1} < b$ we have by (7.45)

$$\|V(b,[\alpha_{k-1},b]) \prod_a^{\alpha_{k-1}} V(t,dt) - Q\| < \varepsilon$$

and by (7.49) we get

$$\|(\prod_a^{\alpha_{k-1}} V(t,dt))^{-1} P(V,D^-) - I\| < 2\varepsilon.$$

Therefore (7.50) yields

$$\|P(V,D) - Q\| < (\varepsilon + \|Q\|)2\varepsilon + \varepsilon = \varepsilon(2\varepsilon + 1 + 2\|Q\|)$$

for an arbitrary δ–fine partition D od $[a,b]$, and this means by definition that the Perron product integral $\prod_a^b V(t,dt)$ exists and its value is Q. □

In a completely analogous way also the following result can be proved.

7.15 Theorem. *Assume that $V : [a,b] \times \mathfrak{J} \to L(\mathbb{R}^n)$ satisfies the condition C and that for every $c \in (a,b]$ the Perron product integral $\prod_c^b V(t,dt)$ exists and has invertible values.*

Let the limit

$$\lim_{c \to a+} \prod_c^b V(t,dt) V(a,[a,c]) = Q$$

exist, where $Q \in L(\mathbb{R}^n)$.

Then $V : [a,b] \times \mathfrak{J} \to L(\mathbb{R}^n)$ is Perron product integrable over $[a,b]$ and

$$\prod_a^b V(t,dt) = Q.$$

Remark. It is not difficult to check that if $V : [a,b] \times \mathfrak{J} \to L(\mathbb{R}^n)$ is Perron product integrable over $[a,b]$, then for every $d \in (a,b]$ we have

$$\lim_{c \to d-} \prod_a^c V(t,dt) = (V_-(d))^{-1} \prod_a^d V(t,dt).$$

and similarly for $d \in [a, b)$

$$\lim_{c \to d+} \prod_c^b V(t, dt) = \prod_d^b V(t, dt)(V_+(d))^{-1}.$$

If $d \in (a, b)$ then

$$\prod_a^b V(t, dt) = \lim_{c \to d+} \prod_c^b V(t, dt) V_+(d) V_-(d) \lim_{c \to d-} \prod_a^c V(t, dt).$$

The proof of the following theorem was given in the paper [65].

7.16 Theorem. *Assume that $W : [a, b] \to \mathbb{R}^n$ is such that the inverse $(W(\tau))^{-1}$ exists for every $\tau \in [a, b]$ and*

$$\max\{\|W(\tau)\|, \|(W(\tau))^{-1}\|\} \le M \tag{7.51}$$

where $M > 0$ is a constant. Let $V : [a, b] \times \mathfrak{J} \to L(\mathbb{R}^n)$ be such that for every $\theta > 0$ there exists a gauge δ on $[a, b]$ such that

$$\sum_{j=1}^k \|V(\tau_j), [\alpha_{j-1}, \alpha_j]) - W(\alpha_j)(W(\alpha_{j-1}))^{-1}\| \le \theta \tag{7.52}$$

provided

$$D = \{\alpha_0, \tau_1, \alpha_1, \dots \alpha_{k-2}, \tau_{k-1}, \alpha_{k-1}, \tau_k, \alpha_k\}$$

is a δ–fine partition of $[a, b]$.

Then the Perron product integral $\prod_a^b V(t, dt)$ exists and is equal to $W(b)(W(a))^{-1}$.

Proof. Let $0 < \theta \le \dfrac{1}{M^2}$. For a δ–fine partition D (δ is the gauge corresponding to θ) define

$$Y_i = (W(\alpha_i))^{-1} V(\tau_i, [\alpha_{i-1}, \alpha_i]) W(\alpha_{i-1}) - I$$

for $i = 1, \dots, k$.

Then

$$\|Y_i\| = \|(W(\alpha_i))^{-1}V(\tau_i, [\alpha_{i-1}, \alpha_i])W(\alpha_{i-1}) - I\| =$$
$$= \|(W(\alpha_i))^{-1}[V(\tau_i, [\alpha_{i-1}, \alpha_i]) -$$
$$-W(\alpha_i)(W(\alpha_{i-1}))^{-1}]W(\alpha_{i-1})\| \leq$$
$$\leq M^2 \|V(\tau_i, [\alpha_{i-1}, \alpha_i]) - W(\alpha_i)(W(\alpha_{i-1}))^{-1}\|$$

for $i = 1, \dots, k$ and therefore by (7.52)

$$\sum_{i=1}^{k} \|Y_i\| \leq$$
$$\leq M^2 \sum_{i=1}^{k} \|V(\tau_i, [\alpha_{i-1}, \alpha_i]) - W(\alpha_i)(W(\alpha_{i-1}))^{-1}\| \leq$$
$$\leq M^2 \theta \leq 1.$$

Take

$$X = (I + Y_k) \dots (I + Y_1) - I =$$
$$= (W(\alpha_k))^{-1}V(\tau_k, [\alpha_{k-1}, \alpha_k])W(\alpha_{k-1}) \dots$$
$$(W(\alpha_1))^{-1}V(\tau_1, [\alpha_0, \alpha_1])W(\alpha_0) - I.$$

Then Lemma 7.13 yields

$$\|X\| = \|(W(b))^{-1}P(V, D)W(a) - I\| \leq 2M^2\theta.$$

Consequently

$$\|P(V, D) - W(b)(W(a))^{-1}\| =$$
$$= \|W(b)[(W(b))^{-1}P(V, D)W(a) - I](W(a))^{-1}\| \leq$$
$$\leq \|W(b)\| \|(W(b))^{-1}P(V, D)W(a) - I\| \|(W(a))^{-1}\| \leq 2M^4\theta$$

and this proves the theorem. □

7.17 Lemma. *Assume that $L \geq 1$ is a constant such that for every matrix $Z \in L(\mathbb{R}^n)$, $Z = (z_{ik})_{i,k=1,\ldots,n}$ the inequality*

$$\frac{1}{L} \max_{i,k=1,\ldots,n} |z_{ik}| \leq \|Z\| \leq L \max_{i,k=1,\ldots,n} |z_{ik}| \tag{7.53}$$

holds. Let

$$0 < \theta < \frac{1}{9L^4}, \tag{7.54}$$

$Z_1, Z_2, \ldots, Z_r \in L(\mathbb{R}^n)$ and assume that for every p–tuple $\{j_1, j_2, \ldots, j_p\} \subset \{1, 2, \ldots, r\}$ the inequality

$$\|(I + Z_{j_p})(I + Z_{j_{p-1}}) \ldots (I + Z_{j_1}) - I\| \leq \theta \tag{7.55}$$

holds. Then

$$\sum_{j=1}^{r} \|Z_j\| \leq M\theta, \tag{7.56}$$

where $M = 4n^2L^2$.

Proof. Let us denote $Z_j = (z_{ik}^{(j)})_{i,k=1,\ldots,n}$, i.e. $z_{ik}^{(j)}$ is the element belonging to the i–th row and k–th column of the matrix Z_j.

Denote further by $J(l,m)$ the set of such values $j \in \{1, \ldots, r\}$ for which

$$\max_{i,k=1,\ldots,n} |z_{ik}^{(j)}| = |z_{lm}^{(j)}|,$$

i.e. $J(l,m)$ is the set of values $j \in \{1, \ldots, r\}$ for which the absolute value of the elements of the matrix Z_j reaches its maximum in the l–th row and m–th column. It is clear that for a given pair (l,m) the set $J(l,m)$ can be empty.

Assume that the estimate (7.56) is not valid, i.e. that

$$4n^2L^2\theta < \sum_{j=1}^{r} \|Z_j\| = \sum_{l=1}^{n} \sum_{m=1}^{n} \sum_{j \in J(l,m)} \|Z_j\|$$

holds. Then evidently there is a pair (l,m) such that

$$4L^2\theta < \sum_{j\in J(l,m)} \|Z_j\| \le \sum_{j\in J(l,m)} L \max_{i,k=1,\dots,n} |z_{ik}^{(j)}| =$$

$$= L \sum_{j\in J(l,m)} |z_{lm}^{(j)}|,$$

where (7.53) was used. Denote now

$$J^+ = \{j \in J(l,m), z_{lm}^{(j)} \ge 0\}$$

and

$$J^- = J(l,m) \setminus J^+ = \{j \in J(l,m), z_{lm}^{(j)} < 0\}.$$

Using the last inequality given above we can conclude that at least one of the inequalities

$$\sum_{j\in J^+} z_{lm}^{(j)} > 2L\theta, \quad -\sum_{j\in J^-} z_{lm}^{(j)} > 2L\theta$$

holds. Assume that the first of these inequalities is satisfied. The case when the second one holds can be treated similarly.

By the assumption (7.55) we have

$$\|Z_j\| \le \theta, \quad j = 1,2,\dots,r,$$

and by (7.53) also

$$z_{lm}^{(j)} \le L\theta \text{ for } j \in J^+$$

and therefore we have a set $\tilde{J}^+ \subset J^+$ such that

$$2L\theta < \sum_{j\in\tilde{J}^+} z_{lm}^{(j)} \le 3L\theta \tag{7.57}$$

and consequently by (7.53) and (7.54) we have

$$\sum_{j\in\widetilde{J}+} \|Z_j\| \le L \sum_{j\in\widetilde{J}+} z_{lm}^{(j)} \le 3L^2\theta < 1.$$

From this inequality we can see that the matrices Z_j with $j \in \widetilde{J}^+$ satisfy the assumptions of Lemma 7.13 and by this lemma we obtain

$$\prod_{j\in\widetilde{J}+} (I + Z_j) - I = \sum_{j\in\widetilde{J}+} Z_j + X$$

and

$$\|X\| \le (\sum_{j\in\widetilde{J}+} \|Z_j\|)^2 \le 9L^4\theta^2. \tag{7.58}$$

Hence by (7.55) we obtain

$$\begin{aligned} \|\sum_{j\in\widetilde{J}+} Z_j\| = \|\prod_{j\in\widetilde{J}+} (I + Z_j) - I - X\| &\le \\ \le \|\prod_{j\in\widetilde{J}+} (I + Z_j) - I\| + \|X\| &\le \\ \le \theta + 9L^4\theta^2. \end{aligned}$$

On the other hand, by (7.57) we have

$$2\theta < \frac{1}{L} \sum_{j\in\widetilde{J}+} z_{lm}^{(j)} \le \frac{1}{L} \max_{i,k=1,\dots,n} |\sum_{j\in\widetilde{J}+} z_{ik}^{(j)}| \le \|\sum_{j\in\widetilde{J}+} Z_j\|$$

and this together with the previous inequality leads to

$$2\theta < \theta + 9L^4\theta^2,$$

that is

$$\frac{1}{9L^4} < \theta$$

which contradicts (7.54) and therefore the estimate (7.56) holds.
□

7.18 Theorem. *Let $V : [a,b] \times \mathfrak{J} \to L(\mathbb{R}^n)$ be Perron product integrable over $[a,b]$ with $\prod_a^b V(t,dt) = Q$ where $Q \in L(\mathbb{R}^n)$ is invertible and assume that the condition C is satisfied.*

Let us define $\Phi : [a,b] \to L(\mathbb{R}^n)$ by the relations

$$\Phi(a) = I,$$

$$\Phi(s) = \prod_a^s V(t,dt) \text{ for } s \in (a,b].$$

Let $\varepsilon \in (0, \frac{1}{9L}\|(\Phi(b))^{-1}\|^{-1}$ where L is the constant given in (7.51) and let $\delta : [,b] \to (0,+\infty)$ be a gauge on $[a,b]$ such that

$$\|P(V,D) - \Phi(b)\| < \varepsilon$$

for every δ–fine partition D of $[a,b]$.

If

$$a \le \beta_1 \le \xi_1 \le \gamma_1 \le \beta_2 \le \xi_2 \le \gamma_2 \le \dots$$
$$\dots \le \beta_m \le \xi_m \le \gamma_m \le b$$

is a δ–fine system $\{(\xi_j,[\beta_j,\gamma_j]),\ j = 1,\dots,m\}$, i.e.

$$\xi_j \in [\beta_j,\gamma_j] \subset [\xi_j - \delta(\xi_j), \xi_j + \delta(\xi_j)], \quad j = 1,\dots,m$$

then

$$\sum_{j=1}^m \|(\Phi(\gamma_j))^{-1} V(\xi_j,[\beta_j,\gamma_j])\Phi(\beta_j) - I\| \le M\|(\Phi(b))^{-1}\|\varepsilon \quad (7.59)$$

where M is the constant from Lemma 7.16 and

$$\sum_{j=1}^m \|V(\xi_j,[\beta_j,\gamma_j]) - \prod_{\beta_j}^{\gamma_j} V(t,dt)\| \le K^2 M\|(\Phi(b))^{-1}\|\varepsilon \quad (7.60)$$

where K is the constant from Theorem 7.8.

Proof. Let us set

$$Z_j = (\Phi(\gamma_j))^{-1} V(\xi_j, [\beta_j, \gamma_j])\Phi(\beta_j) - I$$

for $j = 1, 2, \dots, m$. Since all assumptions of Lemma 7.9 are satisfied, we obtain by (7.28) the inequalities

$$\|(I + Z_{j_p})(I + Z_{j_{p-1}}) \dots (I + Z_{j_1}) - I\| \le \|(\Phi(b))^{-1}\|\varepsilon$$

for every p–tuple $\{j_1, j_2, \dots, j_p\} \subset \{1, 2, \dots, m\}$, $j_1 < j_2 < \dots < j_p$, and by the choice of $\varepsilon > 0$ we also have $\|(\Phi(b))^{-1}\|\varepsilon < \dfrac{1}{9L}$. Hence Lemma 7.17 yields

$$\sum_{j=1}^{m} \|Z_j\| \le M\|(\Phi(b))^{-1}\|\varepsilon$$

and (7.59) holds.

Since

$$\prod_{\beta_j}^{\gamma_j} V(t, dt) = \Phi(\gamma_j)(\Phi(\beta_j))^{-1}$$

for $j = 1, \dots, m$ and therefore also

$$V(\xi_j, [\beta_j, \gamma_j]) - \prod_{\beta_j}^{\gamma_j} V(t, dt) =$$

$$= \Phi(\gamma_j)[(\Phi(\gamma_j))^{-1} V(\xi_j, [\beta_j, \gamma_j])\Phi(\beta_j)](\Phi(\beta_j))^{-1} =$$

$$= \Phi(\gamma_j)(\Phi(\beta_j))^{-1}$$

for $j = 1, \dots, m$, we obtain by Theorem 7.8 the inequality

$$\|V(\xi_j, [\beta_j, \gamma_j]) - \prod_{\beta_j}^{\gamma_j} V(t, dt)\| \le K^2 \|Z_j\|$$

for $j = 1, \dots, m$, which together with (7.59) implies (7.60). □

Remark. Lemma 7.17 and its proof is strictly based on the structure of $n \times n$–matrices which represent linear operators from $L(\mathbb{R}^n)$. All statements given before Lemma 7.17 are independent of the structure of $n \times n$–matrices and they can be formulated, including the definition of the Perron product integral, in terms of general bounded linear operators from the space $L(X)$ of all bounded linear operators on a Banach space X.

It can be shown that if X is a Banach space which is not finite–dimensional then a statement analogous to Lemma 7.17 is not true. A counterexample demonstrating this fact is given in [148].

7.19 Definition. Two functions $V_1, V_2 : [a, b] \times \mathfrak{J} \to L(\mathbb{R}^n)$ are called *equivalent* if for every $\theta > 0$ there is a gauge δ on $[a, b]$ such that

$$\sum_{j=1}^{k} \|V_1(\tau_j, [\alpha_{j-1}, \alpha_j]) - V_2(\tau_j, [\alpha_{j-1}, \alpha_j])\| < \theta$$

for every δ–fine partition

$$D = \{(\tau_j, [\alpha_{j-1}, \alpha_j]), j = 1, \dots, k\}$$

of the interval $[a, b]$.

7.20 Theorem. *Assume that the functions $V_1, V_2 : [a,b] \times \mathfrak{J} \to L(\mathbb{R}^n)$ are equivalent.*

If the function V_1 satisfies condition C and $\prod_a^b V_1(t,dt) = Q$ exists and $Q \in L(\mathbb{R}^)$ is invertible then also $\prod_a^b V_2(t,dt)$ exists and

$$\prod_a^b V_2(t,dt) = \prod_a^b V_1(t,dt).$$

Proof. Assume that $\theta > 0$ is given. By (7.60) from Theorem 7.18 there is a gauge δ on $[a,b]$ such that for every δ–fine partition D we have

$$\sum_{j=1}^k \|V_1(\tau_j, [\alpha_{j-1}, \alpha_j]) - \Phi_1(\alpha_j)(\Phi_1(\alpha_{j-1}))^{-1}\| < \theta$$

where $\Phi_1(s) = \prod_a^s V_1(t,dt)$ for $s \in (a,b]$ and $\Phi_1(a) = I$.

By the equivalence of V_1 and V_2 the gauge δ can be given in such a way that

$$\sum_{j=1}^k \|V_1(\tau_j, [\alpha_{j-1}, \alpha_j]) - V_2(\tau_j, [\alpha_{j-1}, \alpha_j])\| < \theta$$

for every δ–fine partition D.

Hence

$$\sum_{j=1}^k \|V_2(\tau_j, [\alpha_{j-1}, \alpha_j]) - \Phi_1(\alpha_j)(\Phi_1(\alpha_{j-1}))^{-1}\| \le$$

$$\le \sum_{j=1}^k \|V_2(\tau_j, [\alpha_{j-1}, \alpha_j]) - V_1(\tau_j, [\alpha_{j-1}, \alpha_j])\| +$$

$$+\sum_{j=1}^{k} \|V_1(\tau_j,[\alpha_{j-1},\alpha_j]) - V_2(\tau_j,[\alpha_{j-1},\alpha_j])\| < 2\theta$$

and by Theorem 7.16 the Perron product integral $\prod_a^b V_2(t,dt)$ exists and

$$\prod_a^b V_2(t,dt) = \Phi_1(b)(\Phi_1(a))^{-1} =$$

$$= \Phi_1(b) = \prod_a^b V_1(t,dt).$$

□

Let us now introduce another condition for functions $V : [a,b]\times \mathfrak{J} \to L(\mathbb{R}^n)$.

7.21 Definition. A function $V : [a,b] \times \mathfrak{J} \to L(\mathbb{R}^n)$ *satisfies the condition* C^+ if there exists a nondecreasing function $g : [a,b] \to \mathbb{R}$ such that for every $\tau \in [a,b]$ there is a $\rho = \rho(\tau) > 0$ such that

$$\|V(\tau,[\alpha,\beta]) - I\| \le g(\beta) - g(\alpha) \tag{7.61}$$

for all $\alpha, \beta, \tau \in [a,b]$ with $\tau - \rho < \alpha \le \tau \le \beta < \tau + \rho$.

7.22 Lemma. *Assume that a function $V : [a,b] \times \mathfrak{J} \to L(\mathbb{R}^n)$ is Perron product integrable over $[a,b]$ with an invertible value of the product integral and that the conditions C and C^+ are satisfied.*

Then the function $\Phi : [a,b] \to L(\mathbb{R}^n)$ given by

$$\Phi(a) = I, \quad \Phi(s) = \prod_a^s V(t,dt), \ s \in (a,b] \tag{7.62}$$

is of bounded variation on $[a,b]$, i.e. $\Phi \in BV([a,b];L(\mathbb{R}^n))$ and also $\Phi^{-1} \in BV([a,b];L(\mathbb{R}^n))$.

Proof. Assume that $\alpha, \beta \in [a,b]$, $\alpha \leq \beta$. Then if $\tau \in [\alpha, \beta]$, we have

$$\Phi(\beta) - \Phi(\alpha) = [\Phi(\beta)(\Phi(\alpha))^{-1} - I]\Phi(\alpha) =$$

$$= [\prod_{\alpha}^{\beta} V(t,dt) - I]\Phi(\alpha) =$$

$$= [\prod_{\alpha}^{\beta} V(t,dt) - V(\tau,[\alpha,\beta])]\Phi(\alpha) + [V(\tau,[\alpha,\beta]) - I]\Phi(\alpha).$$

By Theorem 7.8 and by the condition C^+ we have

$$\|\Phi(\beta) - \Phi(\alpha)\| \leq$$

$$\leq K[\| \prod_{\alpha}^{\beta} V(t,dt) - V(\tau,[\alpha,\beta])\| + g(\beta) - g(\alpha)] \tag{7.63}$$

provided $\tau - \rho(\tau) < \alpha \leq \tau \leq \beta < \tau + \rho(\tau)$ and ρ is the function from Definition 7.21.

Assume further that $\varepsilon > 0$ is given and that $\delta : [a,b] \to (0,+\infty)$ is such a gauge on $[a,b]$ that

$$\|P(V,D) - \Phi(b)\| < \varepsilon$$

for every δ–fine partition D of $[a,b]$ and that $\delta(\tau) < \rho(\tau)$ for $\tau \in [a,b]$ where $\rho(\tau) > 0$ is given in condition C^+.

Let now $a = s_0 < s_1 < \cdots < s_m = b$ be given and let

$$D^p = \{\alpha_0^p, \tau_1^p, \alpha_1^p, \dots, \tau_{k_p}^p, \alpha_{k_p}^p\}$$

be an arbitrary δ–fine partition of $[s_{p-1}, s_p]$, $p = 1, \dots, m$. Then by (7.63) we have

$$\|\Phi(s_p) - \Phi(s_{p-1})\| \leq \sum_{j=1}^{k_p} \|\Phi(\alpha_j^p) - \Phi(\alpha_{j-1}^p)\| \leq$$

$$\leq K \sum_{j=1}^{k_p} [\| \prod_{\alpha_{j-1}^p}^{\alpha_j^p} V(t,dt) - V(\tau_j^p, [\alpha_{j-1}^p, \alpha_j^p])\| +$$

$$+ g(\alpha_j^p) - g(\alpha_{j-1}^p)] =$$

$$= K \sum_{j=1}^{k_p} [\| \prod_{\alpha_{j-1}^p}^{\alpha_j^p} V(t,dt) - V(\tau_j^p, [\alpha_{j-1}^p, \alpha_j^p])\| + K[g(s_p) - g(s_{p-1})]$$

for every $p = 1, \dots, m$ and henceforth

$$\sum_{p=1}^{m} \|\Phi(s_p) - \Phi(s_{p-1})\| \leq$$

$$\leq K \sum_{p=1}^{m} \sum_{j=1}^{k_p} [\| \prod_{\alpha_{j-1}^p}^{\alpha_j^p} V(t,dt) - V(\tau_j^p, [\alpha_{j-1}^p, \alpha_j^p])\| + K[g(b) - g(a)]. \tag{7.64}$$

Using Theorem 7.18 we obtain the estimate

$$\sum_{p=1}^{m} \sum_{j=1}^{k_p} \| \prod_{\alpha_{j-1}^p}^{\alpha_j^p} V(t,dt) - V(\tau_j^p, [\alpha_{j-1}^p, \alpha_j^p])\| \leq K^2 M \|(\Phi(b))^{-1}\| \varepsilon$$

because evidently $D = D^1 \circ D^2 \circ \cdots \circ D^m$ is a δ–fine partition of $[a,b]$. By (7.64) we therefore have

$$\sum_{p=1}^{m} \|\Phi(s_p) - \Phi(s_{p-1})\| \leq$$

$$\leq K^3 M \|(\Phi(b))^{-1}\| \varepsilon + K[g(b) - g(a)]$$

for an arbitrary choice of points $a = s_0 < s_1 < \cdots < s_m = b$, and consequently

$$\operatorname{var}_a^b \Phi \leq K^3 M \|(\Phi(b))^{-1}\| \varepsilon + K[g(b) - g(a)] < \infty, \tag{7.65}$$

i.e. $\Phi \in BV([a,b]; L(\mathbb{R}^n))$.

It is easy to observe that (7.65) yields the inequality

$$\operatorname{var}_a^b \Phi \leq K[g(b) - g(a)]$$

because $\varepsilon > 0$ in (7.65) can be taken arbitrarily small.

Since we have $(\Phi(s))^{-1} = (\Phi(b))^{-1} \prod_s^b V(t, dt)$, the boundedness of $\operatorname{var}_a^b \Phi^{-1}$ can be shown similarly. □

7.23 Lemma. *Assume that a function $V : [a,b] \times \mathfrak{J} \to L(\mathbb{R}^n)$ is Perron product integrable over $[a,b]$ with an invertible value of the product integral and that the conditions C and C^+ are satisfied.*

Then for every $t \in [a,b]$ the Perron–Stieltjes integral

$$\int_a^t d[\Phi(r)](\Phi(r))^{-1} = \widetilde{A}(t) \in L(\mathbb{R}^n) \tag{7.66}$$

exists.

For $\widetilde{A} : [a,b] \to L(\mathbb{R}^n)$ we have $\widetilde{A} \in BV([a,b]; L(\mathbb{R}^n))$ and the inverses $[I - \Delta^- \widetilde{A}(t)]^{-1}$, $[I + \Delta^+ \widetilde{A}(t)]^{-1}$ exist for $t \in (a,b]$, $t \in [a,b)$, respectively.

Proof. By Lemma 7.22 the $n \times n$–matrix valued functions Φ and Φ^1 are of bounded variation on $[a,b]$. Therefore the Perron–Stieltjes integral in (7.66) exists by Corollary 1.34. The fact that $\widetilde{A}$ from (7.66) is of bounded variation easily follows from Corollary 1.36 (see also Remark 1.37).

By the definition of $\widetilde{A}$ we have for any $\eta > 0$ the equality

$$\widetilde{A}(t) - \widetilde{A}(t-\eta) = \int_{t-\eta}^{t} d[\Phi(r)](\Phi(r))^{-1}$$

and therefore by Theorem 1.16 we have

$$\Delta^- \widetilde{A}(t) = \lim_{\eta \to 0+} \int_{t-\eta}^{t} d[\Phi(r)](\Phi(r))^{-1} =$$

$$= \lim_{\eta \to 0+} [\Phi(t) - \Phi(t-\eta)](\Phi(t))^{-1} = [\Phi(t) - \Phi(t-)](\Phi(t))^{-1} =$$

$$= I - \Phi(t-)(\Phi(t))^{-1}.$$

By (7.37) in Lemma 7.12 we have $\Phi(t-) = (V_-(t))^{-1}\Phi(t)$, i.e.

$$I - \Delta^- \widetilde{A}(t) = \Phi(t-)(\Phi(t))^{-1} =$$

$$= (V_-(t))^{-1}\Phi(t)(\Phi(t))^{-1} = (V_-(t))^{-1}$$

for $t \in (a, b]$ and $V_-(t)$ is invertible by (7.4–).

In a completely analogous way we obtain also

$$I + \Delta^+ \widetilde{A}(t) = V_+(t)$$

for $t \in [a, b)$, where $V_+(t)$ is invertible by (7.4+). □

7.24 Theorem. *Assume that a function $V : [a, b] \times \mathfrak{J} \to L(\mathbb{R}^n)$ is Perron product integrable over $[a, b]$ with an invertible value of the product integral and that the conditions C and C^+ are satisfied.*

Then for every $s \in [a, b]$ the relation

$$\Phi(s) = \Phi(a) + \int_a^s d[\widetilde{A}(r)]\Phi(r) \tag{7.67}$$

holds, where $\Phi : [a,b] \to L(\mathbb{R}^n)$ is given by (7.62) and $\widetilde{A} : [a,b] \to L(\mathbb{R}^n)$ is defined by (7.66) in Lemma 7.23.

Proof. Let us consider the integral

$$\int_a^s d[\widetilde{A}(r)]\Phi(r) = \int_a^s d[\int_a^r d[\Phi(t)](\Phi(t))^{-1}]\Phi(r).$$

To the right hand side the substitution theorem for Perron–Stieltjes integrals can be applied (see e.g. [155 ,I.4.25]) to obtain

$$\int_a^s d[\widetilde{A}(r)]\Phi(r) = \int_a^s d[\Phi(r)](\Phi(r))^{-1}\Phi(r) =$$

$$= \int_a^s d[\Phi(r)] = \Phi(s) - \Phi(a)$$

for every $s \in [a,b]$, i.e. (7.67) holds. □

Looking at the concepts of Chap. 6 we can reformulate the results of Theorem 7.24 and Lemma 7.23 as follows.

7.25 Theorem. *Assume that a function $V : [a,b] \times \mathfrak{J} \to L(\mathbb{R}^n)$ is Perron product integrable over $[a,b]$ with an invertible value and that the conditions C and C^+ are satisfied.*

Then the function $\Phi : [a,b] \to L(\mathbb{R}^n)$ given by

$$\Phi(a) = I, \quad \Phi(s) = \prod_a^s V(t,dt), \ s \in (a,b]$$

is a fundamental matrix of the generalized linear differential equation

$$\frac{dx}{d\tau} = D[\widetilde{A}(t)x],$$

where $\widetilde{A} : [a,b] \to L(\mathbb{R}^n)$ is defined by (7.66) in Lemma 7.23. The matrix valued function $\widetilde{A}$ satisfies the condition (6.13).

This result shows that for a $V : [a,b] \times \mathfrak{J} \to L(\mathbb{R}^n)$ the "indefinite" Perron product integral $\prod_a^s V(t,dt)$ is a fundamental matrix corresponding to a well–behaved generalized linear ordinary differential equation. The condition (6.13) described and used in Chap. 6 is a necessary and sufficient condition for the existence of a unique solution of the corresponding nonhomogeneous equation for every right hand side and every initial condition.

Theorem 7.25 naturally suggests the following problem.

Given $A : [a,b] \to L(\mathbb{R}^n)$ such that

$$A \text{ is of bounded variation on } [a,b], \tag{7.68}$$

$$[I + \Delta^+ A(t)]^{-1} \text{ exists for every } t \in [a,b), \tag{7.69–}$$

$$[I - \Delta^- A(t)]^{-1} \text{ exists for every } t \in (a,b] \tag{7.69+}$$

holds.

Construct a function $V : [a,b] \times \mathfrak{J} \to L(\mathbb{R}^n)$ which is Perron product integrable over $[a,b]$ with an invertible value of its Perron product integral for which the conditions C and C^+ are satisfied and such that for the function $\Phi : [a,b] \to L(\mathbb{R}^n)$ given by

$$\Phi(a) = I, \quad \Phi(s) = \prod_a^s V(t,dt), \ s \in (a,b]$$

the equality

$$\Phi(s) = \Phi(a) + \int_a^s d[A(r)]\Phi(r)$$

holds.

In other words, we are looking for the representation of the fundamental matrix of the generalized linear ordinary differential equation

$$\frac{dx}{d\tau} = DA(t)x$$

for the case when this linear equation satisfies the necessary and sufficient conditions for the existence of a unique solution on $[a, b]$ for all possible initial conditions (see e.g. Theorem 6.14 and the results given in Chap. 6).

Assume that (7.68) and (7.69) hold for $A : [a, b] \to L(\mathbb{R}^n)$. For $\alpha, \beta, \tau \in [a, b]$, $\alpha \le \tau \le \beta$ define

$$W(\tau, [\alpha, \beta]) = [I + A(\beta) - A(\tau)][I + A(\alpha) - A(\tau)]^{-1}. \quad (7.70)$$

Since A satisfies (7.68) we have $\|\Delta^- A(t)\| < \frac{1}{2}$ except a finite set of points $t_1, t_2, \ldots, t_l \in [a, b]$. For $t \notin \{t_1, t_2, \ldots, t_l\}$ we then have

$$[I - \Delta^- A(t)]^{-1} = \sum_{k=0}^{\infty} (\Delta^- A(t))^k$$

and

$$\|[I - \Delta^- A(t)]^{-1}\| \le \sum_{k=0}^{\infty} \|\Delta^- A(t)\|^k < 2.$$

Taking $K_- = \max\{2, \|[I - \Delta^- A(t_1)]^{-1}\|, \ldots, \|[I - \Delta^- A(t_l)]^{-1}\|\}$ we have

$$\|[I - \Delta^- A(t)]^{-1}\| \le K_-$$

for every $t \in (a, b]$. Similarly it can be also shown that

$$\|[I + \Delta^+ A(t)]^{-1}\| \le K_+$$

for every $t \in [a, b)$ where K_+ is a constant.

Since the onesided limits of A exist in $[a,b]$ we can easily state that there is a constant $L > 0$ such that for every $\tau \in [a,b]$ there is a $\delta_1(\tau) > 0$ such that

$$[I + A(\alpha) - A(\tau)]^{-1}, [I + A(\beta) - A(\tau)]^{-1}$$

exist and

$$\|[I + A(\alpha) - A(\tau)]^{-1}\| \leq L, \|[I + A(\beta) - A(\tau)]^{-1}\| \leq L \quad (7.71)$$

whenever $\alpha, \beta \in [a,b]$, $\tau - \delta_1(\tau) < \alpha \leq \tau \leq \beta < \tau + \delta_1(\tau)$.

For $W : [a,b] \times \mathfrak{J} \to L(\mathbb{R}^n)$ given by (7.70) the following identities hold:

$$W(\tau, [\tau, \tau]) = I, \quad \tau \in [a,b],$$

$$W(\tau, [\alpha, \tau]) = [I + A(\alpha) - A(\tau)]^{-1},$$

$$W(\tau, [\tau, \beta]) = [I + A(\beta) - A(\tau)]$$

and therefore

$$W(\tau, [\alpha, \beta]) = W(\tau, [\tau, \beta]) W(\tau, [\alpha, \tau])$$

provided $\alpha, \beta \in [a,b]$, $\tau - \delta_1(\tau) < \alpha \leq \tau \leq \beta < \tau + \delta_1(\tau)$.

Finally, we have

$$\lim_{\beta \to \tau+} W(\tau, [\tau, \beta]) = \lim_{\beta \to \tau+} [I + A(\beta) - A(\tau)] = I + \Delta^+ A(\tau)$$

for $\tau \in [a,b)$ and by Lemma 7.11 also

$$\lim_{\alpha \to \tau-} W(\tau, [\alpha, \tau]) = \lim_{\alpha \to \tau-} [I + A(\alpha) - A(\tau)]^{-1} = [I - \Delta^- A(\tau)]^{-1}$$

for $\tau \in (a,b]$. In this way we have verified that W given by (7.70) satisfies the condition C from Definition 7.4.

Moreover, by (7.71) we have

$$\|W(\tau,[\alpha,\beta])-I\| = \|[I+A(\beta)-A(\tau)][I+A(\alpha)-A(\tau)]^{-1}-I\| =$$

$$= \|[I+A(\beta)-A(\tau)-(I+A(\alpha)-A(\tau))][I+A(\alpha)-A(\tau)]^{-1}\| \leq$$

$$\leq \|A(\beta)-A(\alpha)\|L \leq L(\operatorname{var}_a^\beta A - \operatorname{var}_a^\alpha A)$$

whenever $\alpha,\beta\in[a,b]$, $\tau-\delta_1(\tau)<\alpha\leq\tau\leq\beta<\tau+\delta_1(\tau)$, and this yields that W satisfies the condition C^+ given in Definition 7.21 with the nondecreasing function $g:[a,b]\to\mathbb{R}$ given by $g(s)=L\operatorname{var}_a^s A$ for $s\in[a,b]$.

Now let $\Psi:[a,b]\to L(\mathbb{R}^n)$ be the fundamental matrix of the generalized linear ordinary differential equation

$$\frac{dx}{d\tau} = DA(t)x \tag{7.72}$$

with $\Psi(a)=I$ (see Theorem 6.11). Since the Perron–Stieltjes integral $\int_a^b d[A(r)]\Psi(r)$ exists, the Saks–Henstock lemma 1.13 yields

for every $\varepsilon>0$ *there is a gauge* δ_2 *on* $[a,b]$, $\delta_2(\tau)\leq\delta_1(\tau)$, $\tau\in[a,b]$ *such that if*

$$a\leq\beta_1\leq\xi_1\leq\gamma_1\leq\beta_2\leq\xi_2\leq\gamma_2\leq\dots$$

$$\dots\leq\beta_m\leq\xi_m\leq\gamma_m\leq b$$

is a δ*-fine system* $\{(\xi_j,[\beta_j,\gamma_j]),\ j=1,\dots,m\}$, *i.e.*

$$\xi_j\in[\beta_j,\gamma_j]\subset[\xi_j-\delta(\xi_j),\xi_j+\delta(\xi_j)],\quad j=1,\dots,m$$

then we have

$$\sum_{j=1}^m \|[A(\gamma_j)-A(\beta_{j-1})]\Psi(\xi_j)-\int_{\beta_j}^{\gamma_j} d[A(r)]\Psi(r)\| < \varepsilon. \tag{7.73}$$

7.26 Lemma. *Assume that $A : [a,b] \to L(\mathbb{R}^n)$ satisfies (7.68) and (7.69).*

Let $\Psi : [a,b] \to L(\mathbb{R}^n)$ be the fundamental matrix of the generalized linear ordinary differential equation (7.72) with $\Psi(a) = I$.

Then for every $\theta > 0$ there is a gauge δ on $[a,b]$ such that

$$\sum_{j=1}^{k} \|W(\tau_j, [\alpha_{j-1}, \alpha_j]) - \Psi(\alpha_j)(\Psi(\alpha_{j-1}))^{-1}\| < \theta \qquad (7.74)$$

for every δ–fine partition $D = \{\alpha_0, \tau_1, \alpha_1, \ldots, \alpha_{k-1}, \tau_k, \alpha_k\}$ of $[a,b]$ where $W[a,b] \times \mathfrak{J} \to L(\mathbb{R}^n)$ is given by (7.70).

Proof. Let $\varepsilon > 0$ be arbitrary and let δ be a gauge on $[a,b]$ such that $\delta(\tau) < \delta_2(\tau)$ for $\tau \in [a,b]$ where δ_2 is given in (7.73). If D is a δ–fine partition of $[a,b]$ then $W(\tau_j, [\alpha_{j-1}, \alpha_j])$ is well defined (cf. (7.71)) for $j = 1, \ldots, k$. By definition and (7.71) we have

$$\|W(\tau_j, [\alpha_{j-1}, \alpha_j]) - \Psi(\alpha_j)(\Psi(\alpha_{j-1}))^{-1}\| =$$

$$= \|[I + A(\alpha_j) - A(\tau_j)][I + A(\alpha_{j-1}) - A(\tau_j)]^{-1} - \Psi(\alpha_j).$$

$$.(\Psi(\alpha_{j-1}))^{-1}\| =$$

$$= \|[I + A(\alpha_j) - A(\tau_j) - \Psi(\alpha_j)(\Psi(\tau_j))^{-1}][I + A(\alpha_{j-1}) - A(\tau_j)]^{-1} +$$

$$+\Psi(\alpha_j)(\Psi(\tau_j))^{-1}]([I + A(\alpha_{j-1}) - A(\tau_j)]^{-1} - \Psi(\tau_j).$$

$$(\Psi(\alpha_{j-1}))^{-1})\| \le$$

$$\le L\|I + A(\alpha_j) - A(\tau_j) - \Psi(\alpha_j)(\Psi(\tau_j))^{-1}\| +$$

$$+\|\Psi(\alpha_j)(\Psi(\tau_j))^{-1}\| \|[I + A(\alpha_{j-1}) - A(\tau_j)]^{-1}\|.$$

$$.\|[\Psi(\alpha_{j-1})(\Psi(\tau_j))^{-1} - [I + A(\alpha_{j-1}) - A(\tau_j)]]\Psi(\tau_j)(\Psi(\alpha_{j-1}))^{-1})\|.$$

By Theorem 6.16 there is a constant $M \geq 0$ such that

$$\|\Psi(t)(\Psi(s))^{-1}\| \leq M$$

for all $s, t \in [a, b]$ and therefore we get from the inequality derived above the following one (let us mention that $\|\Psi(s)\| \leq M$ and $\|(\Psi(s))^{-1}\| \leq M$, because we have $\Psi(a) = I$):

$$\|W(\tau_j, [\alpha_{j-1}, \alpha_j]) - \Psi(\alpha_j)(\Psi(\alpha_{j-1}))^{-1}\| \leq$$

$$\leq L\|\Psi(\tau_j) + (A(\alpha_j) - A(\tau_j))\Psi(\tau_j) - \Psi(\alpha_j)\|\|(\Psi(\tau_j))^{-1}\|+$$

$$+LM\|[\Psi(\alpha_{j-1}) - \Psi(\tau_j) - (A(\alpha_{j-1}) - A(\tau_j))\Psi(\tau_j)](\Psi(\tau_j))^{-1}\Psi(\tau_j).$$

$$.(\Psi(\alpha_{j-1}))^{-1})\| \leq$$

$$\leq LM\|(A(\alpha_j) - A(\tau_j))\Psi(\tau_j) - \int_{\tau_j}^{\alpha_j} d[A(r)]\Psi(r)\|+$$

$$+M^2L\|(A(\tau_j) - A(\alpha_{j-1}))\Psi(\tau_j) - \int_{\alpha_{j-1}}^{\tau_j} d[A(r)]\Psi(r)\|$$

for every $j = 1, \ldots, k$.

Using (7.73) in this situation we finally obtain

$$\sum_{j=1}^{k} \|W(\tau_j, [\alpha_{j-1}, \alpha_j]) - \Psi(\alpha_j)(\Psi(\alpha_{j-1}))^{-1}\| \leq$$

$$\leq LM \sum_{j=1}^{k} \|(A(\alpha_j) - A(\tau_j))\Psi(\tau_j) - \int_{\tau_j}^{\alpha_j} d[A(r)]\Psi(r)\|+$$

$$+M^2L \sum_{j=1}^{k} \|(A(\tau_j) - A(\alpha_{j-1}))\Psi(\tau_j) - \int_{\alpha_{j-1}}^{\tau_j} d[A(r)]\Psi(r)\| <$$

$$< \varepsilon LM(M+1).$$

Taking $0 < \varepsilon < \dfrac{\theta}{LM(M+1)+1}$ for an arbitrary $\theta > 0$ we obtain (7.74) for partitions D which are δ–fine with δ corresponding by (7.73) to this choice of ε. □

By Lemma 7.26 and by Theorem 7.16 we now immediately obtain

7.27 Theorem. *Assume that $A : [a,b] \to L(\mathbb{R}^n)$ satisfies (7.68) and (7.69).*

Let $\Psi : [a,b] \to L(\mathbb{R}^n)$ be the uniquely determined fundamental matrix of the generalized linear ordinary differential equation (7.72) with $\Psi(a) = I$.

Then the function $W[a,b] \times \mathfrak{J} \to L(\mathbb{R}^n)$ given by (7.70) is Perron product integrable over $[a,b]$ with the value $\Psi(b)$ and for every $s \in [a,b]$ we have

$$\prod_a^s W(t,dt) = \Psi(s). \tag{7.75}$$

Remark. Taking into account the result of Theorem 7.27 together with the result of Theorem 7.25 we can see that there is a one-to-one correspondence between the "indefinite" Perron product integrals $\prod_a^s V(t,dt)$ of functions $V[a,b] \times \mathfrak{J} \to L(\mathbb{R}^n)$ which satisfy conditions C and C^+ and the fundamental matrices of generalized linear ordinary differential equations (7.72) with $A : [a,b] \to L(\mathbb{R}^n)$ satisfying (7.68) an (7.69). Let us now turn our attention to the case when $A : [a,b] \to L(\mathbb{R}^n)$ is such that

$$A \text{ is of bounded variation on } [a,b], \tag{7.68}$$

and

$$A \text{ is continuous at every point in } [a,b]. \tag{7.76}$$

For $A : [a,b] \to L(\mathbb{R}^n)$ satisfying (7.68) and (7.76) define $V_1, V_2 : [a,b] \times \mathfrak{J} \to L(\mathbb{R}^n)$ by the relations

$$V_1(\tau, [\alpha, \beta]) = I + A(\beta) - A(\alpha) \tag{7.77}$$

for $\tau \in [\alpha, \beta] \subset [a,b]$ and

$$V_2(\tau, [\alpha, \beta]) = e^{A(\beta)-A(\alpha)} = \sum_{k=0}^{\infty} \frac{[A(\beta) - A(\alpha)]^k}{k!} \tag{7.78}$$

for $\tau \in\in [\alpha, \beta] \subset [a,b]$.

By (7.76) for every $\varepsilon \in (0,1)$ and $\tau \in [a,b]$ there is a $\delta(\tau) > 0$ such that

$$\|A(\beta) - A(\alpha)\| < \varepsilon \tag{7.79}$$

for every $[\alpha, \beta] \subset [a,b]$, $\tau - \delta(\tau) < \alpha \le \tau \le \beta < \tau + \delta(\tau)$.

If the pair $(\tau, [\alpha, \beta])$ is δ–fine then

$$V_1(\tau, [\alpha, \beta]) - W(\tau, [\alpha, \beta]) =$$

$$= I + A(\beta) - A(\alpha) - [I + A(\beta) - A(\tau)][I + A(\alpha) - A(\tau)]^{-1} =$$

$$= \{[I + A(\beta) - A(\alpha)][I + A(\alpha) - A(\tau)] - [I + A(\beta) - A(\tau)]\}.$$

$$.[I + A(\alpha) - A(\tau)]^{-1} =$$

$$= [A(\beta) - A(\alpha)][A(\alpha) - A(\tau)][I + A(\alpha) - A(\tau)]^{-1} \tag{7.80}$$

and

$$V_2(\tau, [\alpha, \beta]) - V_1(\tau, [\alpha, \beta]) =$$

$$= e^{A(\beta)-A(\alpha)} - [I + A(\beta) - A(\alpha)] =$$

$$= \sum_{k=0}^{\infty} \frac{[A(\beta) - A(\alpha)]^k}{k!} - [I + A(\beta) - A(\alpha)] =$$

$$= \sum_{k=2}^{\infty} \frac{[A(\beta) - A(\alpha)]^k}{k!}. \tag{7.81}$$

It is not difficult to see that $\delta(\tau) > 0$ can be choosen so small that if $(\tau, [\alpha, \beta])$ is δ–fine then $\|[I + A(\alpha) - A(\tau)]^{-1}\| \leq L$ where L is a constant (see also (7.71)).

By (7.80) we have

$$\|V_1(\tau, [\alpha, \beta]) - W(\tau, [\alpha, \beta])\| \leq \varepsilon L \|A(\beta) - A(\alpha)\| \tag{7.82}$$

while (7.82) leads to

$$\|V_2(\tau, [\alpha, \beta]) - V_1(\tau, [\alpha, \beta])\| \leq$$

$$\sum_{k=2}^{\infty} \frac{\|A(\beta) - A(\alpha)\|^k}{k!} =$$

$$= e^{\|A(\beta) - A(\alpha)\|} - 1 - \|A(\beta) - A(\alpha)\| \leq$$

$$\leq \|A(\beta) - A(\alpha)\|^2 \leq \varepsilon \|A(\beta) - A(\alpha)\| \tag{7.83}$$

whenever the the pair $(\tau, [\alpha, \beta])$ is δ–fine.

Assume now that $D = \{(\tau_j, [\alpha_{j-1}, \alpha_j]), j = 1, \dots, k\}$ is an arbitrary δ–fine partition of $[a, b]$. Then by (7.82) we obtain

$$\sum_{j=1}^{k} \|V_1(\tau_j, [\alpha_{j-1}, \alpha_j]) - W(\tau_j, [\alpha_{j-1}, \alpha_j])\| \leq$$

$$\leq \varepsilon L \sum_{j=1}^{k} \|A(\alpha_j) - A(\alpha_{j-1})\| \leq \varepsilon L \operatorname{var}_a^b A$$

and (7.83) yields

$$\sum_{j=1}^{k} \|V_1(\tau_j, [\alpha_{j-1}, \alpha_j]) - V_2(\tau_j, [\alpha_{j-1}, \alpha_j])\| \leq$$

$$\leq \varepsilon \sum_{j=1}^{k} \|A(\alpha_j) - A(\alpha_{j-1})\| \leq \varepsilon \operatorname{var}_a^b A.$$

Using this two inequalities we can conclude now that

the functions $W, V_1, V_2 : [a,b] \times \mathfrak{J} \to L(\mathbb{R}^n)$ given by the relations (7.70), (7.77), (7.78), respectively, are equivalent in the sense of Definition 7.19.

Using Theorems 7.20 and 7.27 we arrive at the following statement.

7.28 Proposition. *Assume that $A : [a,b] \to L(\mathbb{R}^n)$ satisfies (7.68) and (7.76). Then the functions $W, V_1, V_2 : [a,b] \times \mathfrak{J} \to L(\mathbb{R}^n)$ given by (7.70), (7.77), (7.78), respectively, are Perron product integrable over $[a,b]$ to the value $\Psi(b)$ and for every $s \in [a,b]$ we have*

$$\prod_a^s W(t,dt) = \prod_a^s V_1(t,dt) = \prod_a^s V_2(t,dt) = \Psi(s). \tag{7.84}$$

CHAPTER VIII

CONTINUOUS DEPENDENCE ON PARAMETERS FOR GENERALIZED ORDINARY DIFFERENTIAL EQUATIONS

Let us assume that $G = B \times (a,b)$ where $B = B_c = \{x \in \mathbb{R}^n; \|x\| < c\}$, $c > 0$, $(a,b) \subset \mathbb{R}$, $-\infty < a < b < +\infty$ and $h : [a,b] \to \mathbb{R}$ is a nondecreasing function defined on $[a,b]$, $\omega : [0,+\infty) \to \mathbb{R}$ is a continuous, increasing function with $\omega(0) = 0$.

8.1 Lemma. *Assume that $F_k : G \to \mathbb{R}^n$ belongs to the class $\mathcal{F}(G,h,\omega)$ for $k = 0,1,\dots$ and that*

$$\lim_{k\to\infty} F_k(x,t) = F_0(x,t) \tag{8.1}$$

for $(x,t) \in G$.

If $x : [c,d] \to B$ is a function of bounded variation on $[c,d] \subset (a,b)$ then

$$\lim_{k\to\infty} \int_c^d DF_k(x(\tau),t) = \int_c^d DF_0(x(\tau),t). \tag{8.2}$$

Proof. Let $\varepsilon > 0$ be given. Assume that $\eta > 0$ is such that $\omega(\eta) \leq \dfrac{\varepsilon}{2(h(d)-h(c)+1)}$. Since $x : [c,d] \to B_c$ is a function of bounded variation on $[c,d]$, for every $\eta > 0$ there is a finite step function $\varphi : [c,d] \to B_c$ such that

$$\|x(\tau) - \varphi(\tau)\| \leq \eta \quad \text{for } \tau \in [c,d]. \tag{8.3}$$

Therefore

$$\|F_k(x(\tau),t_2) - F_k(x(\tau),t_1) - F_k(\varphi(\tau),t_2) + F_k(\varphi(\tau),t_1)\| \leq$$

$$\leq \omega(\|x(\tau) - \varphi(\tau)\|)|h(t_2) - h(t_1)| \leq$$

$$\leq \omega(\eta)|h(t_2) - h(t_1)|$$

for $\tau \in [c,d]$, $t_1, t_2 \in [c,d]$ and $k = 0,1,\ldots$ because $F_k \in \mathcal{F}(G,h,\omega)$ (see Definition 3.8). The integrals $\int_c^d DF_k(x(\tau),t)$ $\int_c^d DF_k(\varphi(\tau),t)$ evidently exist by Corollary 3.16, and Corollary 1.35 yields the estimate

$$\|\int_c^d D[F_k(x(\tau),t) - F_k(\varphi(\tau),t)]\| \leq$$

$$\leq \int_c^d \omega(\eta)\, dh(s) = \omega(\eta)(h(d) - h(c)) \tag{8.4}$$

for every $k = 0,1,\ldots$.

Since $F_k \in \mathcal{F}(G,h,\omega)$ we have

$$\|F_k(x,t_2) - F_k(x,t_1)\| \leq |h(t_2) - h(t_1)|$$

for every $x \in B_c$ and $t_1, t_2 \in (a,b)$ and this leads to the conclusion that $\lim_{\rho\to 0+} F_k(x,t+\rho) = F_k(x,t+)$ and $\lim_{\rho\to 0+} F_k(x,t-\rho) = F_k(x,t-)$ for every $(x,t) \in G$ uniformly with respect to $k = 0,1,\ldots$. Hence by (8.1) we obtain

$$\lim_{k\to\infty} F_k(x,t+) = \lim_{k\to\infty}\lim_{\rho\to 0+} F_k(x,t+\rho) =$$

$$= \lim_{\rho\to 0+}\lim_{k\to\infty} F_k(x,t+\rho) =$$

$$= \lim_{\rho\to 0+} F_0(x,t+\rho) = F_0(x,t+)$$

and similarly also

$$\lim_{k\to\infty} F_k(x,t-) = F_0(x,t-)$$

provided $(x,t) \in G$. Using these inequalities and assuming that $\varphi(\tau) = c_j$ for $\tau \in (s_{j-1}, s_j)$, $(s_{j-1}, s_j) \subset [c,d]$ we obtain (see the proof of Corollary 3.15)

$$\int_{s_{j-1}}^{s_j} DF_k(\varphi(\tau),t) =$$

$$= F_k(c_j, s_j-) - F_k(c_j,\sigma_0) - F_k(\varphi(s_j), s_j-) + F_k(\varphi(s_j), s_j)+$$
$$+F_k(c_j,\sigma_0) - F_k(c_j, s_{j-1}+)+$$
$$+F_k(\varphi(s_{j-1}), s_{j-1}+) - F_k(\varphi(s_{j-1}), s_{j-1}) =$$
$$= F_k(c_j, s_j-) - F_k(c_j, s_{j-1}+) + F_k(\varphi(s_{j-1}), s_{j-1}+)-$$
$$-F_k(\varphi(s_{j-1}), s_{j-1}) - F_k(\varphi(s_j), s_j-) + F_k(\varphi(s_j), s_j).$$

and therefore

$$\lim_{k\to\infty} \int_{s_{j-1}}^{s_j} D[F_k(\varphi(\tau),t) - F_0(\varphi(\tau),t)] = 0. \tag{8.5}$$

Since φ is a finite step function we obtain from (8.5) using the additivity of the integral the relation

$$\lim_{k\to\infty} \int_{c}^{d} D[F_k(\varphi(\tau),t) - F_0(\varphi(\tau),t)] = 0. \tag{8.6}$$

By (8.4) we get

$$\| \int_c^d DF_k(x(\tau),t) - \int_c^d DF_0(x(\tau),t)\| \leq$$

$$\leq \| \int_c^d DF_k(x(\tau),t) - \int_c^d DF_k(\varphi(\tau),t)\| +$$

$$+\| \int_c^d DF_0(x(\tau),t) - \int_c^d DF_0(\varphi(\tau),t)\| +$$

$$+\| \int_c^d DF_k(\varphi(\tau),t) - \int_c^d DF_0(\varphi(\tau),t)\| \leq$$

$$\leq 2\omega(\eta)(h(d)-h(c)) + \| \int_c^d DF_k(\varphi(\tau),t) - \int_c^d DF_0(\varphi(\tau),t)\| \leq$$

$$\leq \varepsilon + \| \int_c^d DF_k(\varphi(\tau),t) - \int_c^d DF_0(\varphi(\tau),t)\|$$

by the choice of η. Taking $k \to \infty$ on both sides of this inequality we obtain

$$\lim_{k\to\infty} \| \int_c^d DF_k(x(\tau),t) - \int_c^d DF_0(x(\tau),t)\| \leq$$

$$\leq \varepsilon + \lim_{k\to\infty} \| \int_c^d DF_k(\varphi(\tau),t) - \int_c^d DF_0(\varphi(\tau),t)\|,$$

and since ε can be taken arbitrarily small we obtain the result

$$\lim_{k\to\infty} \int_c^d DF_k(x(\tau),t) = \int_c^d DF_0(x(\tau),t).$$

□

8.2 Theorem. *Assume that $F_k : G \to \mathbb{R}^n$ belongs to the class $\mathcal{F}(G,h,\omega)$ for $k = 0,1,\dots$ and that*

$$\lim_{k\to\infty} F_k(x,t) = F_0(x,t) \tag{8.7}$$

for $(x,t) \in G$.

Let $x_k : [\alpha,\beta] \to \mathbb{R}^n$, $k = 1,2,\dots$ *be solutions of the generalized differential equation*

$$\frac{dx}{d\tau} = DF_k(x,t) \tag{8.8}$$

on $[\alpha,\beta] \subset (a,b)$ *such that*

$$\lim_{k\to\infty} x_k(s) = x(s), \quad s \in [\alpha,\beta], \tag{8.9}$$

and $(x(s),s) \in G$ *for* $s \in [\alpha,\beta]$.

Then $x : [\alpha,\beta] \to \mathbb{R}^n$ *is of bounded variation on* $[\alpha,\beta]$ *and it is a solution of the generalized differential equation*

$$\frac{dx}{d\tau} = DF_0(x,t) \tag{8.10}$$

on $[\alpha,\beta]$.

Proof. By Lemma 3.10 we have

$$\|x_k(s_2) - x_k(s_1)\| \le |h(s_2) - h(s_1)|$$

for every $k = 1,2,\dots$ and $s_2, s_1 \in [\alpha,\beta]$.

Hence

$$\|x_k(s)\| \le \|x_k(\alpha)\| + h(s) - h(\alpha) \le \|x_k(\alpha)\| + h(\beta) - h(\alpha)$$

and

$$\mathrm{var}_\alpha^\beta\, x_k \le h(\beta) - h(\alpha). \tag{8.11}$$

By (8.9) we have $x_k(\alpha) \to x(\alpha)$ for $k \to \infty$ and therefore the sequence (x_k) of functions on $[\alpha,\beta]$ is equibounded and by (8.11) of uniformly bounded variation on $[\alpha,\beta]$. By Helly's Choice

Theorem (see [104]) there exists a subsequence of (x_k) which converges pointwise to a function of bounded variation on $[\alpha, \beta]$. Hence we conclude by (8.9) that the function $x : [\alpha, \beta] \to \mathbb{R}^n$ is of bounded variation on $[\alpha, \beta]$, and Corollary 3.16 leads to the conclusion that the integral $\int_\alpha^\beta DF_0(x(\tau), t)$ exists.

By definition of a solution of the generalized differential equation (8.8) we have

$$x_k(s_2) - x_k(s_1) = \int_{s_1}^{s_2} DF_k(x_k(\tau), t) \tag{8.12}$$

for every $s_1, s_2 \in [\alpha, \beta]$ and $k = 1, 2, \ldots$. Our aim is to show that

$$\lim_{k\to\infty} \int_{s_1}^{s_2} DF_k(x_k(\tau), t) = \int_{s_1}^{s_2} DF_0(x(\tau), t) \tag{8.13}$$

for any $s_1, s_2 \in [\alpha, \beta]$ because passing to the limit $k \to \infty$ in (8.12) we obtain

$$x(s_2) - x(s_1) = \int_{s_1}^{s_2} DF_0(x(\tau), t)$$

for every $s_1, s_2 \in [\alpha, \beta]$ provided (8.13) is true, and this means that $x : [\alpha, \beta] \to \mathbb{R}^n$ is a solution of (8.10) on the interval $[\alpha, \beta]$.

To prove (8.13) let us consider the difference

$$\int_{s_1}^{s_2} DF_k(x_k(\tau), t) - \int_{s_1}^{s_2} DF_0(x(\tau), t) =$$

$$= \int_{s_1}^{s_2} D[F_k(x_k(\tau), t) - F_k(x(\tau), t)] +$$

$$+ \int_{s_1}^{s_2} D[F_k(x(\tau), t) - F_0(x(\tau), t)]$$

for $\alpha \leq s_1 < s_2 \leq \beta$.

By Lemma 8.1 we have

$$\lim_{k\to\infty} \int_{s_1}^{s_2} D[F_k(x(\tau),t) - F_0(x(\tau),t)] = 0. \tag{8.14}$$

Since $F_k : G \to \mathbb{R}^n$ belongs to the class $\mathcal{F}(G,h,\omega)$ for $k = 0,1,\dots$ we have

$$\|F_k(x_k(\tau),t_2) - F_k(x_k(\tau),t_1) - F_k(x(\tau),t_2) + F_k(x(\tau),t_1)\| \leq$$
$$\leq \omega(\|x_k(\tau) - x(\tau)\|)|h(t_2) - h(t_1)| \tag{8.15}$$

for $\tau, t_1, t_2 \in [\alpha,\beta]$.

The functions $x_k - x$, $k = 1,2,\dots$ are of bounded variation on $[\alpha,\beta]$ and therefore the functions $\omega(\|x_k(\tau) - x(\tau)\|)$ are regulated nonnegative real functions because they have onesided limits at every point $[\alpha,\beta]$. Hence by Corollary 1.34 the integrals

$$\int_{s_1}^{s_2} \omega(\|x_k(s) - x(s)\|)\, dh(s)$$

exist for every $k = 1,2,\dots$. Using (8.15) it can be shown that the assumptions of Corollary 3.15 hold and by this corollary we obtain the inequality

$$\|\int_{s_1}^{s_2} D[F_k(x_k(\tau),t) - F_k(x(\tau),t)]\| \leq$$
$$\leq \int_{s_1}^{s_2} \omega(\|x_k(s) - x(s)\|)\, dh(s) \tag{8.16}$$

for every $s_1, s_2 \in [\alpha,\beta]$ and $k = 1,2,\dots$.

Moreover, (8.9) implies

$$\lim_{k\to\infty} \omega(\|x_k(s) - x(s)\|) = 0, \quad s \in [\alpha,\beta]$$

and we evidently have also

$$0 \leq \omega(\|x_k(s) - x(s)\|) \leq K = \text{ const.}$$

for $s \in [\alpha, \beta]$. Hence by Corollary 1.32 of the convergence theorem we obtain

$$\lim_{k\to\infty} \int_{s_1}^{s_2} \omega(\|x_k(s) - x(s)\|)\, dh(s) = 0$$

and by (8.16) also

$$\lim_{k\to\infty} \int_{s_1}^{s_2} D[F_k(x_k(\tau), t) - F_k(x(\tau), t)] = 0.$$

This relation together with (8.14) yields (8.13) and concludes the proof. □

8.3 *Examples.*

1. For $k = 1, 2, \ldots$ let us define functions $h_k : [-1, 1] \to \mathbb{R}$ as follows:

$$h_k(t) = t \qquad \text{for } t \in [-1, 0],$$
$$h_k(t) = t + \frac{1}{k} \qquad \text{for } t \in (0, 1]$$

and set $h_0(t) = t$ for $t \in [-1, 1]$. Now define

$$F_k(x, t) = x h_k(t), \quad |x| < 1,\ t \in [-1, 1],\ k = 0, 1, \ldots .$$

For the functions $F_k : (-1, 1) \times [-1, 1] \to \mathbb{R}$ we have

$$|F_k(x, t_2) - F_k(x, t_1)| \leq (h_k(t_2) - h_k(t_1))|x| \leq h_k(t_2) - h_k(t_1)$$

and

$$|F_k(x,t_2) - F_k(x,t_1) - F_k(y,t_2) + F_k(y,t_1)| \leq$$
$$\leq |x-y|(h_k(t_2) - h_k(t_1))$$

provided $x, y \in (-1,1)$ and $-1 \leq t_1 \leq t_2 \leq 1$. Let us set

$$h(t) = t \qquad \text{for } t \in [-1,0],$$
$$h(t) = t+1 \qquad \text{for } t \in (0,1].$$

It is easy to see that we have $F_k \in \mathcal{F}(G,h,\omega)$ with $\omega(r) = r$ for $r \geq 0$, $G = (-1,1) \times [-1,1]$ and also

$$\lim_{k\to\infty} F_k(x,t) = x \lim_{k\to\infty} h_k(t) = xh_0(t) = F_0(x,t).$$

Let us consider the generalized differential equation

$$\frac{dx}{d\tau} = DF_k(x,t) = D[xh_k(t)] \tag{8.17}$$

with the initial condition $x(-1) = x_0$ with $|x_0| < \dfrac{1}{2e^2}$. A solution $x_k : [-1,1] \to (-1,1) \subset \mathbb{R}$ of (8.17) exists on $[-1,1]$ and satisfies by definition the integral equation

$$x(s) = x_0 + \int_{-1}^{s} D[x(\tau)h_k(t)] = x_0 + \int_{-1}^{s} x(\tau)\,dh_k(\tau).$$

It is easy to check that the function $x_k : [-1,1] \to (-1,1)$ given by

$$x_k(t) = e^{t+1}x_0 \qquad \text{for } t \in [-1,0],$$
$$x_k(t) = e^{t+1}(1+\frac{1}{k})x_0 \qquad \text{for } t \in (0,1]$$

is a solution of (8.17) with $x_k(-1) = x_0$ and

$$\lim_{k\to\infty} x_k(t) = e^{t+1}x_0 \text{ for } t \in [-1,1].$$

The limit function $x_0(t) = e^{t+1}x_0$, $t \in [-1,1]$ is evidently a solution of the classical initial value problem

$$\dot{x} = x, \quad x(-1) = x_0$$

and by Theorem 8.2 also of the corresponding generalized form

$$\frac{dx}{d\tau} = DF_0(x,t) = D[xt], \quad x(-1) = x_0. \tag{8.18}$$

2. For $k = 1, 2, \dots$ let us define functions $h_k : [-1,1] \to \mathbb{R}$ as follows:

$$h_k(t) = t \qquad \text{for } t \in [-1, \frac{1}{k}],$$
$$h_k(t) = t + \frac{1}{k} \quad \text{for } t \in (\frac{1}{k}, 1]$$

and set $h_0(t) = t$ for $t \in [-1,1]$. Define as in the first example

$$F_k(x,t) = xh_k(t), \quad |x| < 1,\, t \in [-1,1],\, k = 0, 1, \dots .$$

For the functions $F_k : (-1,1) \times [-1,1] \to \mathbb{R}$ we have

$$|F_k(x,t_2) - F_k(x,t_1)| \le (h_k(t_2) - h_k(t_1))|x| \le h_k(t_2) - h_k(t_1)$$

and

$$|F_k(x,t_2) - F_k(x,t_1) - F_k(y,t_2) + F_k(y,t_1)| \le$$
$$\le |x-y|(h_k(t_2) - h_k(t_1))$$

provided $x, y \in (-1,1)$ and $-1 \le t_1 \le t_2 \le 1$.

Moreover, we again have

$$\lim_{k\to\infty} F_k(x,t) = x \lim_{k\to\infty} h_k(t) = xh_0(t) = F_0(x,t)$$

provided $x \in (-1,1)$ and $-1 \le t \le 1$.

It is easy to check that the function $x_k : [-1,1] \to (-1,1)$ given by

$$x_k(t) = e^{t+1}x_0 \qquad \text{for } t \in [-1, \frac{1}{k}],$$
$$x_k(t) = e^{t+1}(1+\frac{1}{k})x_0 \quad \text{for } t \in (\frac{1}{k}, 1]$$

is a solution of (8.17) with $x_k(-1) = x_0$ and

$$\lim_{k\to\infty} x_k(t) = e^{t+1}x_0 \quad \text{for } t \in [-1,1].$$

The limit function $x_0(t) = e^{t+1}x_0$, $t \in [-1,1]$ is again a solution of the initial value problem

$$\dot{x} = x, \quad x(-1) = x_0$$

and also of the corresponding generalized form

$$\frac{dx}{d\tau} = DF_0(x,t) = D[xt], \quad x(-1) = x_0. \tag{8.18}$$

Nevertheless, this evident convergence result cannot be explained by Theorem 8.2 because it is not possible to have a nondecreasing function $h : [-1,1] \to \mathbb{R}$ such that $F_k \in \mathcal{F}(G,h,\omega)$ for all $k = 0,1,\ldots$. Indeed, for the functions $F_k : (-1,1) \times [-1,1] \to \mathbb{R}$ we have

$$|F_k(x,t_2) - F_k(x,t_1)| = (h_k(t_2) - h_k(t_1))|x|$$

for $x \in (-1,1)$ and $-1 \le t_1 \le t_2 \le 1$. If $F_k \in \mathcal{F}(G,h,\omega)$were valid for all $k = 0,1,\dots$, then we should have

$$|F_k(x,t_2) - F_k(x,t_1)| = (h_k(t_2) - h_k(t_1))|x| \le h(t_2) - h(t_1)$$

for $x \in (-1,1)$ and $-1 \le t_1 \le t_2 \le 1$.

Hence e.g. for $x = \frac{1}{2}$ the above inequality implies

$$\frac{1}{2k} \le \frac{1}{2}\left[h_k(\frac{2k-1}{2k(k-1)}) - h_k(\frac{2k+1}{2k(k+1)})\right] \le$$

$$\le h(\frac{2k-1}{2k(k-1)}) - h(\frac{2k+1}{2k(k+1)})$$

for every $k > 1$ and this means that for an arbitrary positive constant $C > 0$ we obtain the inequality

$$h(1) - h(-1) > C$$

which contradicts the fact that h is a finite function. Therefore we cannot have a nondecreasing $h : [-1,1] \to \mathbb{R}$ such that $F_k \in \mathcal{F}(G,h,\omega)$ for all $k = 0,1,\dots$.

Looking at the second example we can see that in this case we have

$$\limsup_{k\to\infty}(h_k(t_2) - h_k(t_1)) \le h_0(t_2) - h_0(t_1)$$

for $-1 \le t_1 \le t_2 \le 1$, and we notice that the function h_0 is continuous in $[-1,1]$. This example motivates our subsequent considerations.

Let us introduce some notation. Let $\omega : [0,+\infty) \to [0,+\infty)$ be continuous, nondecreasing and $\omega(0) = 0$. Assume that $h_0 : [a,b] \to \mathbb{R}$ is a nondecreasing function which is continuous from the left on $[a, b]$ and that an arbitrary $\eta > 0$ is given. If $[\alpha, \beta] \subset [a, b]$ then denote

$$D = D(\eta, h_0, [\alpha, \beta]) = \{t \in [\alpha, \beta]; \omega(h_0(t+) - h_0(t)) > \eta\}.$$

It is clear that for every $\eta > 0$ the set D is finite. The existence of the onesided limits of the function h_0 at every point in $[\alpha, \beta]$ and the continuity of ω yields the existence of a function $\delta_1 : [\alpha, \beta] \to (0, +\infty)$ with the following properties:

If $t \notin D$ then

$$\omega(h_0(t + \delta_1(t)) - h_0(t - \delta_1(t))) < \eta \tag{8.19}$$

provided $t \in (\alpha, \beta)$ and

$$\omega(h_0(\alpha + \delta_1(\alpha)) - h_0(\alpha)) < \eta, \quad \omega(h_0(\beta) - h_0(\beta - \delta_1(\beta))) < \eta \tag{8.20}$$

if $\alpha \notin D, \beta \notin D$, respectively.

If $t \in D$ then

$$\begin{aligned} \omega(h_0(t + \delta_1(t)) - h_0(t+)) < \eta, \quad & t \in [\alpha, \beta), \\ \omega(h_0(t) - h_0(t - \delta_1(t))) < \eta, \quad & t \in (\alpha, \beta]. \end{aligned} \tag{8.21}$$

If $t \in D$ is one of the endpoints of the interval $[\alpha, \beta]$ then only the respective one of these inequalities is required.

Further, define $\delta_2 : [\alpha, \beta] \to (0, +\infty)$ by the relations

$$\delta_2(\tau) = \text{dist}(\tau, D) \text{ for } \tau \notin D, \quad \delta_2(\tau) = 1 \text{ for } \tau \in D \tag{8.22}$$

and take

$$\delta_{\eta, h_0}(\tau) = \min(\delta_1(\tau), \delta_2(\tau)) \text{ for } \tau \in [\alpha, \beta]. \tag{8.23}$$

The function $\delta_{\eta,h_0} : [\alpha,\beta] \to (0,+\infty)$ is evidently a gauge on $[\alpha,\beta]$ and by Cousin's Lemma 1.4 the set of δ_{η,h_0}–fine partitions of $[\alpha,\beta]$ is nonempty. Assume that

$$A = \{\alpha_0, \tau_1, \alpha_1, \dots, \alpha_{k-1}, \tau_k, \alpha_k\}$$

is a δ_{η,h_0}–fine partition of $[\alpha,\beta]$. Then

$$D \subset \{\tau_1, \tau_2, \dots, \tau_k\} \tag{8.24}$$

because $\delta_{\eta,h_0}(\tau) < \delta_1(\tau)$ for $\tau \in [\alpha,\beta]$ by (8.23) and (8.22) holds. This can be shown in the same way as in the proof of Theorem 1.11. Moreover, by virtue of the properties of the gauge δ_{η,h_0} the partition A has the following properties:

$$\text{if } \tau_m \notin D \text{ then } \omega(h_0(\alpha_m) - h_0(\alpha_{m-1})) < \eta \tag{8.25}$$

and

$$\begin{aligned} \text{if } \tau_m \in D \text{ then } &\omega(h_0(\alpha_m) - h_0(\tau_m+)) < \eta \text{ and} \\ &\omega(h_0(\tau_m) - h_0(\alpha_{m-1})) < \eta. \end{aligned} \tag{8.26}$$

The property (8.25) easily follows from (8.19) and (8.20) while (8.26) is a consequence of (8.21).

Since the right hand side limits of h_0 exist at every point of $[\alpha,\beta)$ we can see that the δ_{η,h_0}–fine partition A of $[\alpha,\beta]$ can be chosen such that

$$\alpha_m > \tau_m \text{ if } \tau_m \in D \cap [\alpha,\beta). \tag{8.27}$$

Concluding these preparatory considerations we state that for every nondecreasing function $h_0 : [\alpha,\beta] \to \mathbb{R}$ which is continuous from the left and for every $\eta > 0$ there exists a partition of the interval $[\alpha,\beta]$ wich satisfies (8.24), (8.25), (8.26) and (8.27).

Since the set of points of continuity of h_0 is dense in $[\alpha,\beta]$ and all the inequalities in (8.25) – (8.27) are sharp, the partition can be chosen in such a way that the points $\alpha_1, \dots, \alpha_{k-1}$ are points of continuity of h_0.

8.4 Lemma. *Assume that the functions $h, h_0 : [a,b] \to \mathbb{R}$ are nondecreasing and continuous from the left. Let $F \in \mathcal{F}(G,h,\omega)$ and let $x : [\alpha,\beta] \to \mathbb{R}^n$, $[\alpha,\beta] \subset [a,b]$ be such that $(x(\tau),\tau) \in G$ for $\tau \in [\alpha,\beta]$ and*

$$\|x(t_2) - x(t_1)\| \leq h_0(t_2) - h_0(t_1), \quad \alpha \leq t_1 \leq t_2 \leq \beta. \tag{8.28}$$

Then the integral $\int_\alpha^\beta DF(x(\tau),t)$ exists and if $\eta > 0$ is arbitrary and $A = \{\alpha_0, \tau_1, \alpha_1, \dots, \alpha_{k-1}, \tau_k, \alpha_k\}$ is a partition of $[\alpha,\beta]$ which satisfies (8.24), (8.25), (8.26) and (8.27) then

$$\| \int_\alpha^\beta DF(x(\tau),t) - \sum_{m=1}^{k} S_m \| \leq \eta(h(\beta) - h(\alpha) + h_0(\beta) - h_0(\alpha)) \tag{8.29}$$

where

$$S_m = F(x(\tau_m),\alpha_m) - F(x(\tau_m),\alpha_{m-1})$$

if $\tau_m \notin D$ and

$$\begin{aligned} S_m =& F(x(\tau_m+),\alpha_m) - F(x(\tau_m+),\tau_m+) + \\ &+ F(x(\tau_m),\tau_m+) - F(x(\tau_m),\alpha_{m-1}) \end{aligned}$$

provided $\tau \in D$.

Proof. The existence of the integral $\int_\alpha^\beta DF(x(\tau),t)$ is clear by Corollary 3.16 because by (8.28) the function x is of bounded variation on $[\alpha,\beta]$.

Let $\eta > 0$ be given and assume that the partition A satisfies the conditions given in the statement. The integral $\int_{\alpha_{m-1}}^{\alpha_m} DF(x(\tau),t)$ exists for every $m = 1,\dots,k$ (see Theorem 1.10). Let $m \in \{1,\dots,k\}$ be fixed and assume $\tau \in D$. Then for every $\varepsilon > 0$ there exist gauges δ^-, δ^+ on $[\alpha_{m-1},\tau_m]$, $[\tau_m,\alpha_m]$, respectively, such that

$$\| \int_{\alpha_{m-1}}^{\tau_m} DF(x(\tau),t) - S(A^-)\| \leq \frac{\varepsilon}{2}$$

for every δ^-–fine partition A^- of $[\alpha_{m-1}, \tau_m]$ where $S(A^-)$ is the integral sum corresponding to A^- and

$$\|\int_{\tau_m}^{\alpha_m} DF(x(\tau), t) - S(A^+)\| \leq \frac{\varepsilon}{2}$$

for every δ^+–fine partition A^- of $[\tau_m, \alpha_m]$ where $S(A^+)$ is the integral sum corresponding to A^+. From these inequalities we immediately get

$$\|\int_{\alpha_{m-1}}^{\alpha_m} DF(x(\tau), t) - S(A^-) - S(A^+)\| \leq \varepsilon. \tag{8.30}$$

Without loss of generality we assume that the value $\delta^+(\tau_m) > 0$ is such that

$$h(\alpha) - h(\tau_m+) < \eta \frac{h_0(\tau_m+) - h_0(\tau_m)}{\omega(h_0(\tau_m+) - h_0(\tau_m))}$$

for every $\alpha \in (\tau_m, \tau_m + \delta^+(\tau_m))$.

Assume further that

$$A^- = \{\tilde{\alpha}_0, \tilde{\tau}_1, \tilde{\alpha}_1, \ldots, \tilde{\alpha}_{l-1}, \tilde{\tau}_l, \tilde{\alpha}_l\}$$

is a δ^-–fine partition of $[\alpha_{m-1}, \tau_m]$ with

$$\alpha_{m-1} = \tilde{\alpha}_0 \leq \tilde{\tau}_1 \leq \tilde{\alpha}_1 \leq \cdots \leq \tilde{\alpha}_{l-1} \leq \tilde{\tau}_l = \tilde{\alpha}_l = \tau_m$$

and that

$$A^+ = \{\tilde{\alpha}_l, \tilde{\tau}_{l+1}, \tilde{\alpha}_{l+1}, \ldots, \tilde{\alpha}_{l+s-1}, \tilde{\tau}_{l+s}, \tilde{\alpha}_{l+s}\}$$

is a δ^+–fine partition of $[\tau_m, \alpha_m]$ with

$$\begin{aligned} \tau_m = &\tilde{\alpha}_l \leq \tilde{\tau}_{l+1} \leq \tilde{\alpha}_{l+1} \leq \tilde{\tau}_{l+2} \leq \\ &\leq \cdots \leq \tilde{\alpha}_{l+s-1} \leq \tilde{\tau}_{l+s} = \tilde{\alpha}_{l+s} = \alpha_m. \end{aligned}$$

We give an estimate for $\|S(A^-) + S(A^+) - S_m\|$ using the properties of the class $\mathcal{F}(G, h, \omega)$. We have

$$\|S(A^-) + S(A^+) - S_m\| \leq$$

$$\leq \|\sum_{j=1}^{l}[F(x(\tilde{\tau}_j), \tilde{\alpha}_j) - F(x(\tilde{\tau}_j), \tilde{\alpha}_{j-1})]-$$

$$-(F(x(\tau_m), \tau_m+) - F(x(\tau_m), \alpha_{m-1}))+$$

$$+\sum_{j=l+1}^{l+s}[F(x(\tilde{\tau}_j), \tilde{\alpha}_j) - F(x(\tilde{\tau}_j), \tilde{\alpha}_{j-1})]-$$

$$-(F(x(\tau_m+), \alpha_m) - F(x(\tau_m+), \tau_m+))\| =$$

$$= \|\sum_{j=1}^{l}[F(x(\tilde{\tau}_j), \tilde{\alpha}_j) - F(x(\tilde{\tau}_j), \tilde{\alpha}_{j-1})-$$

$$-F(x(\tau_m), \tilde{\alpha}_j) + F(x(\tau_m), \tilde{\alpha}_{j-1})]-$$

$$-F(x(\tau_m), \tau_m+) - F(x(\tau_m), \tilde{\alpha}_l)+$$

$$+F(x(\tilde{\tau}_{l+1}), \tilde{\alpha}_{l+1}) - F(x(\tilde{\tau}_{l+1}), \tilde{\alpha}_l)-$$

$$-F(x(\tau_m+), \tilde{\alpha}_{l+1}) + F(x(\tau_m+), \tau_m+)+$$

$$+\sum_{j=l+2}^{l+s}[F(x(\tilde{\tau}_j), \tilde{\alpha}_j) - F(x(\tilde{\tau}_j), \tilde{\alpha}_{j-1})-$$

$$-F(x(\tau_m+), \tilde{\alpha}_j) + F(x(\tau_m+), \tilde{\alpha}_{j-1})]\| \leq$$

$$\leq \sum_{j=1}^{l}\|F(x(\tilde{\tau}_j), \tilde{\alpha}_j) - F(x(\tilde{\tau}_j), \tilde{\alpha}_{j-1})-$$

$$-F(x(\tau_m), \tilde{\alpha}_j) + F(x(\tau_m), \tilde{\alpha}_{j-1})\|+$$

$$+\|F(x(\tau_m),\tilde{\alpha}_{l+1})-F(x(\tau_m),\tau_m+)-$$

$$-F(x(\tau_m+),\tilde{\alpha}_{l+1})+F(x(\tau_m+),\tau_m+)\|+$$

$$+\sum_{j=l+2}^{l+s}\|F(x(\tilde{\tau}_j),\tilde{\alpha}_j)-F(x(\tilde{\tau}_j),\tilde{\alpha}_{j-1})-$$

$$-F(x(\tau_m+),\tilde{\alpha}_j)+F(x(\tau_m+),\tilde{\alpha}_{j-1})\|\leq$$

$$\leq\sum_{j=1}^{l}\omega(\|x(\tilde{\tau}_j)-x(\tau_m)\|)(h(\tilde{\alpha}_j)-h(\tilde{\alpha}_{j-1}))+$$

$$+\omega(\|x(\tau_m+)-x(\tau_m)\|)(h(\tilde{\alpha}_{l+1})-h(\tau_m+))+$$

$$+\sum_{j=l+2}^{l+s}\omega(\|x(\tilde{\tau}_j)-x(\tau_m+)\|)(h(\tilde{\alpha}_j)-h(\tilde{\alpha}_{j-1}))\leq$$

$$\leq\eta(h(\tau_m)-h(\alpha_{m-1}))+$$

$$+\omega(h_0(\tau_m+)-h_0(\tau_m))\eta\frac{h_0(\tau_m+)-h_0(\tau_m)}{\omega(h_0(\tau_m+)-h_0(\tau_m))}+$$

$$+\eta(h(\alpha_m)-h(\tilde{\alpha}_{l+1}))\leq$$

$$\leq\eta(h(\alpha_m)-h(\alpha_{m-1})+h_0(\tau_m+)-h_0(\tau_m))\leq$$

$$\leq\eta(h(\alpha_m)-h(\alpha_{m-1})+h_0(\alpha_m)-h_0(\alpha_{m-1})).$$

This estimate together with (8.30) leads to the inequality

$$\|\int_{\alpha_{m-1}}^{\alpha_m}DF(x(\tau),t)-S_m\|\leq\varepsilon+\|S(A^-)+S(A^+)-S_m\|<$$

$$<\varepsilon+\eta(h(\alpha_m)-h(\alpha_{m-1})+h_0(\alpha_m)-h_0(\alpha_{m-1})).$$

Since $\varepsilon > 0$ can be chosen arbitrarily small this inequality yields

$$\| \int_{\alpha_{m-1}}^{\alpha_m} DF(x(\tau), t) - S_m \| \leq$$

$$\leq \eta(h(\alpha_m) - h(\alpha_{m-1}) + h_0(\alpha_m) - h_0(\alpha_{m-1})). \tag{8.31}$$

For the case when $\tau \notin D$ the same inequality can be proved with less effort. Using (8.31) and the inequality

$$\| \int_{\alpha}^{\beta} DF(x(\tau), t) - \sum_{m=1}^{k} S_m \| \leq \sum_{m=1}^{k} \| \int_{\alpha_{m-1}}^{\alpha_m} DF(x(\tau), t) - S_m \|$$

we can see immediately that (8.29) holds. □

Let us now derive another theorem on continuous dependence on a parameter which is based on the above Lemma 8.3. The structure of the statement is close to that of Theorem 8.2.

8.5 Theorem. *Assume that $F_k : G \to \mathbb{R}^n$ belongs to the class $\mathcal{F}(G, h_k, \omega)$ for $k = 0, 1, \dots$ where $h_k : [a, b] \to \mathbb{R}$ are nondecreasing functions which are continuous from the left when $k = 1, 2, \dots$ and the function $h_0 : [a, b] \to \mathbb{R}$ is nondecreasing and continuous on $[a, b]$.*

Assume further that

$$\limsup_{k \to \infty} (h_k(t_2) - h_k(t_1)) \leq h_0(t_2) - h_0(t_1) \tag{8.32}$$

for every $a \leq t_1 \leq t_2 \leq b$.

Suppose that

$$\lim_{k \to \infty} F_k(x, t) = F_0(x, t) \tag{8.7}$$

for $(x, t) \in G$.

Let $x_k : [\alpha, \beta] \to \mathbb{R}^n$, $k = 1, 2, \dots$ *be a sequence of solutions of the generalized differential equation*

$$\frac{dx}{d\tau} = DF_k(x, t) \tag{8.8}$$

on $[\alpha, \beta] \subset (a, b)$ *such that*

$$\lim_{k \to \infty} x_k(s) = x(s), \quad s \in [\alpha, \beta], \tag{8.9}$$

and $(x(s), s) \in G$ *for* $s \in [\alpha, \beta]$.

Then $x : [\alpha, \beta] \to \mathbb{R}^n$ *is a continuous function of bounded variation on* $[\alpha, \beta]$, *and it is a solution of the generalized differential equation*

$$\frac{dx}{d\tau} = DF_0(x, t) \tag{8.10}$$

on the interval $[\alpha, \beta]$.

Proof. Since x_k is a solution of (8.8) we have by Lemma 3.10 $\|x_k(t_2) - x_k(t_1)\| \le h_k(t_2) - h_k(t_1)$ for $k = 1, 2, \dots$ and $t_1, t_2 \in [\alpha, \beta]$, $t_1 \le t_2$. Therefore by (8.9) and (8.32) we have also

$$\|x(t_2) - x(t_1)\| \le h_0(t_2) - h_0(t_1)$$

for $t_1, t_2 \in [\alpha, \beta]$, $t_1 \le t_2$. Hence the function x is continuous on $[\alpha, \beta]$ and the integral $\int_\alpha^\beta DF_0(x(\tau), t)$ exists by Corollary 3.18.

We will show that $x : [\alpha, \beta] \to \mathbb{R}^n$ is a solution of (8.10). To do this, it is sufficient to prove that for every $s_1, s_2 \in [\alpha, \beta]$, $s_1 \le s_2$ we have

$$\lim_{k \to \infty} \int_{s_1}^{s_2} DF_k(x_k(\tau), t) = \int_{s_1}^{s_2} DF_0(x(\tau), t). \tag{8.33}$$

Suppose that $s_1, s_2 \in [\alpha, \beta]$, $s_1 < s_2$ and choose an arbitrary $\eta > 0$. Determine a partition

$$A = \{\alpha_0, \tau_1, \alpha_1, \dots, \alpha_{l-1}, \tau_l, \alpha_l\}$$

of the interval $[s_1, s_2]$ for which

$$\omega(h_0(\alpha_m) - h_0(\alpha_{m-1})) < \eta, \quad m = 1, \dots, l.$$

Since the function h_0 is continuous, the set

$$D = D(\eta, h_0, [\alpha, \beta]) = \{t \in [\alpha, \beta]; \omega(h_0(t+) - h_0(t)) > \eta\}$$

is empty and (8.25), (8.26) is automatically satisfied. The assumption (8.32) implies that there is a $k_0 \in \mathbb{N}$ such that for $k > k_0$ we have

$$\omega(h_k(\alpha_m) - h_k(\alpha_{m-1})) < \eta, \quad m = 1, \dots, l$$

and we have also $\omega(h_k(\tau_m+) - h_k(\tau_m)) < \eta$ for $m = 1, \dots, l$ because $h_k(\tau_m+) - h_k(\tau_m) \leq h_k(\alpha_m) - h_k(\alpha_{m-1})$ and this means that $\tau_m \notin D(\eta, h_k, [\alpha, \beta])$ for $m = 1, \dots, l$ and $k > k_0$. Hence for all sufficiently large $k \in \mathbb{N}$ the partition A satisfies (8.25), (8.26) and (8.27) also with the functions h_k instead of h_0.

Due to (8.9) it can be also assumed that

$$\omega(\|x_k(\tau_m) - x(\tau_m)\|) < \eta$$

for all $m = 1, \dots, l$ and $k > k_0$.

By Lemma 8.4 we obtain for $k > k_0$ the inequality

$$\| \int_{s_1}^{s_2} DF_k(x_k(\tau), t) - \sum_{m=1}^{l} S_{m,k} \| \leq 2\eta(h_k(s_2) - h_k(s_1))$$

where $S_{m,k} = F_k(x_k(\tau_m), \alpha_m) - F_k(x_k(\tau_m), \alpha_{m-1})$, and also

$$\| \int_{s_1}^{s_2} DF_0(x(\tau), t) - \sum_{m=1}^{l} S_{m,0} \| \leq 2\eta(h_0(s_2) - h_0(s_1))$$

where $S_{m,0} = F_0(x(\tau_m), \alpha_m) - F_0(x(\tau_m), \alpha_{m-1})$.

Hence

$$\| \int_{s_1}^{s_2} DF_k(x_k(\tau), t) - \int_{s_1}^{s_2} DF_0(x(\tau), t) \| \leq$$

$$\leq \| \sum_{m=1}^{l} (S_{m,k} - S_{m,0}) \| + 2\eta(h_k(s_2) - h_k(s_1) + h_0(s_2) - h_0(s_1)). \tag{8.34}$$

Let us consider the first term on the right hand side of (8.34). We have

$$\| \sum_{m=1}^{l} (S_{m,k} - S_{m,0}) \| \leq$$

$$\leq \sum_{m=1}^{l} \| F_k(x_k(\tau_m), \alpha_m) - F_k(x_k(\tau_m), \alpha_{m-1}) -$$

$$- F_0(x(\tau_m), \alpha_m) + F_0(x(\tau_m), \alpha_{m-1}) \| \leq$$

$$\leq \sum_{m=1}^{l} \| F_k(x_k(\tau_m), \alpha_m) - F_k(x_k(\tau_m), \alpha_{m-1}) -$$

$$- F_k(x(\tau_m), \alpha_m) + F_k(x(\tau_m), \alpha_{m-1}) \| +$$

$$+ \sum_{m=1}^{l} \| F_k(x(\tau_m), \alpha_m) - F_k(x(\tau_m), \alpha_{m-1}) -$$

$$- F_0(x(\tau_m), \alpha_m) + F_0(x(\tau_m), \alpha_{m-1}) \| \leq$$

$$\leq \sum_{m=1}^{l} \omega(\| x_k(\tau_m) - x(\tau_m) \|)(h_k(\alpha_m) - h_k(\alpha_{m-1})) +$$

$$+ \sum_{m=1}^{l} [\| F_k(x(\tau_m), \alpha_m) - F_0(x(\tau_m), \alpha_m) \| +$$

$$+\|F_k(x(\tau_m),\alpha_{m-1})-F_0(x(\tau_m),\alpha_{m-1})\|]\leq$$

$$\leq\sum_{m=1}^{l}\eta(h_k(\alpha_m)-h_k(\alpha_{m-1}))+$$

$$+\sum_{m=1}^{l}[\|F_k(x(\tau_m),\alpha_m)-F_0(x(\tau_m),\alpha_m)\|+$$

$$+\|F_k(x(\tau_m),\alpha_{m-1})-F_0(x(\tau_m),\alpha_{m-1})\|].$$

By (8.7) the sum on the right hand side of this inequality converges to zero for $k\to\infty$. Using the fact that $\eta>0$ can be taken arbitrarily small we obtain by (8.34) that for every $\varepsilon>0$ there is a $k_1\in\mathbb{N}$ such that for $k>k_1$ we have

$$\|\int_{s_1}^{s_2}DF_k(x_k(\tau),t)-\int_{s_1}^{s_2}DF_0(x(\tau),t)\|<\varepsilon,$$

i.e. (8.33) holds and the theorem is proved. □

Theorems 8.2 and 8.5 are in a certain sense weak forms of continuous dependence results for generalized ordinary differential equations. The most important assumption is the relation (8.7) which ensures that if a sequence $x_k:[\alpha,\beta]\to\mathbb{R}^n$ of solutions of (8.8), $k=1,2,\dots$ converges pointwise to a certain function $x:[\alpha,\beta]\to\mathbb{R}^n$ then this limit is a solution of the equation (8.10). There are different additional conditions on the right hand sides F_k of (8.8) and F_0 of (8.10) in Theorems 8.2 and 8.5.

Now we present results with an additional uniqueness condition for the "limit" equation (8.10).

8.6 Theorem. *Assume that $F_k:G\to\mathbb{R}^n$ belongs to the class $\mathcal{F}(G,h,\omega)$ for $k=0,1,\dots$ where $h:[a,b]\to\mathbb{R}$ is nondecreasing, continuous from the left, and that*

$$\lim_{k\to\infty}F_k(x,t)=F_0(x,t) \tag{8.7}$$

for $(x,t) \in G$.

Let $x : [\alpha, \beta] \to \mathbb{R}^n$, $[\alpha, \beta] \subset [a, b]$ *be a solution of the generalized differential equation*

$$\frac{dx}{d\tau} = DF_0(x,t) \tag{8.10}$$

on $[\alpha, \beta]$ *which has the following uniqueness property:*

if $x : [\alpha, \gamma] \to \mathbb{R}^n$, $[\alpha, \gamma] \subset [\alpha, \beta]$ *is a solution of (8.10) such that* $y(\alpha) = x(\alpha)$ *then* $y(t) = x(t)$ *for every* $t \in [\alpha, \gamma]$.

Assume further that there is a $\varrho > 0$ *such that if* $s \in [\alpha, \beta]$ *and* $\|y - x(s)\| < \varrho$ *then* $(y, s) \in G = B_c \times [\alpha, \beta]$, *and let* $y_k \in \mathbb{R}^n$, $k = 1, 2, \ldots$ *satisfy*

$$\lim_{k \to \infty} y_k = x(\alpha).$$

Then for sufficiently large $k \in \mathbb{N}$ *there exists a solution* x_k *of the generalized differential equation*

$$\frac{dx}{d\tau} = DF_k(x,t) \tag{8.8}$$

on $[\alpha, \beta]$ *with* $x_k(\alpha) = y_k$ *and*

$$\lim_{k \to \infty} x_k(s) = x(s), \quad s \in [\alpha, \beta].$$

Proof. By assumption we have $(y, \alpha) \in G$ provided

$$\|y - x(\alpha)\| < \frac{\varrho}{2}$$

or

$$\|y - x(\alpha+)\| = \|y - x(\alpha) - F_0(x(\alpha), \alpha+) + F_0(x(\alpha), \alpha)\| < \frac{\varrho}{2}.$$

Since $y_k \to x(\alpha)$ for $k \to \infty$, we have by (8.7) also

$$y_k + F_k(y_k, \alpha+) - F_k(y_k, \alpha) \to x(\alpha) + F_0(x(\alpha), \alpha+) - F_0(x(\alpha), \alpha)$$

for $k \to \infty$ because

$$\|F_k(y_k, \alpha+) - F_k(x(\alpha), \alpha+) - F_k(y_k, \alpha) + F_k(x(\alpha), \alpha)\| \leq$$

$$\leq \omega(\|y_k - x(\alpha)\|)(h(\alpha+) - h(\alpha))$$

and

$$F_k(x(\alpha), \alpha+) - F_k(x(\alpha), \alpha) - F_0(x(\alpha), \alpha+) + F_0(x(\alpha), \alpha) \to 0$$

for $k \to \infty$. Hence we can conclude that there is a $k_1 \in \mathbb{N}$ such that for $k > k_1$ we have $(y_k, \alpha) \in G$ as well as

$$(y_k + F_k(y_k, \alpha+) - F_k(y_k, \alpha), \alpha) \in G.$$

Since the set B_c is open there exists $d > \alpha$ such that if $t \in [\alpha, d]$ and

$$\|x - (y_k + F_k(y_k, \alpha+) - F_k(y_k, \alpha))\| \leq h(t) - h(\alpha+)$$

then $(x, t) \in G$ for $k > k_1$. Using Theorem 4.2 on local existence for generalized ordinary differential equations we obtain that for $k > k_1$ there exists a solution $x_k : [\alpha, d] \to \mathbb{R}^n$ of the generalized differential equation (8.8) on $[\alpha, d]$ such that $x_k(\alpha) = y_k$, $k > k_1$. We claim that $\lim_{k\to\infty} x_k(t) = x(t)$ for $t \in [\alpha, d]$.

Let us mention that the solutions x_k of (8.8) exist on the interval $[\alpha, d]$ and that this interval is the same for all $k > k_1$. Indeed, looking at the proof of Theorem 4.2 it is easy to check that the value $d > \alpha$ depends only on the function h which is common for all right hand sides F_k of (8.8).

By Theorem 8.2, if the sequence (x_k) contains a pointwise convergent subsequence on $[\alpha, d]$ then the limit of this subsequence is necessarily $x(t)$ for $t \in [\alpha, d]$ by the uniqueness assumption on the solution x of (8.10).

By Lemma 3.10 the sequence (x_k), $k > k_1$ of functions on $[\alpha, d]$ is equibounded and of uniformly bounded variation on $[\alpha, d]$. Therefore by Helly's Choice Theorem (see [104]) the sequence (x_k) contains a pointwise convergent subsequence and $x(t)$ is therefore the only accumulation point of the sequence $x_k(t)$ for every $t \in [\alpha, d]$, i.e. $\lim_{k\to\infty} x_k(t) = x(t)$ for $t \in [\alpha, d]$.

In this way we have shown that the theorem holds on $[\alpha, d]$, $d > \alpha$. Let us assume that the convergence result does not hold on the whole interval $[\alpha, \beta]$. Then there exists $d^\star \in (\alpha, \beta)$ such that for every $d < d^\star$ there is a solution x_k of (8.8) with $x_k(\alpha) = y_k$ on $[\alpha, d]$ provided $k \in \mathbb{N}$ is sufficiently large and $\lim_{k\to\infty} x_k(t) = x(t)$ for $t \in [\alpha, d]$ but this does not hold on $[\alpha, d]$ for $d > d^\star$. By Lemma 3.10 we have

$$\|x_k(t_2) - x_k(t_1)\| \le |h(t_2) - h(t_1)|, \quad t_2, t_1 \in [\alpha, d^\star)$$

for $k \in \mathbb{N}$ sufficiently large. Therefore the limits $x_k(d^\star-)$ exist and we also have

$$\lim_{k\to\infty} x_k(d^\star-) = x(d^\star-) = x(d^\star)$$

since the solution x is continuous from the left. Defining $x_k(d^\star) = x_k(d^\star-)$ we obtain $\lim_{k\to\infty} x_k(d^\star) = x(d^\star)$ and this means that Theorem 8.6 holds on the closed interval $[\alpha, d^\star]$, too. Using now $d^\star < \beta$ as the starting point we can show in the same way as above that the theorem holds also on the interval $[d^\star, d^\star + \Delta]$ with some $\Delta > 0$ and this contradicts our assumption. Therefore the theorem holds on the whole interval $[\alpha, \beta]$. □

8.7 *Remark.* Let us mention that in Theorem 8.6 the continuity from the left of the function h is assumed. This assumption is made here to avoid technical problems of the same kind as in Chap. IV. In this case we have to take care for onesided (right) discontinuities of a solution only. In the general case the idea is the same but some additional reasoning has to be done for the possible discontinuities from the left.

Theorem 8.6 is derived from the result given in Theorem 8.2. Now we give a similar result based on Theorem 8.5. Assuming again the uniqueness of a solution of the "limit" equation we obtain a result which is stronger than the result of the previous Theorem 8.6.

8.8 Theorem. *Assume that $F_k : G \to \mathbb{R}^n$ belongs to the class $\mathcal{F}(G, h_k, \omega)$ for $k = 0, 1, \ldots$ where $h_k : [a, b] \to \mathbb{R}$ are nondecreasing functions which are continuous from the left when $k = 1, 2, \ldots$ and the function $h_0 : [a, b] \to \mathbb{R}$ is nondecreasing and continuous on $[a, b]$.*

Assume further that

$$\limsup_{k \to \infty}(h_k(t_2) - h_k(t_1)) \leq h_0(t_2) - h_0(t_1) \tag{8.32}$$

for every $a \leq t_1 \leq t_2 \leq b$.

Suppose that

$$\lim_{k \to \infty} F_k(x, t) = F_0(x, t) \tag{8.7}$$

for $(x, t) \in G$.

Let $x : [\alpha, \beta] \to \mathbb{R}^n$, $[\alpha, \beta] \subset [a, b]$ be a solution of the generalized differential equation

$$\frac{dx}{d\tau} = DF_0(x, t) \tag{8.10}$$

on $[\alpha, \beta]$ which has the following uniqueness property:

if $y : [\alpha, \gamma] \to \mathbb{R}^n$, $[\alpha, \gamma] \subset [\alpha, \beta]$ is a solution of (8.10) such that $y(\alpha) = x(\alpha)$ then $y(t) = x(t)$ for every $t \in [\alpha, \gamma]$.

Assume further that there is a $\varrho > 0$ such that if $s \in [\alpha, \beta]$ and $\|y - x(s)\| < \varrho$ then $(y, s) \in G = B_c \times [\alpha, \beta]$, and let $y_k \in \mathbb{R}^n$, $k = 1, 2, \ldots$ satisfy

$$\lim_{k \to \infty} y_k = x(\alpha).$$

Then for every $\mu > 0$ there exists a $k_\star \in \mathbb{N}$ such that for $k \in \mathbb{N}$, $k > k_\star$ there exists a solution x_k of the generalized differential equation

$$\frac{dx}{d\tau} = DF_k(x, t) \tag{8.8}$$

on $[\alpha, \beta]$ with $x_k(\alpha) = y_k$ and

$$\|x_k(s) - x(s)\| < \mu, \quad s \in [\alpha, \beta] \tag{8.35}$$

holds.

Proof. The existence of the solutions x_k of the equation (8.8) for sufficiently large $k \in \mathbb{N}$ and the pointwise convergence

$$\lim_{k \to \infty} x_k(s) = x(s), \quad s \in [\alpha, \beta]$$

can be shown in the same way as in Theorem 8.6 with minor changes arising from the assumption (8.32).

For showing (8.35) let us consider the difference $x_k(s) - x(s)$ for sufficiently large $k \in \mathbb{N}$ and for $s \in [\alpha, \beta]$. By the definition of a solution we have

$$x_k(s) - x(s) =$$

$$= y_k - x(\alpha) + \int_\alpha^s D[F_k(x_k(\tau), t) - F_0(x(\tau), t)] \tag{8.36}$$

for every $s \in [\alpha, \beta]$.

Let an arbitrary $\eta > 0$ be given. Choose a $\kappa > 0$ such that $\kappa < \eta$, $\omega(\kappa) < \eta$ and determine a partition

$$A = \{\alpha_0, \tau_1, \alpha_1, \ldots, \alpha_{l-1}, \tau_l, \alpha_l\}$$

of $[\alpha, \beta]$ such that

$$h_0(\alpha_m) - h_0(\alpha_{m-1}) < \kappa, \alpha_{m-1} = \tau_m < \alpha_m,$$
$$m = 1, 2, \ldots, l. \tag{8.37}$$

It is evident that such a partition A exists because the function h_0 is uniformly continuous on $[\alpha, \beta]$. The choice of κ gives also

$$\omega(h_0(\alpha_m) - h_0(\alpha_{m-1})) < \omega(\kappa) < \eta,$$
$$m = 1, 2, \ldots, l. \tag{8.38}$$

By the continuity of h_0 we have $D = D(\eta, h_0, [\alpha, \beta]) = \emptyset$ and the partition A satisfies (8.25), (8.26) and (8.27) . Since (8.32) is assumed, there exists a $k_0 \in \mathbb{N}$ such that for $k > k_0$, $k \in \mathbb{N}$ we have

$$h_k(\alpha_m) - h_k(\alpha_{m-1}) < \kappa, \; m = 1, 2, \ldots, l,$$

and consequently also

$$\omega(h_k(\alpha_m) - h_k(\alpha_{m-1})) < \eta, \; m = 1, 2, \ldots, l. \tag{8.39}$$

Further, $\tau_m = \alpha_{m-1}$, $m = 1, 2, \ldots, l$ and the inequality

$$h_k(\alpha_{m-1}+) - h_k(\alpha_{m-1}) < h_k(\alpha_m) - h_k(\alpha_{m-1}) < \kappa$$

yields

$$\omega(h_k(\tau_m+) - h_k(\tau_m)) < \omega(\kappa) < \eta$$

and therefore $\tau_m = \alpha_{m-1} \notin D(\eta, h_k, [\alpha, \beta])$ for $m = 1, 2, \ldots, l$. Hence for $k > k_0$ the partition A satisfies (8.25), (8.26) and (8.27) with h_k instead of h_0.

Due to the pointwise convergence of solutions x_k to x there is a $k_1 > k_0$, $k_1 \in \mathbb{N}$ such that for every $k > k_1$ and $m = 1, 2, \ldots, l$ we have

$$\|x_k(\tau_m) - x(\tau_m)\| < \kappa$$

and therefore also

$$\omega(\|x_k(\tau_m) - x(\tau_m)\|) < \omega(\kappa) < \eta. \tag{8.40}$$

Assume now that $s \in (\alpha, \beta]$ is given. Then there is a $p \in \{1, 2, \ldots, l\}$ such that $s \in (\alpha_{p-1}, \alpha_p]$. Denoting

$$A_s = \{\alpha_0, \tau_1, \alpha_1, \ldots, \alpha_{p-1}, \tau_p, \alpha_p = s\}$$

we obtain a partition of $[\alpha, s]$ and this partition satisfies (8.25), (8.26) and (8.27) and also the same relations with h_k instead of h_0 provided $k > k_1$.

We will consider the difference

$$\int_\alpha^s D[F_k(x_k(\tau), t) - F_0(x(\tau), t)].$$

Since by the properties of a solution we have

$$\|x_k(t_2) - x_k(t_1)\| \leq h_k(t_2) - h_k(t_1)$$

and

$$\|x(t_2) - x(t_1)\| \leq h_0(t_2) - h_0(t_1)$$

for $\alpha \leq t_1 \leq t_2 \leq s$ and $k > k_1$, Lemma 8.4 yields the inequality

$$\| \int_\alpha^s DF_k(x_k(\tau), t) - \sum_{m=1}^{p} S_{m,k}\| \leq 2\eta(h_k(s) - h_k(\alpha))$$

for $k > k_1$ where

$$S_{m,k} = F_k(x_k(\tau_m), \alpha_m) - F_k(x_k(\tau_m), \alpha_{m-1}) =$$

$$= F_k(x_k(\alpha_{m-1}), \alpha_m) - F_k(x_k(\alpha_{m-1}), \alpha_{m-1})$$

for $m = 1, 2, \dots, p-1$ and

$$S_{p,k} = F_k(x_k(\alpha_{p-1}), s) - F_k(x_k(\alpha_{p-1}), \alpha_{p-1}),$$

and also

$$\| \int_\alpha^s DF_0(x(\tau), t) - \sum_{m=1}^p S_{m,0} \| \leq 2\eta(h_0(s) - h_0(\alpha))$$

where

$$S_{m,0} = F_0(x(\tau_m), \alpha_m) - F_0(x(\tau_m), \alpha_{m-1}) =$$

$$= F_0(x(\alpha_{m-1}), \alpha_m) - F_0(x(\alpha_{m-1}), \alpha_{m-1})$$

for $m = 1, 2, \dots, p-1$ and

$$S_{p,0} = F_0(x(\alpha_{p-1}), s) - F_0(x(\alpha_{p-1}), \alpha_{p-1}).$$

For $k > k_1$ this yields

$$\| \int_\alpha^s DF_k(x_k(\tau), t) - \int_\alpha^s DF_0(x(\tau), t) \| \leq$$

$$\leq \sum_{m=1}^p \| S_{m,k} - S_{m,0} \| + 2\eta(h_k(s) - h_k(\alpha) + h_0(s) - h_0(\alpha)) =$$

$$= \sum_{m=1}^{p-1} \| F_k(x_k(\alpha_{m-1}), \alpha_m) - F_k(x_k(\alpha_{m-1}), \alpha_{m-1}) -$$

$$- F_0(x(\alpha_{m-1}), \alpha_m) + F_0(x(\alpha_{m-1}), \alpha_{m-1}) \| +$$

$$+ \| F_k(x_k(\alpha_{p-1}), s) - F_k(x_k(\alpha_{p-1}), \alpha_{p-1}) -$$

$$-F_0(x_k(\alpha_{p-1}), s) + F_k(x_k(\alpha_{p-1}), \alpha_{p-1})\|+$$

$$+2\eta(h_k(s) - h_k(\alpha) + h_0(s) - h_0(\alpha)). \tag{8.41}$$

By the properties of the class $\mathcal{F}(G, h_k, \omega)$ and by (8.40) we obtain for $k > k_1$ and $m = 1, 2, \dots, p-1$

$$\|F_k(x_k(\alpha_{m-1}), \alpha_m) - F_k(x_k(\alpha_{m-1}), \alpha_{m-1})-$$

$$-F_0(x(\alpha_{m-1}), \alpha_m) + F_0(x(\alpha_{m-1}), \alpha_{m-1})\| \leq$$

$$\leq \|F_k(x_k(\alpha_{m-1}), \alpha_m) - F_k(x_k(\alpha_{m-1}), \alpha_{m-1})-$$

$$-F_k(x(\alpha_{m-1}), \alpha_m) + F_k(x(\alpha_{m-1}), \alpha_{m-1})\|+$$

$$+\|F_k(x(\alpha_{m-1}), \alpha_m) - F_0(x(\alpha_{m-1}), \alpha_m)\|+$$

$$+\|F_k(x(\alpha_{m-1}), \alpha_{m-1}) + F_0(x(\alpha_{m-1}), \alpha_{m-1})\| \leq$$

$$\leq \omega(\|x_k(\alpha_{m-1}) - x(\alpha_{m-1})\|)(h_k(\alpha_m) - h_k(\alpha_{m-1}))+$$

$$+\|F_k(x(\alpha_{m-1}), \alpha_m) - F_0(x(\alpha_{m-1}), \alpha_m)\|+$$

$$+\|F_k(x(\alpha_{m-1}), \alpha_{m-1}) + F_0(x(\alpha_{m-1}), \alpha_{m-1})\| \leq$$

$$\leq \eta(h_k(\alpha_m) - h_k(\alpha_{m-1}))+$$

$$+\|F_k(x(\alpha_{m-1}), \alpha_m) - F_0(x(\alpha_{m-1}), \alpha_m)\|+$$

$$+\|F_k(x(\alpha_{m-1}), \alpha_{m-1}) + F_0(x(\alpha_{m-1}), \alpha_{m-1})\| \tag{8.42}$$

and similarly also

$$\|F_k(x_k(\alpha_{p-1}), s) - F_k(x_k(\alpha_{p-1}), \alpha_{p-1})-$$

$$-F_0(x(\alpha_{p-1}), s) + F_0(x(\alpha_{p-1}), \alpha_{p-1})\| \leq$$

$$\leq \omega(\|x_k(\alpha_{p-1}) - x(\alpha_{p-1})\|)(h_k(\alpha_p) - h_k(\alpha_{p-1}))+$$

$$+\|F_k(x(\alpha_{p-1}), \alpha_p) - F_0(x(\alpha_{p-1}), \alpha_p)\|+$$

$$+\|F_k(x(\alpha_{p-1}),\alpha_{p-1}) + F_0(x(\alpha_{p-1}),\alpha_{p-1})\| +$$

$$+h_k(\alpha_p) - h_k(s) + h_0(\alpha_p) - h_0(s) \le$$

$$\le \eta(h_k(\alpha_p) - h_k(\alpha_{p-1})) +$$

$$+\|F_k(x(\alpha_{p-1}),\alpha_p) - F_0(x(\alpha_{p-1}),\alpha_p)\| +$$

$$+\|F_k(x(\alpha_{p-1}),\alpha_{p-1}) + F_0(x(\alpha_{p-1}),\alpha_{p-1})\| +$$

$$+h_k(\alpha_p) - h_k(s) + h_0(\alpha_p) - h_0(s). \tag{8.43}$$

Hence (8.38), (8.39), (8.41), (8.42) and (8.42) yield

$$\|\int_\alpha^s DF_k(x_k(\tau),t) - \int_\alpha^s DF_0(x(\tau),t)\| \le$$

$$\le 2\eta(h_k(s) - h_k(\alpha) + h_0(s) - h_0(\alpha)) +$$

$$+\eta\sum_{m=1}^{p}(h_k(\alpha_m) - h_k(\alpha_{m-1})) +$$

$$+\sum_{m=1}^{p}[\|F_k(x(\alpha_{m-1}),\alpha_m) - F_0(x(\alpha_{m-1}),\alpha_m)\| +$$

$$+\|F_k(x(\alpha_{m-1}),\alpha_{m-1}) + F_0(x(\alpha_{m-1}),\alpha_{m-1})\|] +$$

$$+h_k(\alpha_p) - h_k(\alpha_{p-1}) + h_0(\alpha_p) - h_0(\alpha_{p-1}) \le$$

$$\le 2\eta[1 + h_0(\beta) - h_0(\alpha)] + \eta[h_k(\beta) - h_k(\alpha)] +$$

$$+\sum_{m=1}^{l}[\|F_k(x(\alpha_{m-1}),\alpha_m) - F_0(x(\alpha_{m-1}),\alpha_m)\| +$$

$$+\|F_k(x(\alpha_{m-1}),\alpha_{m-1}) + F_0(x(\alpha_{m-1}),\alpha_{m-1})\|]. \tag{8.44}$$

By the assumption (8.7), for every $\mu > 0$ there is a $k_2 \in \mathbb{N}$, $k_2 > k_1$ such that

$$\sum_{m=1}^{l} [\|F_k(x(\alpha_{m-1}), \alpha_m) - F_0(x(\alpha_{m-1}), \alpha_m)\| +$$

$$+\|F_k(x(\alpha_{m-1}), \alpha_{m-1}) + F_0(x(\alpha_{m-1}), \alpha_{m-1})\|] \leq \frac{\mu}{4}$$

for $k > k_2$ and by (8.32) there is a $k_3 \in \mathbb{N}$, $k_3 > k_2$ such that for $k \in \mathbb{N}$, $k > k_3$ we have

$$h_k(\beta) - h_k(\alpha) \leq h_0(\beta) - h_0(\alpha) + 1.$$

Since $\eta > 0$ can be arbitrary, we choose it so that

$$\eta < \frac{\mu}{20(1 + h_0(\beta) - h_0(\alpha))}.$$

By the inequality (8.44) we then obtain for every $s \in [\alpha, \beta]$

$$\| \int_\alpha^s DF_k(x_k(\tau), t) - \int_\alpha^s DF_0(x(\tau), t)\| \leq$$

$$\leq \frac{\mu}{4} + \frac{5\mu(1 + h_0(\beta) - h_0(\alpha))}{20(1 + h_0(\beta) - h_0(\alpha))} = \frac{\mu}{2}.$$

Hence by (8.36) we have

$$\|x_k(s) - x(s)\| \leq \|y_k - x(\alpha)\| + \frac{\mu}{2}$$

for every $s \in [\alpha, \beta]$ and, finally, we take $k_\star \in \mathbb{N}$ such that $k_\star > k_3$ and $\|y_k - x(\alpha)\| \leq \frac{\mu}{2}$ for $k > k_\star$. Then we obtain

$$\|x_k(s) - x(s)\| < \mu$$

for every $s \in [\alpha, \beta]$, $k > k_\star$ and the theorem is proved. □

Applications to classical ODE's

Now we will use the previous results for deriving continuous dependence results for classical ordinary differential equations. The link between generalized differential equations and the classical Carathéodory concept is given by Theorem 5.14.

Let P be a metric space, $p_0 \in P$ na accumulation point of P.

Let $\overline{B}$ be the closure of the set B.

Assume that a function $f : \overline{B} \times [a,b] \times P \to \mathbb{R}^n$ satisfies the following conditions:

$f(x,\cdot,p)$ is Lebesgue measurable on $[a,b]$ for $(x,p) \in \overline{B} \times P$,

there exists a Lebesgue measurable function $m : [a,b] \to \mathbb{R}$ such that $\int_a^b m(s)\,ds < +\infty$ and

$$\|f(s,x\| \leq m(s)$$

for $(x,s,p) \in \overline{B} \times [a,b] \times P$,
there exists a Lebesgue measurable function $l : [a,b] \to \mathbb{R}$ such that $\int_a^b l(s)\,ds < +\infty$ and

$$\|f(s,x,p) - f(s,y,p)\| \leq l(s)\omega(\|x-y\|)$$

for $(x,s,p),(y,s,p) \in \overline{B} \times [a,b] \times P$.

Let us define

$$F(x,t,p) = \int_c^t f(x,s,p)\ ds \tag{8.45}$$

for $(x,t,p) \in \overline{B} \times [a,b] \times P$.

8.9 Theorem. *Assume that for some $c \in [a,b]$ we have*

$$\lim_{p \to p_0} \int_c^t f(x,s,p)\, ds = \int_c^t f(x,s,p_0)\, ds \tag{8.46}$$

for $(x,t) \in \overline{B} \times [a,b]$.

Let $x(t,p) : [\alpha,\beta] \times P \to \mathbb{R}^n$, $p \neq p_0$ be solutions of

$$\dot{x} = f(x,t,p) \tag{8.47}$$

on $[\alpha,\beta] \subset [a,b]$ such that

$$\lim_{p \to p_0} x(t,p) = y(t) \quad \text{for } t \in [\alpha,\beta], \tag{8.48}$$

and $y(t) \in \overline{B}$ for $t \in [\alpha,\beta]$.

Then $y : [\alpha,\beta] \to \mathbb{R}^n$ is a solution of

$$\dot{x} = f(x,t,p_0) \tag{8.49}$$

on $[\alpha,\beta]$.

Proof. By the assumptions made above, Proposition 5.5 yields that the function F given by (8.45) belongs to $\mathcal{F}(\overline{B} \times [a,b], h, \omega)$ for all $p \in P$ where

$$h(t) = \int_c^t m(s)\, ds + \int_c^t l(s)\, ds, \quad t \in [a,b].$$

The relation (8.46) can be written in the form

$$\lim_{p \to p_0} F(x,t,p) = F(x,t,p_0)$$

when (8.45) is taken into account. By Theorem 5.14 the equation $\dot{x} = f(x,t,p)$ has the same set of solutions as the generalized ordinary differential equation

$$\frac{dx}{d\tau} = DF(x,t,p) \tag{8.50}$$

for all $p \in P$. Consequently, using (8.48) and Theorem 8.2 we obtain that the function $y : [\alpha, \beta] \to \mathbb{R}^n$ is a solution of the generalized differential equation

$$\frac{dx}{d\tau} = DF(x, t, p_0).$$

Therefore again by Theorem 5.14, y is a solution of (8.49) on $[\alpha, \beta]$ and this proves the theorem. □

8.10 Theorem. *Assume that for some $c \in [a, b]$ we have*

$$\lim_{p \to p_0} \int_c^t f(x, s, p)\, ds = \int_c^t f(x, s, p_0)\, ds \tag{8.46}$$

for $(x, t) \in \overline{B} \times [a, b]$.

Let $x : [\alpha, \beta] \to \mathbb{R}^n$, $[\alpha, \beta] \subset [a, b]$ be a solution of the differential equation (8.49) on $[\alpha, \beta]$ which has the following uniqueness property:

if $y : [\alpha, \gamma] \to \mathbb{R}^n$, $[\alpha, \gamma] \subset [\alpha, \beta]$ is a solution of (8.10) such that $y(\alpha) = x(\alpha)$ then $y(t) = x(t)$ for every $t \in [\alpha, \gamma]$.

Assume further that there is a $\varrho > 0$ such that if $s \in [\alpha, \beta]$ and $\|y - x(s)\| < \varrho$ then $(y, s) \in G = B_c \times [\alpha, \beta]$, and let $y_p \in \mathbb{R}^n$, $p \in P$ satisfy

$$\lim_{p \to p_0} y_p = x(\alpha).$$

Then for every $\mu > 0$ there exists an $\eta_\star > 0$ such that for $dist(p, p_0) < \eta_\star$ (by $dist$ the metric in P is denoted) there exists a solution $x(t, p)$ of the differential equation (8.47) on $[\alpha, \beta]$ with $x(\alpha, p) = y_p$ and

$$\|x(s, p) - x(s)\| < \mu, \quad s \in [\alpha, \beta]$$

holds.

Proof. It is easy to check that the function F given by (8.45) satisfies the assumptions of Theorem 8.8 with the parameter $p \in$

P instead of the sequential setting with $k \in \mathbb{N}$ in Theorem 8.8. Theorem 8.8 together with Theorem 5.14 yields the statement of our theorem. We are not going into details here. Looking at the proof of Theorem 8.8 it can be easily seen that it can be repeated for the case of the parameter $p \in P$ with with minor technical changes. □

8.11 *Remark.* Both the previous Theorems 8.9 and 8.10 are corollaries of continuous dependence results for generalized ordinary differential equations. They represent continuous dependence theorems for classical ordinary differential equations under the relatively weak "integral continuity" assumption represented by (8.46). In this context let us mention the results of M.A. Krasnoselskij and S.G. Krejn [66], J. Kurzweil and Z. Vorel [22] as well as of Z. Artstein [1], [2], [3], [4].

Averaging for generalized ordinary differential equations

Now we derive a special result for generalized ordinary differential equations with a small parameter which is analogous to the known results substantiating the method of averaging for classical ordinary differential equations.

8.12 Theorem. *Assume that* $G = B \times [0, +\infty)$, $B = \{x \in \mathbb{R}^n;\ \|x\| < c\}$, $c > 0$ *and that* $F \in \mathcal{F}(G, h, \omega)$ *where* $h : [0, +\infty) \to \mathbb{R}$ *is continuous from the left, nondecreasing, and* $\omega : [0, +\infty) \to [0, +\infty)$ *is continuous, increasing and* $\omega(0) = 0$.

Assume that

$$\limsup_{r \to \infty} \frac{h(r+\alpha) - h(\alpha)}{r} \leq C,\ C = const \tag{8.51}$$

for every $\alpha \geq 0$ *and*

$$\lim_{r \to \infty} \frac{F(x,r)}{r} = F_0(x),\ x \in B. \tag{8.52}$$

Let $y : [0, +\infty) \to \mathbb{R}^n$ be a uniquely determined solution of the autonomous ordinary differential equation

$$\dot{y} = F_0(y) \tag{8.53}$$

which belongs to B together with its ρ–neighbourhood with $\rho > 0$, i.e. there is a $\rho > 0$ such that $\{x \in \mathbb{R}^n; \|x - y(t)\| < \rho\} \subset B$ for every $t \in [0, +\infty)$.

Then for every $\mu > 0$ and $L > 0$ there is an $\varepsilon_0 > 0$ such that for $\varepsilon \in (0, \varepsilon_0)$ the inequality

$$\|x_\varepsilon(t) - \xi_\varepsilon(t)\| < \mu$$

holds for $t \in [0, \frac{L}{\varepsilon}]$ where x_ε is a solution of the generalized ordinary differential equation

$$\frac{dx}{d\tau} = D[\varepsilon F(x, t)] \tag{8.54}$$

on $[0, \frac{L}{\varepsilon}]$ such that $x_\varepsilon(0) = y(0)$, and ξ_ε is a solution of the autonomous ordinary differential equation

$$\dot{x} = \varepsilon F_0(x) \tag{8.55}$$

on $[0, \frac{L}{\varepsilon}]$ such that $\xi_\varepsilon(0) = y(0)$.

Proof. For $y \in B,\ t \in [0, +\infty)$ and $\varepsilon > 0$ define

$$G_\varepsilon(y, t) = \varepsilon F(y, \frac{t}{\varepsilon})$$

and take $h_\varepsilon(t) = \varepsilon h(\frac{t}{\varepsilon})$ for $t \geq 0$. The function h_ε is evidently nondecreasing and continuous from the left on $[0, +\infty)$.

Since $F \in \mathcal{F}(G, h, \omega)$ we obtain by definition

$$\|G_\varepsilon(y, t_2) - G_\varepsilon(y, t_1)\| = \varepsilon \|F(y, \frac{t_2}{\varepsilon}) - F(y, \frac{t_1}{\varepsilon})\| \leq$$

$$\leq \varepsilon |h(\frac{t_2}{\varepsilon}) - h(\frac{t_1}{\varepsilon})| = |h_\varepsilon(t_2) - h_\varepsilon(t_1)|,$$

and similarly also

$$\|G_\varepsilon(y, t_2) - G_\varepsilon(y, t_1) - G_\varepsilon(x, t_2) + G_\varepsilon(x, t_1)\| \leq$$

$$\leq \omega(\|y - x\|)|h_\varepsilon(t_2) - h_\varepsilon(t_1)|$$

whenever $x, y \in B$, $t_1, t_2 \in [0, +\infty)$. These inequalities mean that we have $G_\varepsilon \in \mathcal{F}(G, h_\varepsilon, \omega)$ for $\varepsilon > 0$.

If $y \in B$, then (8.48) yields

$$\lim_{r \to \infty} \frac{F(y, r) - F(y, 0)}{r} = \lim_{r \to \infty} \frac{F(y, r)}{r} = F_0(y)$$

and therefore (8.51) and (8.52) imply that for every $\eta > 0$ there is an $R > 0$ such that for $r > R$ we have

$$\|F_0(y)\| \leq \|F_0(y) - \frac{F(y, r) - F(y, 0)}{r}\| +$$

$$+ \frac{\|F(y, r) - F(y, 0)\|}{r} \leq \eta + \frac{h(r) - h(0)}{r} < 2\eta + C$$

because $F \in \mathcal{F}(G, h, \omega)$ implies $\|F(y, r) - F(y, 0)\| \leq h(r) - h(0)$. Since $\eta > 0$ can be chosen arbitrarily small, we have

$$\|F_0(y)\| \leq C, \quad y \in B. \tag{8.56}$$

Analogously, if $x, y \in B$ then for every $\eta > 0$ there is an $R > 0$ such that for $r > R$ we have

$$\|F_0(x) - F_0(y)\| < \eta + \frac{\|F(y, r) - F(y, 0) - F(x, r) + F(x, 0)\|}{r} \leq$$

$$\leq \eta + \omega(\|y-x\|)\frac{h(r)-h(0)}{r} \leq \eta(1+\omega(\|y-x\|)) + C\omega(\|y-x\|),$$

and again since $\eta > 0$ can be arbitrarily small, we obtain

$$\|F_0(x) - F_0(y)\| \leq C\omega(\|y-x\|) \tag{8.57}$$

provided $x, y \in B$.

For $y \in B,\ t > 0$ we obtain

$$\lim_{\varepsilon \to 0+} G_\varepsilon(y,t) = \lim_{\varepsilon \to 0+} \varepsilon F(y, \frac{t}{\varepsilon}) =$$

$$= \lim_{\varepsilon \to 0+} t\frac{\varepsilon}{t}F(y, \frac{t}{\varepsilon}) = tF_0(y)$$

and also

$$\lim_{\varepsilon \to 0+} G_\varepsilon(y,0) = \lim_{\varepsilon \to 0+} \varepsilon F(0, \frac{t}{\varepsilon}) = 0.$$

Denote $G_0(y,t) = tF_0(y)$ for $y \in B,\ t \geq 0$. Then the relations given above imply

$$\lim_{\varepsilon \to 0+} G_\varepsilon(y,t) = G_0(y,t). \tag{8.58}$$

By (8.56) and (8.57) we have $G_0 \in \mathcal{F}(G,h,\omega)$ where $h_0(t) = Ct,\ t \geq 0$.

Further, for $0 \leq t_1 < t_2 < +\infty$ we obtain by definition

$$h_\varepsilon(t_2) - h_\varepsilon(t_1) = \varepsilon(h(\frac{t_2}{\varepsilon}) - h(\frac{t_1}{\varepsilon})) =$$

$$= (t_2 - t_1)\frac{\varepsilon}{t_2 - t_1}(h(\frac{t_2 - t_1}{\varepsilon} + \frac{t_1}{\varepsilon}) - h(\frac{t_1}{\varepsilon}))$$

and the assumption (8.51) yields

$$\limsup_{\varepsilon \to 0+}(h_\varepsilon(t_2) - h_\varepsilon(t_1)) \leq C(t_2 - t_1) = h_0(t_2) - h_0(t_1) \tag{8.59}$$

because we have

$$\lim_{\varepsilon \to 0+} \frac{t_2 - t_1}{\varepsilon} = +\infty.$$

It is easy to se that (8.59) is satisfied in the case $t_1 = t_2$ as well.

Using the fact that $y : [0, +\infty) \to B$ is a solution of (8.53) we obtain by the properties of the generalized Perron integral the equality

$$y(s_2) - y(s_1) = \int_{s_1}^{s_2} F_0(y(\tau))\, d\tau =$$

$$= \int_{s_1}^{s_2} D[F_0(y(\tau))t] = \int_{s_1}^{s_2} DG_0(y(\tau), t)$$

for $s_1, s_2 \in [0, +\infty)$, i.e. y is a solution of the generalized ordinary differential equation

$$\frac{dy}{d\tau} = G_0(y, t) \tag{8.60}$$

on $[0, +\infty)$, and by the assumption this solution is uniquely determined. In this way we have shown that all assumptions of Theorem 8.8 are satisfied for the case of the continuous parameter $\varepsilon \to 0+$. Using the result of Theorem 8.8 we obtain that for every $\mu > 0$ and $L > 0$ there exists a value $\varepsilon_0 > 0$ such that for $\varepsilon \in (0, \varepsilon_0)$ there is a solution y_ε of the generalized differential equation

$$\frac{dy}{d\tau} = DG_\varepsilon(y, t) \tag{8.61}$$

on the interval $[0, L]$ such that $y_\varepsilon(0) = y(0)$ and

$$\|y_\varepsilon(s) - y(s)\| \le \mu \tag{8.62}$$

for all $s \in [0, L]$.

For the solution $y_\varepsilon : [0, L] \to B$ of (8.61) we have

$$y_\varepsilon(s_2) - y_\varepsilon(s_1) = \int_{s_1}^{s_2} DG_\varepsilon(y_\varepsilon(\tau), t) = \varepsilon \int_{s_1}^{s_2} DF(y_\varepsilon(\tau), \frac{t}{\varepsilon})$$

whenever $s_1, s_2 \in [0, L]$. For $t \in [0, \frac{L}{\varepsilon}]$ denote $x_\varepsilon(t) = y_\varepsilon(\varepsilon t)$. Then

$$x_\varepsilon(t_2) - x_\varepsilon(t_1) = y_\varepsilon(\varepsilon t_2) - y_\varepsilon(\varepsilon t_1) =$$

$$= \varepsilon \int_{\varepsilon t_1}^{\varepsilon t_2} DF(y_\varepsilon(\sigma), \frac{s}{\varepsilon}) = \varepsilon \int_{\varepsilon t_1}^{\varepsilon t_2} DF(x_\varepsilon(\frac{\sigma}{\varepsilon}), \frac{s}{\varepsilon})$$

for $t_1, t_2 \in [0, \frac{L}{\varepsilon}]$. Applying the Substitution Theorem 1.18 with the continuous monotone substitution $\varphi(\sigma) = \frac{\sigma}{\varepsilon}$ we obtain

$$\int_{\varepsilon t_1}^{\varepsilon t_2} DF(x_\varepsilon(\frac{\sigma}{\varepsilon}), \frac{s}{\varepsilon}) = \int_{\varphi(\varepsilon t_1)}^{\varphi(\varepsilon t_2)} DF(x_\varepsilon(\tau), t) = \int_{t_1}^{t_2} DF(x_\varepsilon(\tau), t)$$

for any $t_1, t_2 \in [0, \frac{L}{\varepsilon}]$. This together with the previous equality yields

$$x_\varepsilon(t_2) - x_\varepsilon(t_1) = \varepsilon \int_{t_1}^{t_2} DF(x_\varepsilon(\tau), t)$$

for $t_1, t_2 \in [0, \frac{L}{\varepsilon}]$ and $x_\varepsilon(0) = y_\varepsilon(0) = y(0)$. Therefore the function $x_\varepsilon : [0, \frac{L}{\varepsilon}] \to B$ is a solution of the generalized differential equation (8.54) on $[0, \frac{L}{\varepsilon}]$.

Analogously it can be shown that the function $\xi_\varepsilon : [0, \frac{L}{\varepsilon}] \to B$ given by $\xi_\varepsilon(t) = y(\varepsilon t)$ is a solution of the autonomous ordinary differential equation (8.55) on $[0, \frac{L}{\varepsilon}]$.

Finally, by (8.62) we obtain

$$\|x_\varepsilon(t) - \xi_\varepsilon(t)\| = \|y_\varepsilon(\varepsilon t) - y(\varepsilon t)\| < \mu$$

for every $t \in [0, \frac{L}{\varepsilon}]$ and the theorem is proved. □

8.13 *Remark.* Theorem 8.12 forms an analogue of the known result of N.N. Bogoljubov on the method of averaging. The classical autonomous differential equation (8.55) is the averaged equation for (8.54), where the process of averaging is described by the relation (8.52).

As an application of Theorem 8.12 we give a result for averaging in the case of differential equations with impulses. We use the equivalence of such type of equations with generalized ordinary differential equations as they are described in Chap. V. Results on averaging of differential equations with impulses were given for the first time by A.M. Samojlenko (see especially the monograph [120]).

8.14 Theorem. *Assume that $G = B \times [0, +\infty)$, $B = \{x \in \mathbb{R}^n;\ \|x\| < c\}$, $c > 0$ and $K \geq 0$. Let $f : G \to \mathbb{R}^n$ be such a function that*

$$\|f(x,s)\| \leq K,$$

and

$$\|f(x,s) - f(y,s)\| \leq K\|x - y\|$$

for $x, y \in B$, $s \in [0, +\infty)$. Further, assume that a sequence of points $0 \leq t_1 < t_2 < \cdots$ is given such that

$$\limsup_{r\to\infty} \frac{1}{r} \sum_{\alpha \leq t_i \leq \alpha + r} 1 \leq d$$

for every $\alpha \geq 0$ and that $I_i : B \to \mathbb{R}^n$ $i = 1, 2, \cdots$ is a sequence of functions such that

$$\|I_i(x)\| \leq K,$$

and

$$\|I_i(x) - I_i(y)\| \leq K\|x - y\|$$

for $x, y \subset B$, $i = 1, 2, \cdots$. *Suppose that*

$$\lim_{r\to\infty} \frac{1}{r} \int_0^r f(x,s)\,ds = f_0(x), \;\; x \in B$$

and

$$\lim_{r\to\infty} \frac{1}{r} \sum_{0\le t_i<r} I_i(x) = I_0(x), \;\; x \in B.$$

Let $y : [0, +\infty) \to \mathbb{R}^n$ *be a uniquely determined solution of the autonomous differential equation*

$$\dot{y} = f_0(y) + I_0(y)$$

which belongs to B *together with its* ρ*–neighbourhood with* $\rho > 0$.

Then for every $\mu > 0$ *and* $L > 0$ *there is an* $\varepsilon_0 > 0$ *such that for* $\varepsilon \in (0, \varepsilon_0)$ *the inequality*

$$\|x_\varepsilon(t) - \xi_\varepsilon(t)\| < \mu$$

holds for $t \in [0, \frac{L}{\varepsilon}]$ *where* x_ε *is a solution of the differential equation with impulses*

$$\dot{x} = \varepsilon f(x,t), \;\; t \neq t_i \tag{8.63}$$

$$\Delta x|_{t_i} = x(t_i+) - x(t_i) = \varepsilon I_i(x(t_i)), \;\; i = 1, 2, \cdots$$

on $[0, \frac{L}{\varepsilon}]$ *such that* $x_\varepsilon(0) = y(0)$, *and* ξ_ε *is a solution of the "averaged" system of the autonomous ordinary differential equation*

$$\dot{x} = \varepsilon[f_0(x) + I_0(x)] \tag{8.64}$$

on $[0, \frac{L}{\varepsilon}]$ *such that* $\xi_\varepsilon(0) = y(0)$.

Proof. By Theorem 5.20 the system (8.63) is equivalent to

$$\frac{dx}{d\tau} = D\varepsilon F(x,t)$$

where

$$F(x,t) = \int_0^t f(x,s)\,ds + \sum_{i=1}^{\infty} H_{t_i}(t)$$

and $H_\nu(t) = 0$ for $t \in [0,\nu]$, $H_\nu(t) = 0$ for $t > \nu$.By the assumptions it follows that the function F belongs to $\mathcal{F}(G,h,\omega)$ where

$$h(t) = Kt + K\sum_{i=1}^{\infty} H_{t_i}(t)$$

for $t \geq 0$ is evidently continuous from the left and nondecreasing by the properties of the function H_ν and $\omega(r) = r$ for $r \geq 0$.

The assumptions further yield

$$\lim_{r\to\infty} \frac{1}{r} F(x,r) = f_0(x) + I_0(x) = F_0(x)$$

and

$$\limsup_{r\to\infty} \frac{h(r+\alpha) - h(\alpha)}{r} \leq K(1+d)$$

for $x \in B$ and $\alpha \in [0,+\infty)$. All assumptions of Theorem 8.12 being satisfied we obtain the desired result immediately when using the above mentioned equivalence of the system with impulses (8.63) and the corresponding generalized differential equation. □

CHAPTER IX

EMPHATIC CONVERGENCE FOR GENERALIZED ORDINARY DIFFERENTIAL EQUATIONS

Let us consider in this chapter the convergence effect which occurs in the theory of ordinary differential equations when the right hand sides of the equations converge to the Dirac delta function.

We consider these phenomena in the framework of generalized ordinary differential equations and the results presented below are in fact a continuation of the previous Chap. VIII on continuous dependence on a parameter.

Let us start with a simple example which shows the convergence effect mentioned above.

9.1 *Example.* Let $\delta_k : [-1,1] \to \mathbb{R}$, $k = 1,2,\dots$ be the δ–sequence given by

$$\delta_k(t) = k \quad \text{for } t \in (0, \frac{1}{k}],$$

$$\delta_k(t) = 0 \quad \text{for } t \in [-1,1] \setminus (0, \frac{1}{k}].$$

Assume that $A, B \in L(\mathbb{R}^n)$ are given constant $n \times n$– matrices. Let us consider the sequence of linear ordinary differential equations

$$\dot{x} = [A + \delta_k(t)B]x \tag{9.1}$$

where $x \in \mathbb{R}^n$, $t \in [-1,1]$ and the initial condition $x(-1) = \widetilde{x} \in \mathbb{R}^n$ is given. It is easy to check that if for a given $k \in \mathbb{N}$ we define the function $x_k : [-1,1] \to \mathbb{R}^n$ by the relations

$$x_k(t) = e^{A(t+1)}\widetilde{x} \quad \text{for } t \in [-1,0],$$

$$x_k(t) = e^{(A+kB)t}x_k(0) = e^{(A+kB)t}e^{A}\widetilde{x} \quad \text{for } t \in (0,\frac{1}{k}],$$

$$x_k(t) = e^{A(t-\frac{1}{k})}x_k(\frac{1}{k}) = e^{At}e^{B}e^{A}\widetilde{x} \quad \text{for } t \in (\frac{1}{k},1] \tag{9.2}$$

then $x_k : [-1,1] \to \mathbb{R}^n$, $k = 1,2,\dots$ is a solution of (9.1) with the given initial condition $x(-1) = \widetilde{x} \in \mathbb{R}^n$. Passing to the limit $k \to \infty$ we easily get

$$\lim_{k\to\infty} x_k(t) = e^{A(t+1)}\widetilde{x} = e^{At}e^{A}\widetilde{x} \quad \text{for } t \in [-1,0]$$

and

$$\lim_{k\to\infty} x_k(t) = e^{At}e^{B}e^{A}\widetilde{x} \quad \text{for } t \in (0,1].$$

Denoting

$$x(t) = e^{At}e^{A}\widetilde{x} \quad \text{for } t \in [-1,0],$$
$$x(t) = e^{At}e^{B}e^{A}\widetilde{x} \quad \text{for } t \in (0,1] \tag{9.3}$$

we have simply

$$\lim_{k\to\infty} x_k(t) = x(t) \quad \text{for } t \in [-1,1]$$

and we can ask wether the limit x is a solution of some diffferential equation.

The function $x : [-1,1] \to \mathbb{R}^n$ given by (9.3) is continuous on the intervals $[-1,0]$ and $(0,1]$ and can exhibit a discontinuity at the point $t = 0$ for which

$$x(0+) - x(0) = \lim_{t\to 0+} e^{At}e^{B}e^{A}\widetilde{x} - e^{A}\widetilde{x} =$$

$$= (e^B - I)e^A\widetilde{x} = (e^B - I)x(0),$$

I denotes the $n \times n$ identity matrix.

From this relation we can immediately see that x is continuous at $t = 0$ if either $x(0) = 0$ (equivalently $\widetilde{x} = 0$) or the matrix $e^B - I$ is singular and $x(0) = e^A\widetilde{x}$ belongs to its null space. Otherwise the function x has a discontinuity at $t = 0$ and therefore it cannot be a solution of a classical ordinary differential equation on the interval $[-1, 1]$. Nonetheless, the function $x : [-1, 1] \to \mathbb{R}^n$ given by (9.3) is a solution of a generalized ordinary differential equation which can be easily constructed using the facts given in Chap. V.

It can be easily observed that the function $x : [-1, 1] \to \mathbb{R}^n$ given by (9.3) is a solution of the linear ordinary differential equation

$$\dot{x} = Ax$$

on the intervals $[-1, 0]$ and $(0, 1]$ with

$$x(0+) = x(0) + (e^B - I)x(0).$$

This can be treated as a differential system with an impulse acting at $t = 0$ as in Chap. V. By the results given in Chap. V this system is equivalent to the generalized ordinary differential equation

$$\frac{dx}{d\tau} = D[At + (e^B - I)H(t)]x \tag{9.4}$$

where $H(t) = 0$ for $t \leq 0$ and $H(t) = 1$ for $t > 0$. Using the result stated in Theorem 5.20 we obtain that the limit x of the sequence (x_k) of solutions of the linear ordinary differential equations (9.1) is a solution of the generalized ordinary differential equation (9.4).

The ordinary differential equation (9.1) is for every $k \in \mathbb{N}$ equivalent to a generalized ordinary differential equation by virtue of the results given in Theorem 5.14. For the construction of this generalized ordinary differential equation let us define

$$F_k(x,t) = \int_{-1}^{t} [A + \delta_k(s)B]x\, ds = [At + H_k(t)B]x$$

where $H_k(t) = \int_{-1}^{t} \delta_k(s)\, ds$ and by definition we have

$$H_k(t) = 0 \quad \text{for } t \in [-1,0],$$

$$H_k(t) = \int_0^t \delta_k(s)\, ds = \int_0^t k\, ds = kt \quad \text{for } t \in (0, \frac{1}{k}],$$

$$H_k(t) = 1 \quad \text{for } t \in (\frac{1}{k}, 0].$$

By Theorem 5.14 the solutions $x_k : [-1,1] \to \mathbb{R}^n$, $k = 1,2,\dots$ of (9.1) on the interval $[-1,1]$ are also solutions of the generalized ordinary differential equations

$$\frac{dx}{d\tau} = D[At + H_k(t)B]x \tag{9.5}$$

on the interval $[-1,1]$ for every $k \in \mathbb{N}$.

In (9.5) a sequance of generalized ordinary differential equations is given and the righjt hand sides of this sequence of equations obviously satisfy

$$\lim_{k\to\infty} [At + H_k(t)B]x = [At + H(t)B]x. \tag{9.6}$$

In spite of this fact, having in mind the convergence results for generalized ordinary differential equations desribed in Chap. VIII one would expect that the solutions $x_k : [-1,1] \to \mathbb{R}^n$ of the initial value problem

$$\frac{dx}{d\tau} = D[At + H_k(t)B]x, \quad x(-1) = \widetilde{x} \tag{9.7}$$

should converge to a solution of the "limit" problem

$$\frac{dx}{d\tau} = D[At + H(t)B]x, \quad x(-1) = \widetilde{x} \tag{9.8}$$

on the interval [-1,1].

It is easy to check that the solution of the initial value problem (9.8) is the function $x : [-1,1] \to \mathbb{R}^n$ given by

$$x(t) = e^{A(t+1)}\widetilde{x} \quad \text{for } t \in [-1,0],$$

$$x(t) = e^{At}(I + B)e^{A}\widetilde{x} \quad \text{for } t \in (0,1].$$

Comparing this with the actual limit of the sequence of solutions of (9.7) which is given by the relations (9.3), we can see that they are different unless the exceptional case $I + B = e^B$ occurs.

The discrepancy is caused by the discontinuity of the limit function $[At + H(t)B]x$ of the right hand sides $[At + H_k(t)B]x$ of the generalized ordinary differential equations (9.5) at the point $t = 0$. Nevertheless, the structure of the equations (9.4) and (9.8) is similar.

The true and effective limit equation to the sequence of equations (9.5) is the generalized ordinary differential equation (9.4) and the convergence theorems 8.2 and 8.5 given in Chap. VIII cannot be applied to the case described in this example.

The discontinuity of the solution of the limit equation (9.4) is in the example given by the limit

$$\lim_{k\to\infty} [x_k(\frac{1}{k}) - x_k(0)].$$

Since we are treating linear equations in the example, this limit can be calculated exactly. In the general nonlinear case such a calculation cannot be carried out, nevertheless the idea works in this case, too.

We will use the concept of *emphatic convergence* which was introduced by J. Kurzweil in [69].

9.2 Definition. Assume that $G = B_c \times (a, b)$, $B_c = \{x \in \mathbb{R}^n; \|x\| < c\}$, $c > 0$ and $F_k : G \to \mathbb{R}^n$, $k = 1, 2, \dots$. The sequence of functions F_k, $k = 1, 2, \dots$ *converges emphatically to* F_0 for $k \to \infty$ if the following conditions are satisfied:

(i) there exist an increasing continuous function $\omega : [0, +\infty) \to [0, +\infty)$, $\omega(0) = 0$ and functions $h_k : [a, b] \to \mathbb{R}$, $k = 0, 1, 2, \dots$ which are nondecreasing and continuous from the left, such that

$$F_k \in \mathcal{F}(G, h_k, \omega), \quad k = 0, 1, 2, \dots,$$

(ii)

$$\limsup_{k \to \infty} (h_k(t_2) - h_k(t_1)) \le h_0(t_2) - h_0(t_1)$$

provided h_0 is continuous at the points t_1 and t_2, $a < t_1 \le t_2 < b$,

(iii)

$$\lim_{k \to \infty} F_k(x, t) = F_0(x, t) + F_*(x, t)$$

if $(x, t) \in G$, t is a point of continuity of the function h_0 and $F_* : G \to \mathbb{R}^n$ is such that

$$\|F_*(x, t_2) - F_*(x, t_1)\| \le |h_*(t_2) - h_*(t_1)|$$

for $t_1, t_2 \in (a, b)$ where $h_* : (a, b) \to \mathbb{R}^n$ is the break function corresponding to h_0,

(iv) $x + F_0(x, t+) - F_0(x, t) \in B_c$ for every $x \in B_c$, $t \in (a, b)$,

(v) if $h_0(t_0+) > h_0(t_0)$, $(x_0, t_0) \in G$ then for every $\varepsilon > 0$ there is a $\delta > 0$ such that for each $\delta' \in (0, \delta)$ there is a $k_0 \in \mathbb{N}$ with the following property:
if $y : [t_0 - \delta', t_0 + \delta'] \to \mathbb{R}^n$ is a solution of the generalized ordinary differential equation

$$\frac{dx}{d\tau} = DF_k(x, t)$$

on $[t_0 - \delta', t_0 + \delta']$, $k > k_0$ and $\|y(t_0 - \delta') - x_0\| \le \delta$ then

$$\|y(t_0 + \delta') - y(t_0 - \delta') - [F_0(x_0, t_0+) - F_0(x_0, t_0)]\| < \varepsilon.$$

9.3 Theorem. *Assume that $G = B_c \times (a, b)$. Let $h_k : (a, b) \to \mathbb{R}$, $k = 0, 1, 2, \ldots$ be nondecreasing functions continuous from the left. Let $d \in (a, b)$ be such that $h_0(t+) = h_(t)$ for $t \neq d$. Assume further that $F_k \in \mathcal{F}(G, h_k, \omega)$, $k = 0, 1, 2, \ldots$ and that the sequence (F_k) converges emphatically to F_0 for $k \to \infty$.*

Let $x_k : [\alpha, \beta] \to \mathbb{R}^n$ be solutions of

$$\frac{dx}{d\tau} = DF_k(x, t) \tag{9.9}$$

on an interval $[\alpha, \beta] \subset (a, b)$, $k = 1, 2, \ldots$ such that

$$\lim_{k\to\infty} x_k(t) = z(t) \tag{9.10}$$

for $t \in [\alpha, \beta]$, $t \neq d$ and $d \in (\alpha, \beta)$. Then the function $x : [\alpha, \beta] \to \mathbb{R}^n$ defined by $x(t) = z(t)$ for $t \in [\alpha, \beta]$, $t \neq d$ and $x(d) = \lim_{s \to d-} x(s)$ is a function of bounded variation on $[\alpha, \beta]$, and it is a solution of the generalized ordinary differential equation

$$\frac{dx}{d\tau} = DF_0(x, t) \tag{9.11}$$

on the interval $[\alpha, \beta]$.

Proof. By Lemma 3.10 we have

$$\|x_k(s_2) - x_k(s_1)\| \le |h_k(s_2) - h_k(s_1)|$$

for $s_1, s_2 \in [\alpha, \beta]$ and (ii) from Definition 9.2 gives for $k \to \infty$ the inequality

$$\|z(s_2) - z(s_1)\| \le |h_0(s_2) - h_0(s_1)|$$

for $s_1, s_2 \in [\alpha, \beta]$, $s_1, s_2 \neq d$. This yields the existence of the onesided limits $\lim_{s\to d-} z(s) = x(d)$ and $\lim_{s\to d+} z(s) = x(d+)$. Therefore $x : [\alpha, \beta] \to \mathbb{R}^n$ is of bounded variation on $[\alpha, \beta]$.

Assume that $\alpha \le s_1 < d < s_2 \le \beta$. By (iii) in Definition 9.2 and by Theorem 8.5 we obtain that for any $\Delta > 0$ the limit function x is a solution of the generalized ordinary differential equation

$$\frac{dx}{d\tau} = D[F_0(x,t) + F_*(x,t)]$$

on the intervals $[\alpha, d-\Delta]$ and $[d+\Delta, \beta]$. Therefore for any $\Delta > 0$ with $\Delta \le \min(s_2 - d, d - s_1)$ we have

$$x(d-\Delta) - x(s_1) = \int_{s_1}^{d-\Delta} D[F_0(x(\tau),t) + F_*(x(\tau),t)] =$$

$$= \int_{s_1}^{d-\Delta} DF_0(x(\tau),t) \tag{9.12}$$

and

$$x(s_2) - x(d+\Delta) = \int_{d+\Delta}^{s_2} D[F_0(x(\tau),t) + F_*(x(\tau),t)] =$$

$$= \int_{d+\Delta}^{s21} DF_0(x(\tau),t) \tag{9.13}$$

because evidently

$$\int_{s_1}^{d-\Delta} D[F_*(x(\tau),t) = \int_{d+\Delta}^{s21} D[F_*(x(\tau),t) = 0$$

by the properties of the function $F_*(x,t)$ given in Definition 9.2.

For a given $\varepsilon > 0$ there is a $\delta_1 > 0$ such that

$$|h_0(d) - h_0(d-\rho)| < \frac{\varepsilon}{2}, \ |h_0(d+\rho) - h_0(d+)| < \frac{\varepsilon}{2} \tag{9.14}$$

for every $\rho \in (0, \delta_1)$. Assume that $\delta \in (0, \delta_1)$ corresponds to ε by the requirement (v) from Definition 9.2. From the existence of the limit $\lim_{s \to d-} z(s) = x(d)$ we obtain that there is a $\Delta \in (0, \delta)$ such that

$$\|x(d - \Delta) - x(d)\| < \frac{\delta}{2}$$

and by (9.10) there is a $k_1 \in \mathbb{N}$, $k_1 > k_0$ such that

$$\|x_k(d - \Delta) - x(d - \Delta)\| < \frac{\delta}{2}$$

for $k > k_1$. Hence for $k > k_1$ we have

$$\|x_k(d - \Delta) - x(d)\| < \delta.$$

Using (v) from Definition 9.2 we obtain

$$\|x_k(d + \Delta) - x(d - \Delta) - \int_{d-\Delta}^{d+\Delta} DF_0(x(\tau), t)\| \leq$$

$$\leq \|x_k(d + \Delta) - x(d - \Delta) - (F_0(x(d), d+) - F_0(x(d), d))\| +$$

$$+ \|F_0(x(d), d+) - F_0(x(d), d) - \int_{d-\Delta}^{d+\Delta} DF_0(x(\tau), t)\| \leq$$

$$\leq \varepsilon + \| \int_{d-\Delta}^{d} DF_0(x(\tau), t)\| +$$

$$+ \|F_0(x(d), d+) - F_0(x(d), d) - \int_{d}^{d+\Delta} DF_0(x(\tau), t)\| \leq$$

$$\leq \varepsilon + |h_0(d) - h_0(d - \Delta)| +$$

$$+ \|F_0(x(d), d+) - F_0(x(d), d) - \int_{d}^{d+\Delta} DF_0(x(\tau), t)\| \leq$$

$$\leq \varepsilon + \frac{\varepsilon}{2} + \|F_0(x(d), d+) - F_0(x(d), d) - \int_{d-\Delta}^{d+\Delta} DF_0(x(\tau), t)\|.$$

By Theorem 1.16 we obtain

$$\lim_{s \to d+} \int_s^{d+\Delta} DF_0(x(\tau), t) =$$

$$= F_0(x(d), d+) - F_0(x(d), d) - \int_d^{d+\Delta} DF_0(x(\tau), t)$$

and because

$$\| \int_s^{d+\Delta} DF_0(x(\tau), t)\| \leq |h_0(d + \Delta) - h_0(s)|$$

we obtain by (9.14) the inequality

$$\|F_0(x(d), d+) - F_0(x(d), d) - \int_{d-\Delta}^{d+\Delta} DF_0(x(\tau), t)\| \leq$$

$$\leq h_0(d + \Delta) - h_0(d+) < \frac{\varepsilon}{2}$$

because we also have $\Delta < \delta_1$.

Hence for every $k > k_1$, $k \in \mathbb{N}$ we get

$$\|x_k(d + \Delta) - x_k(d - \Delta) - \int_{d-\Delta}^{d+\Delta} DF_0(x(\tau), t)\| < 2\varepsilon.$$

By (9.10) there exists $k_2 \in \mathbb{N}$, $k_2 > k_1$ such that

$$\|x_k(d + \Delta) - x(d + \Delta)\| < \frac{\varepsilon}{2},$$

$$\|x_k(d - \Delta) - x(d - \Delta)\| < \frac{\varepsilon}{2}$$

whenever $k > k_1$ and therefore

$$\|x(d+\Delta) - x(d-\Delta) - \int_{d-\Delta}^{d+\Delta} DF_0(x(\tau), t)\| < 3\varepsilon,$$

i.e.

$$x(d+\Delta) - x(d-\Delta) = \int_{d-\Delta}^{d+\Delta} DF_0(x(\tau), t) \tag{9.15}$$

since $\varepsilon > 0$ was given arbitrarily.

Using (9.12), (9.13) and (9.15) we finally obtain

$$x(s_2) - x(s_1) = \int_{s_1}^{s_2} DF_0(x(\tau), t). \tag{9.16}$$

The case when $\alpha \leq s_1 = d < s_2 \leq \beta$ can be treated similarly. The remaining cases of possible positions of s_1, s_2 in the interval $[\alpha, \beta]$ are covered directly by Theorem 8.5 and we obtain that (9.16) holds for every $s_1, s_2 \in [\alpha, \beta]$, which proves the theorem. □

9.4 *Remark.* Theorem 9.3 represents a continuous dependence result for generalized ordinary differential equations provided the right hand sides converge emphatically to a function $F_0 \in \mathcal{F}(G, h_0, \omega)$, in the case when the function h_0 exhibits a discontinuity at a single point $d \in (a, b)$. The idea of the proof of the general case of a nondecreasing function h_0 with the possibly infinite number of discontinuities is essentially the same but the proof is technically rather complicated. The general result reads as follows:

Assume that $G = B_c \times (a, b)$. Let $h_k : (a, b) \to \mathbb{R}$, $k = 0, 1, 2, \ldots$ be nondecreasing functions continuous from the left. Assume further that $F_k \in \mathcal{F}(G, h_k, \omega)$, $k = 0, 1, 2, \ldots$ and that the sequence (F_k) converges emphatically to F_0 for $k \to \infty$.

Let $x_k : [\alpha, \beta] \to \mathbb{R}^n$ *be solutions of*

$$\frac{dx}{d\tau} = DF_k(x,t)$$

on an interval $[\alpha, \beta] \subset (a,b)$, $k = 1, 2, \ldots$ *such that*

$$\lim_{k \to \infty} x_k(t) = z(t)$$

for $t \in [\alpha, \beta]$ *whenever* $h_0(t+) = h_0(t)$. *Then* $x : [\alpha, \beta] \to \mathbb{R}^n$ *defined by* $x(t) = z(t)$ *for* $t \in [\alpha, \beta]$ *with* $h_0(t+) = h_0(t)$ *and continuous from the left in* $[\alpha, \beta]$ *is a function of bounded variation on* $[\alpha, \beta]$, *and it is a solution of the generalized ordinary differential equation*

$$\frac{dx}{d\tau} = DF_0(x,t)$$

on the interval $[\alpha, \beta]$.

The main tool for the proof of this result is Lemma 8.4 and the fact that the set of points at which the jump $h_0(t+) - h_0(t)$ is greater than a given positive constant is finite. See also Theorem 4.1 in [69] where a detailed proof is given.

Looking at the special case described in Theorem 9.3 and at Definition 9.2 we can see that we have to determine the value of the difference $F_0(y, d+) - F_0(y, d)$ which determines the jump $x(d+) - x(d)$ of the function x at a point d provided $x(d) = y$ (see Lemma 3.12). This value can be determined via the requirement (v) in Definition 9.2, i.e. we have to know the limit (for $k \to \infty$) of increments of the solutions of (9.9) on a short interval containig the point d. In the general case, to obtain such an information can be complicated.

Looking at Example 9.1 we can see that the jump of the limit equation at the point $d = 0$ was evaluated by taking the limit of the difference $x_k(\frac{1}{k}) - x_k(0)$ for $k \to \infty$. In Example 9.1 we have

$$F_k(x,t) = [At + H_k(t)B]x,$$

$$F_0(x,t) + F_*(x,t) = [At + H(t)B]x,$$

$$\lim_{k\to\infty} F_k(x,t) = F_0(x,t) + F_*(x,t)$$

and

$$F_0(x,t) = [At + H(t)(e^B - I)]x$$

because according to Theorem 9.3 and (v) from Definition 9.2 obviously

$$F_0(y, d+) - F_0(y,d) = \lim_{k\to\infty} (x_k(\frac{1}{k}) - x_k(0)) = (e^B - I)y$$

for a solution x_k of (9.1) for which $x_k(0) = y$.

9.5 *Example.* Let $G = \mathbb{R}^n \times [-1,1]$. Assume that $F \in \mathcal{F}(G, h, \omega)$ where $h : [-1,1] \to \mathbb{R}$ is nondecreasing and continuous in $[-1,1]$.

For $k = 1, 2, \ldots$ let us have $\alpha_k \in [-1,0]$, $\beta_k \in [0,1]$ where $\alpha_k < \beta_k$ and $\lim_{k\to\infty} \alpha_k = \lim_{k\to\infty} \beta_k = 0$. Let $\Phi_k : [-1,1] \to [0,1]$ be continuous on $[-1,1]$, increasing on $[\alpha_k, \beta_k]$ and such that

$$\Phi_k(t) = 0 \quad \text{for} \quad t \in [-1, \alpha_k],$$

$$\Phi_k(t) = 1 \quad \text{for} \quad t \in [\beta_k, 1].$$

For the restriction $\Phi_k : [\alpha_k, \beta_k] \to [0,1]$ let us denote by $\Phi_k^{-1} : [0,1] \to [\alpha_k, \beta_k]$ the inverse function to Φ_k. The function Φ_k^{-1} is continuous and increasing on $[0,1]$.

Define

$$H(t) = 0 \text{ for } t \le 0, \ H(t) = 1 \text{ for } t > 0.$$

Then

$$\lim_{k \to \infty} \Phi_k(t) = H(t)$$

for every $t \in [-1, 1]$, $t \neq 0$.

Let a function $g : \mathbb{R}^n \to \mathbb{R}^n$ be given such that

$$\|g(x)\| \le K, \quad \|g(x) - g(y)\| \le L\|x - y\| \tag{9.17}$$

for $x, y \in \mathbb{R}^n$.

For $(x, t) \in G$ define

$$F_k(x, t) = F(x, t) + g(x)\Phi_k(t), \quad k = 1, 2, \dots .$$

Then by definition we have $F_k \in \mathcal{F}(G, h_k, \Omega)$, where $h_k(t) = h(t) + K\Phi_k(t)$ and $\Omega(\rho) = \omega(\rho) + L\rho$. By definition we further have

$$\lim_{k \to \infty} F_k(x, t) = F(x, t) + g(x)H(t) \tag{9.18}$$

for every $(x, t) \in G$ with $t \neq 0$, and the function $F(x, t) + g(x)H(t)$ clerly belongs to $\mathcal{F}(G, h + KH, \Omega)$. Moreover, we have

$$\limsup_{k \to \infty} (h_k(t_2) - h_k(t_1)) \le h_0(t_2) - h_0(t_1)$$

for $t_1, t_2 \in [-1, 0) \cup (0, 1]$ where $h_0(t) = h(t) + KH(t)$, $t \in [-1, 1]$.

We consider the problem whether there is a function $F_0(x, t)$ to which the sequence $F_k(x, t)$ converges emphatically for $k \to \infty$.

Let us consider the autonomous ordinary differential equation

$$\dot{y} = g(y) \tag{9.19}$$

and assume that for $x \in \mathbb{R}^n$ the function $v(s, x)$, $s \in [0, 1]$ is the uniquely determined solution of (9.19) on $[0, 1]$ for which we have $v(0, x) = x$.

Since the function h is continuous at 0, for every $\eta > 0$ there exists $\delta > 0$ such that

$$h(\delta) - h(-\delta) < \eta, \tag{9.20}$$

and of course also $h(\beta) - h(\alpha) < \eta$ for every interval $[\alpha, \beta] \subset [-\delta, \delta]$. Let $\delta' \in (0, \delta)$ be given and let $k_0 \in \mathbb{N}$ be such that for $k > k_0$ we have

$$[\alpha_k, \beta_k] \subset [-\delta', \delta'].$$

Assume that $x_0 \in \mathbb{R}^n$ is given and that $y : [-\delta', \delta'] \to \mathbb{R}^n$ is a solution of the generalized ordinary differential equation

$$\frac{dy}{d\tau} = DF_k(y, t) = D[F(y, t) + g(y)\Phi_k(t)] \tag{9.21}$$

such that $\|y(-\delta') - x_0\| < \delta$. Then by the definition of a solution we have

$$y(r) = y(-\delta') + \int_{-\delta'}^{r} DF(y(\tau), t) + \int_{-\delta'}^{r} D[g(y(\tau))\Phi_k(t)] =$$

$$= y(-\delta') + \int_{-\delta'}^{r} DF(y(\tau), t) + \int_{-\delta'}^{r} g(y(\tau))\, d\Phi_k(\tau) =$$

$$= y(-\delta') + \int_{-\delta'}^{r} DF(y(\tau), t) + \int_{\alpha_k}^{r} g(y(\tau))\, d\Phi_k(\tau)$$

for every $r \in [-\delta', \delta']$ since $\Phi_k(t) = 0$ for $t \leq \alpha_k$. We use the notation of the Stieltjes integral in the second integral.

If now $s \in [0, 1]$ then $\Phi_k^{-1}(s) \in [\alpha_k, \beta_k] \subset [-\delta', \delta']$ and we have

$$y(\Phi_k^{-1}(s)) =$$

$$= y(-\delta') + \int_{-\delta'}^{\Phi_k^{-1}(s)} DF(y(\tau), t) + \int_{\alpha_k}^{\Phi_k^{-1}(s)} g(y(\tau))\, d\Phi_k(\tau) =$$

$$= y(-\delta') + \int_{-\delta'}^{\Phi_k^{-1}(s)} DF(y(\tau), t) + \int_{\Phi_k^{-1}(0)}^{\Phi_k^{-1}(s)} g(y(\tau))\, d\Phi_k(\tau).$$

Applying the Substitution Theorem 1.18 to the last integral we obtain

$$y(\Phi_k^{-1}(s)) = y(-\delta') + \int_{-\delta'}^{\Phi_k^{-1}(s)} DF(y(\tau), t) + \int_0^s g(y(\Phi_k^{-1}(\sigma)))\, d\sigma. \tag{9.22}$$

Since $v(s, x_0)$ is a solution of (9.19) on $[0, 1]$, we have

$$v(s, x_0) = x_0 + \int_0^s g(v(\sigma, x_0))\, d\sigma \tag{9.23}$$

for every $s \in [0, 1]$. (9.22) and (9.23) yield further

$$y(\Phi_k^{-1}(s)) - v(s, x_0) =$$

$$= y(-\delta') - x_0 + \int_{-\delta'}^{\Phi_k^{-1}(s)} DF(y(\tau), t) +$$

$$+ \int_0^s g(y(\Phi_k^{-1}(\sigma)))\, d\sigma - \int_0^s g(v(\sigma, x_0))\, d\sigma$$

and therefore

$$\|y(\Phi_k^{-1}(s)) - v(s, x_0)\| \leq$$

$$\leq \|y(-\delta') - x_0\| + \|\int_{-\delta'}^{\Phi_k^{-1}(s)} DF(y(\tau), t)\| +$$

$$+ \int_0^s \|g(y(\Phi_k^{-1}(\sigma))) - g(v(\sigma, x_0))\|\, d\sigma$$

for all $s \in [0,1]$. Consequently, taking into account (9.17) and (9.20) we obtain

$$\|y(\Phi_k^{-1}(s)) - v(s,x_0)\| \leq$$

$$\leq \delta + h(\Phi_k^{-1}(s)) - h(-\delta) +$$

$$+L\int_0^s \|y(\Phi_k^{-1}(\sigma)) - v(\sigma,x_0)\|\, d\sigma \leq$$

$$\leq \delta + \eta + L\int_0^s \|y(\Phi_k^{-1}(\sigma)) - v(\sigma,x_0)\|\, d\sigma.$$

Using the Gronwall lemma (see also Corollary 1.43) we obtain from the last inequality the estimate

$$\|y(\Phi_k^{-1}(s)) - v(s,x_0)\| \leq (\delta+\eta)e^{Ls}, \quad s \in [0,1]$$

and for $s = 1$ also

$$\|y(\beta_k) - v(1,x_0)\| = \|y(\Phi_k^{-1}(1)) - v(1,x_0)\| \leq (\delta+\eta)e^{L}.$$

Further we have

$$y(\delta') - y(\beta_k) = \int_{\beta_k}^{\delta'} D[F(y(\tau),t) + g(y(\tau))\Phi_k(t)] =$$

$$= \int_{\beta_k}^{\delta'} DF(y(\tau),t)$$

because $\Phi_k(t) = 1$ for $t \geq \beta_k$, and consequently, by (9.20),

$$\|y(\delta') - y(\beta_k)\| \leq \|\int_{\beta_k}^{\delta'} DF(y(\tau),t)\| \leq h(\delta') - h(\beta_k) < \eta.$$

Hence

$$\|y(\delta') - v(1, x_0)\| \leq \|y(\delta') - y(\beta_k)\| +$$

$$+ \|y(\beta_k) - v(1, x_0)\| \leq \eta + (\delta + \eta)e^L$$

and

$$\|y(\delta') - y(-\delta') - (v(1, x_0) - v(0, x_0))\| \leq$$

$$\leq \|y(\delta') - v(1, x_0)\| + \|y(-\delta') - x_0\| \leq$$

$$\leq \eta + \delta + (\delta + \eta)e^L = (\delta + \eta)(1 + e^L).$$

For a given $\varepsilon > 0$ the values of $\eta > 0$ and $\delta > 0$ can be taken so small that $(\delta + \eta)(1 + e^L) < \varepsilon$ and we can easily conclude that for every $\varepsilon > 0$ there is a $\delta > 0$ such that if $\delta' \in (0, \delta)$ and $k > k_0$ then for every solution $y : [-\delta', \delta'] \to \mathbb{R}^n$ of (9.21) on the interval $[-\delta', \delta']$ such that $\|y(-\delta') - x_0\| \leq \delta$ the inequality

$$\|y(\delta') - y(-\delta') - (v(1, x_0) - x_0)\| < \varepsilon \tag{9.24}$$

holds.

For $(x, t) \in G$ define

$$F_0(x, t) = F(x, t) + (v(1, x) - x)H(t)$$

where $v(\cdot, x)$ is the uniquely determined solution of (9.19) with $v(0, x) = x$. Then

$$F_0(x, 0+) - F_0(x, 0) = v(1, x) - x.$$

It is easy to see that (iv) from Definition 9.2 holds and using the definition of F_0 we can write (9.24) in the form

$$\|y(\delta') - y(-\delta') - (F_0(x_0, 0+) - F_0(x_0, 0))\| < \varepsilon$$

and the results presented above show that (v) from Definition 9.2 is fulfilled. The remaining parts of Definition 9.2 are easy to check with

$$F_*(x,t) = [g(x) - (v(1,x) - x)]H(t)$$

for $(x,t) \in G$ and finally it can be concluded that the functions F_k converge emphatically to F_0 for $k \to \infty$. Therefore the continuous dependence result given in Theorem 9.3 can be used in this situation.

The case described by this example is applicable for a sequence of ordinary differential equations

$$\dot{x} = f(x,t) + g(x)\varphi_k(t),\ k = 1,2,\dots \tag{9.25}$$

where $\varphi_k : [-1,1] \to \mathbb{R}^n$, $k = 1,2,\dots$ is a sequence of Lebesgue integrable functions which tends positively to the Dirac function, i.e.

$$\varphi_k(t) \geq 0, \quad t \in [-1,1],$$

and the sequence of functions $\Phi_k : [-1,1] \to \mathbb{R}$ given by

$$\Phi_k(t) = \int_{-1}^{t} \varphi_k(s)\,ds,\ k = 1,2,\dots,\ t \in [-1,1]$$

satisfies the assumptions given at the beginning of this example. For $g : \mathbb{R}^n \to \mathbb{R}^n$ we assume (9.17).

If $f : G \to \mathbb{R}^n$ satisfies the Carathéodory conditions and

$$\|f(x,t)\| \leq r(t), \quad t \in [-1,1]$$

where r is a Lebesgue integrable function in $[-1,1]$ then we can define

$$F(x,t) = \int_{-1}^{t} f(x,s)\,ds$$

and by Theorem 5.14 the ordinary differential equation (9.25) is equivalent to the generalized ordinary differential equation

$$\frac{dx}{d\tau} = D[F(x,t) + g(x)\Phi_k(t)], \quad k = 1, 2, \dots .$$

The right hand sides of this sequence of generalized ordinary differential equations emphatically converge to the function

$$F(x,t) + (v(1,x) - x)H(t)$$

where $v(t,x)$ is the solution of (9.19) defined on $[0,1]$ and such that $v(0,x) = x$.

Let us define a function $x : [-1,1] \to \mathbb{R}^n$ as follows:

Let $u : [-1,0] \to \mathbb{R}^n$ be a unique (for increasing values of t) solution of the ordinary differential equation

$$\dot{x} = f(x,t) \tag{9.26}$$

on $[-1,0]$. Let $v(t,u(0))$ be the unique (for increasing values of t) solution of (9.19) defined on $[0,1]$ such that $v(0,u(0)) = u(0)$. Let further $w : [0,1] \to \mathbb{R}^n$ be a unique (for increasing values of t) solution of (9.26) on the interval $[0,1]$ for which $w(0) = v(1,u(0))$. Let us set

$$x(t) = u(t) \quad \text{for } t \in [-1,0],$$
$$x(t) = w(t) \quad \text{for } t \in (0,1].$$

Then $x : [-1,1] \to \mathbb{R}^n$ is a solution of the generalized ordinary differential equation

$$\frac{dx}{d\tau} = D[F(x,t) + (v(1,x) - x)H(t)]$$

(see Theorem 5.20). It can be further shown that if $y_k \to x(-1)$ for $k \to \infty$ then for sufficiently large $k \in \mathbb{N}$ there exists a solution $x_k : [-1,1] \to \mathbb{R}^n$ of (9.25) on $[-1,1]$ and

$$\lim_{k\to\infty} x_k(t) = x(t)$$

for $t \in [-1, 1]$, $t \neq 0$.

This convergence phenomenon expresses the fact that the dynamics of the system (9.25) in a small neighbourhood of 0 is emphatically forced by the large term $g(x)\varphi_k(t)$ which influences the system in a short time in the same way as the term $g(x)$ does in a time interval the length of which is close to the integral of φ_k, i.e. close to 1.

Remark. At the end of Chapter V the work of D. Fraňková on generalized ordinary differential equations with a substitution was mentioned. Her approach is a good tool for explaining emphatic convergence effects of the type described by Definition 9.2. The corresponding convergence results are given in the paper [26].

CHAPTER X

VARIATIONAL STABILITY FOR GENERALIZED ORDINARY DIFFERENTIAL EQUATIONS

Concepts of stability

Let us assume that $G = B_c \times (a,b)$ where $B_c = \{x \in \mathbb{R}^n;\ \|x\| < c\}$, $c > 0$. Let $F : B_c \times [0,+\infty) \to \mathbb{R}^n$ be given.

We assume further that $h : [0,+\infty) \to \mathbb{R}$ is a nondecreasing function defined on $[0,+\infty)$, $\omega : [0,+\infty) \to \mathbb{R}$ is a continuous,increasing function with $\omega(0) = 0$ and $F \in \mathcal{F}(G,h,\omega)$ where $G = B_c \times [0,+\infty)$.

In addition to these usual conditions we assume that

$$F(0,t_2) - F(0,t_1) = 0 \quad \text{for every } t_1, t_2 \geq 0. \tag{10.1}$$

This assumption evidently implies that

$$\int_{s_1}^{s_2} DF(0,t) = F(0,s_2) - F(0,s_1) = 0, \quad s_1, s_2 \in [0,+\infty)$$

and therefore the function x given by $x(s) = 0$ for $s \geq 0$ is a solution of the generalized ordinary differential equation

$$\frac{dx}{d\tau} = DF(x,t) \tag{10.2}$$

on the whole half–axis $[0,+\infty)$.

Let us introduce some concepts of stability of the trivial solution $x(s) = 0,\ s \in [0,+\infty)$ of the equation (10.2).

10.1 Definition. The solution $x \equiv 0$ of (10.2) is called *variationally stable* if for every $\varepsilon > 0$ there exists $\delta = \delta(\varepsilon) > 0$ such that if $y : [t_0, t_1] \to B_c$, $0 \leq t_0 < t_1 < +\infty$ is a function of bounded variation on $[t_0, t_1]$, continuous from the left on $(t_0, t_1]$ with

$$\|y(t_0)\| < \delta$$

and

$$\mathrm{var}_{t_0}^{t_1}(y(s) - \int_{t_0}^{s} DF(y(\tau), t)) < \delta$$

then we have

$$\|y(t)\| < \varepsilon \quad \text{for } t \in [t_0, t_1].$$

10.2 Definition. The solution $x \equiv 0$ of (10.2) is called *variationally attracting* if there exists $\delta_0 > 0$ and for every $\varepsilon > 0$ there is a $T = T(\varepsilon) \geq 0$ and $\gamma = \gamma(\varepsilon) > 0$ such that if $y : [t_0, t_1] \to B_c$, $0 \leq t_0 < t_1 < +\infty$ is a function of bounded variation on $[t_0, t_1]$, continuous from the left on $(t_0, t_1]$ with

$$\|y(t_0)\| < \delta_0$$

and

$$\mathrm{var}_{t_0}^{t_1}(y(s) - \int_{t_0}^{s} DF(y(\tau), t)) < \gamma$$

then

$$\|y(t)\| < \varepsilon$$

for all $t \in [t_0, t_1] \cap [t_0 + T(\varepsilon), +\infty)$ and $t_0 \geq 0$.

10.3 Definition. The solution $x \equiv 0$ of (10.2) is called *variationally–asymptotically stable* if it is variationally stable and variationally attracting.

Following the same lines we now give definitions, which introduce the concept of stability of the trivial solution $x \equiv 0$ of (10.2) with respect to perturbations of the equation (10.2) which are in some sense small.

Together with (10.2) we consider the perturbed equation

$$\frac{dx}{d\tau} = D[F(x,t) + P(t)] \tag{10.3}$$

where $P : [0, +\infty) \to \mathbb{R}^n$ is a function which is locally of bounded variation and continuous from the left on $[0, +\infty)$.

Clearly we have

$$\|F(x,t_2) + P(t_2) - F(x,t_1) - P(t_1)\| \leq$$
$$\leq |h(t_2) + \operatorname{var}_0^{t_2} P - h(t_1) - \operatorname{var}_0^{t_1} P|$$

for $x \in B_c$ and $t_1, t_2 \in [0, +\infty)$ and

$$\|F(x,t_2) + P(t_2) - F(x,t_1) - P(t_1) -$$
$$-(F(y,t_2) + P(t_2) - F(y,t_1) - P(t_1))\| \leq$$
$$\leq \omega(\|x - y\|)|h(t_2) - h(t_1)| \leq$$
$$\leq \omega(\|x - y\|)|h(t_2) + \operatorname{var}_0^{t_2} P - h(t_1) - \operatorname{var}_0^{t_1} P|$$

and therefore the right hand side $F(x,t) + P(t)$ of the generalized ordinary differential equation (10.3) belongs to the class $\mathcal{F}(G, \widetilde{h}, \omega)$ where $\widetilde{h}(t) = h(t) + \operatorname{var}_0^t P$, and all fundamental results (e.g. the local existence of a solution) hold for the equation (10.3).

10.4 Definition. The solution $x \equiv 0$ of (10.2) is called *stable with respect to perturbations* if for every $\varepsilon > 0$ there exists $\delta = \delta(\varepsilon) > 0$ such that if $\|y_0\| < \delta$, $y_0 \in \mathbb{R}^n$ and $P \in BV([t_0, t_1])$ is a function of bounded variation on $[t_0, t_1]$, continuous from the left on $(t_0, t_1]$ and such that

$$\operatorname{var}_{t_0}^{t_1} P < \delta$$

then we have

$$\|y(t, t_0, y_0)\| < \varepsilon \quad \text{for } t \in [t_0, t_1]$$

where $y(t, t_0, y_0)$ is a solution of (10.3) with $y(t_0, t_0, y_0) = y_0$.

10.5 Definition. The solution $x \equiv 0$ of (10.2) is called *attractive with respect to perturbations* if there exists $\delta_0 > 0$ and for every $\varepsilon > 0$ there is a $T = T(\varepsilon) \geq 0$ and $\gamma = \gamma(\varepsilon) > 0$ such that if

$$\|y_0\| < \delta_0, \ y_0 \in \mathbb{R}^n$$

and $P \in BV([t_0, t_1])$ is a function of bounded variation on $[t_0, t_1]$, continuous from the left on $(t_0, t_1]$ and such that

$$\operatorname{var}_{t_0}^{t_1} P < \gamma$$

then

$$\|y(t, t_0, y_0)\| < \varepsilon$$

for all $t \in [t_0, t_1] \cap [t_0 + T(\varepsilon), +\infty)$ and $t_0 \geq 0$ where $y(t, t_0, y_0)$ is a solution of (10.3) with $y(t_0, t_0, y_0) = y_0$.

10.6 Definition. The solution $x \equiv 0$ of (10.2) is called *asymptotically stable with respect to perturbations* if it is stable and attractive with respect to perturbations.

10.7 *Remark.* The concept of variational stability comes from the following intuitive idea: if a certain function given on some $[t_0, t_1] \subset [0, +\infty)$ is such that the initial value $y(t_0)$ is close to zero and on the interval $[t_0, t_1]$ the function y is "almost" a solution of (10.2), i.e. the variation of the function

$$y(s) - y(t_0) - \int_{t_0}^{s} DF(y(\tau), t)$$

on $[t_0, t_1]$ is small enough, then y is close to zero on the interval $[t_0, t_1]$.

On the other hand, the stability with respect to perturbations is motivated by the desire that the solutions of the perturbed equation (10.3) be close to zero on a certain $[t_0, t_1]$ whenever the value $y(t_0)$ is close to zero and the perturbing term P in (10.3) is small in the sense that $\mathrm{var}_{t_0}^{t_1} P$ is small.

The following result shows that both these natural concepts of stability are equivalent.

10.8 Theorem. *The trivial solution $x \equiv 0$ of (10.2) is variationally stable if and only if it is stable with respect to perturbations.*

The trivial solution $x \equiv 0$ of (10.2) is variationally attracting if and only if it is attracting with respect to perturbations.

Proof. Let us prove the first statement.

1) Assume that $x \equiv 0$ is variationally stable. For a given $\varepsilon > 0$ let $\delta > 0$ be given by Definition 10.1. Assume that $y_0 \in \mathbb{R}^n$, $\|y_0\| < \delta$ and $\mathrm{var}_{t_0}^{t_1} P < \delta$ and that $y(t) = y(t, t_0, y_0)$ is a solution of (10.3) on $[t_0, t_1]$. Then $\|y(t_0)\| = \|y(t_0, t_0, y_0)\| = \|y_0\| < \delta$ and by the definition of a solution we have for any $s_1, s_2 \in [t_0, t_1]$

$$y(s_2) - y(s_1) = \int_{s_1}^{s_2} DF(y(\tau), t) + P(s_2) - P(s_1).$$

Hence

$$y(s_2) - \int_{t_0}^{s_2} DF(y(\tau), t) - y(s_1) + \int_{t_0}^{s_1} DF(y(\tau), t) =$$

$$= P(s_2) - P(s_1)$$

for $s_1, s_2 \in [t_0, t_1]$ and consequently

$$\mathrm{var}_{t_0}^{t_1}\Big(y(s) - \int_{t_0}^{s} DF(y(\tau), t)\Big) = \mathrm{var}_{t_0}^{t_1} P < \delta.$$

The variational stability yields

$$\|y(t)\| = \|y(t, t_0, y_0)\| < \varepsilon, \ t \in [t_0, t_1],$$

and this means that the solution $x \equiv 0$ of (10.2) is stable with respect to perturbations.

2) Assume that the solution $x \equiv 0$ of (10.2) is stable with respect to perturbations. For $\varepsilon > 0$ let $\delta > 0$ be given by Definition 10.4. Let $y : [t_0, t_1] \to \mathbb{R}^n$ be of bounded variation on $[t_0, t_1]$, continuous from the left and such that

$$\|y(t_0)\| < \delta \text{ and } \operatorname{var}_{t_0}^{t_1}(y(s) - \int_{t_0}^{s} DF(y(\tau), t)) < \delta.$$

For $s_1, s_2 \in [t_0, t_1]$ we have

$$y(s_2) - y(s_1) = \int_{s_1}^{s_2} DF(y(\tau), t) =$$

$$= \int_{s_1}^{s_2} DF(y(\tau), t) + y(s_2) - \int_{t_0}^{s_2} DF(y(\tau), t) - y(s_1) +$$

$$+ \int_{t_0}^{s_1} DF(y(\tau), t) = \int_{s_1}^{s_2} DF(y(\tau), t) + P(s_2) - P(s_1) \quad (10.4)$$

where $P(s) = y(s) - \int_{t_0}^{s} DF(y(\tau), t)$ for $s \in [t_0, t_1]$. By Lemma 3.9 this function P is of bounded variation on $[t_0, t_1]$, continuous from the left, and (10.4) shows that the function y is a solution of (10.3) on $[t_0, t_1]$ with this P and $\|y(t_0)\| < \delta$. Moreover,

$$\operatorname{var}_{t_0}^{t_1} P = \operatorname{var}_{t_0}^{t_1}(y(s) - \int_{t_0}^{s} DF(y(\tau), t)) < \delta.$$

Hence by the assumption of stability with respect to perturbations we get $\|y(t)\| < \varepsilon$ for $t \in [t_0, t_1]$ and $x \equiv 0$ is variationally stable.

Now we prove the second part of the statement.

1) Assume that $x \equiv 0$ is variationally attracting. Then there exists $\delta_0 > 0$ and for a given $\varepsilon > 0$ also $T > 0$ and $\gamma > 0$ by

Definition 10.2. If now $y_0 \in \mathbb{R}^n$ is such that $\|y_0\| < \delta$, P is of bounded variation on $[t_0, t_1]$ and continuous from the left on this interval where $\text{var}_{t_0}^{t_1} P < \delta$ and $y(t) = y(t, t_0, y_0)$ is a solution of (10.3) on $[t_0, t_1]$ then $\|y(t_0)\| = \|y_0\| < \delta_0$ and

$$\text{var}_{t_0}^{t_1}(y(s) - \int_{t_0}^{s} DF(y(\tau), t)) = \text{var}_{t_0}^{t_1} P < \gamma.$$

Hence by Definition 10.2 we have $\|y(t, t_0, y_0)\| = \|y(t)\| < \varepsilon$ for all $t > t_0 + T$ and $t_0 \geq 0$.

2) If $x \equiv 0$ is attractive with respect to perturbations then we can set $P(s) = y(s) - \int_{t_0}^{s} DF(y(\tau), t)$ for $s \in [t_0, t_1]$ to show that $x \equiv 0$ is variationally attractive directly from the properties described in Definition 10.5. □

Theorem 10.8 and Definitions 10.3 and 10.6 simply yield the following result.

10.9 Theorem. *The trivial solution $x \equiv 0$ of (10.2) is variational–asymptotically stable if and only if it is asymptotically stable with respect to perturbations.*

10.10 *Remark.* The concept of variational stability was introduced by H. Okamura. I. Vrkoč considered Carathéodory equations in [172] and pointed out that Okamura's variational stability is equivalent to his concept of integral stability. There is an improvement of Vrkoč's results given by S.–N. Chow and J.A. Yorke [17].

In the case of classical ordinary differential equations variational stability is a somewhat exotic concept but in the case of generalized ordinary differential equations it seems to be very natural because in our setting the solutions of such equations are functions of bounded variation and in the case of variational stability we are measuring the distance of two solutions using the norm in the space BV. Now we turn our attention to the method of Ljapunov functions for the stability concepts described above. First we derive some auxiliary results.

10.11 Proposition. *Assume that* $-\infty < a < b < +\infty$ *and that* $f, g : [a, b] \to \mathbb{R}$ *are functions which are continuous from the left in* $(a, b]$.

If for every $\sigma \in [a, b]$ *there exists* $\delta(\sigma) > 0$ *such that for every* $\eta \in (0, \delta(\sigma))$ *the inequality*

$$f(\sigma + \eta) - f(\sigma) \leq g(\sigma + \eta) - g(\sigma)$$

holds, then

$$f(s) - f(a) \leq g(s) - g(a)$$

for all $s \in [a, b]$.

Proof. Let us denote

$$M = \{s \in [a, b]; f(\sigma) - f(a) \leq g(\sigma) - g(a), \sigma \in [a, s]\}$$

and set $S = \sup M$. Since

$$f(a + \eta) - f(a) \leq g(a + \eta) - g(a)$$

for $\eta \in (0, \delta(a))$ and $\delta(a) > 0$, the set M is nonempty, $S > a$ and $f(s) - f(a) \leq g(s) - g(a)$ for every $s < S$. Using the continuity from the left of the functions f and g we have also

$$f(S) - f(a) \leq g(S) - g(a).$$

If $S < b$ then by assumption we have

$$f(S + \eta) - f(S) \leq g(S + \eta) - g(S)$$

for every $\eta \in (0, \delta(S))$, $\delta(S) > 0$ and therefore also

$$f(S + \eta) - f(a) = f(S + \eta) - f(S) + f(S) - f(a) \leq$$
$$\leq g(S + \eta) - g(S) + g(S) - g(a) = g(S + \eta) - g(a).$$

This implies that $S + \eta \in M$ for $\eta \in (0, \delta(S))$, i.e. $S < \sup M$ and this contradiction yields $S = b$ and $M = [a, b]$. □

For deriving Ljapunov type theorems we need another technical lemma.

10.12 Lemma. *Suppose that $V : [0, +\infty) \times \mathbb{R}^n \to \mathbb{R}$ is such that for every $x \in \mathbb{R}^n$ the function $V(\cdot, x) : [0, +\infty) \to \mathbb{R}$ is continuous from the left in $(0, +\infty)$. Assume that*

$$|V(t,x) - V(t,y)| \leq K\|x - y\| \tag{10.5}$$

for $x, y \in \mathbb{R}^n$, $t \in [0, +\infty)$ with a constant $K > 0$.

Further assume that there is a real function $\Phi : \mathbb{R}^n \to \mathbb{R}$ such that for every solution $x : (\alpha, \beta) \to \mathbb{R}^n$ of the generalized ordinary differential equation (10.2) on $(\alpha, \beta) \subset [0, +\infty)$ we have

$$\limsup_{\eta \to 0+} \frac{V(t+\eta, x(t+\eta)) - V(t, x(t))}{\eta} \leq \Phi(x(t)) \tag{10.6}$$

for $t \in (\alpha, \beta)$.

If $y : [t_0, t_1] \to \mathbb{R}^n$, $0 \leq t_0 < t_1 < +\infty$ is continuous from the left on $(t_0, t_1]$ and of bounded variation on $[t_0, t_1]$, then the inequality

$$V(t_1, x(t_1)) \leq$$

$$\leq V(t_0, x(t_0)) + K \operatorname{var}_{t_0}^{t_1}[y(s) - \int_{t_0}^{s} DF(y(\tau), t)] + M(t_1 - t_0) \tag{10.7}$$

holds, where $M = \sup_{t \in [t_0, t_1]} \Phi(y(t))$.

Proof. Let $y : [t_0, t_1] \to \mathbb{R}^n$ be given and let $\sigma \in [t_0, t_1]$ be an arbitrary point. It is clear that the function $V(t, y(t)) : [t_0, t_1] \to \mathbb{R}$ is continuous from the left on $(t_0, t_1]$.

Assume that $x : [\sigma, \sigma + \eta_1(\sigma)] \to \mathbb{R}^n$ is a solution of (10.2) on the interval $[\sigma, \sigma + \eta_1(\sigma)]$, $\eta_1(\sigma) > 0$ with the initial condition $x(\sigma) = y(\sigma)$. The existence of such a solution is guaranteed by the local existence theorem 4.2. By the assumption (10.5) we then have

$$V(\sigma + \eta, y(\sigma + \eta)) - V(\sigma + \eta, x(\sigma + \eta)) \leq$$

$$\leq K\|y(\sigma+\eta)-x(\sigma+\eta)\| =$$

$$= K\|y(\sigma+\eta)-y(\sigma)-\int_{\sigma}^{\sigma+\eta} DF(x(\tau),t)\|$$

for every $\eta \in [0, \eta_1(\sigma)]$.

By this inequality and by (10.6) we obtain

$$V(\sigma+\eta, y(\sigma+\eta)) - V(\sigma, x(\sigma)) =$$

$$= V(\sigma+\eta, y(\sigma+\eta)) - V(\sigma+\eta, x(\sigma+\eta)) +$$

$$+ V(\sigma+\eta, x(\sigma+\eta)) - V(\sigma, x(\sigma)) \leq$$

$$\leq K\|y(\sigma+\eta)-y(\sigma)-\int_{\sigma}^{\sigma+\eta} DF(x(\tau),t)\| + \eta\Phi(y(\eta)) \leq$$

$$\leq K\|y(\sigma+\eta)-y(\sigma)-\int_{\sigma}^{\sigma+\eta} DF(x(\tau),t)\| + \eta M + \eta\varepsilon$$

where $\varepsilon > 0$ is arbitrary and $\eta \in (0, \eta_2(\sigma))$ with $\eta_2(\sigma) \leq \eta_1(\sigma)$, $\eta_2(\sigma) > 0$ is sufficiently small. Denote

$$P(s) = y(s) - \int_{t_0}^{s} DF(y(\tau),t) \quad \text{for } s \in [t_0, t_1].$$

The function $P : [t_0, t_1] \to \mathbb{R}^n$ is continuous from the left on $(t_0, t_1]$, of bounded variation on $[t_0, t_1]$, and the last inequality can be used to derive

$$V(\sigma+\eta, y(\sigma+\eta)) - V(\sigma, x(\sigma)) \leq$$

$$\leq K\|y(\sigma+\eta)-y(\sigma)-\int_{\sigma}^{\sigma+\eta} DF(y(\tau),t)\| +$$

$$+ K\|\int_{\sigma}^{\sigma+\eta} D[F(y(\tau),t)-F(x(\tau),t)]\| + \eta M + \eta\varepsilon \leq$$

$$\leq K\|P(\sigma+\eta)-P(\sigma)\|+\eta M+\eta\varepsilon+$$

$$+K\|\int_{\sigma}^{\sigma+\eta} D[F(y(\tau),t)-F(x(\tau),t)]\|\leq$$

$$\leq K(\operatorname{var}_{t_0}^{\sigma+\eta} P-\operatorname{var}_{t_0}^{\sigma} P)+\eta M+\eta\varepsilon+$$

$$+K\|\int_{\sigma}^{\sigma+\eta} D[F(y(\tau),t)-F(x(\tau),t)]\| \tag{10.8}$$

for every $\eta \in (0,\eta_2(\sigma))$.

Let us consider the last term in (10.8). Since $F \in \mathcal{F}(G,h,\omega)$ we obtain by Theorems 1.16 and 1.35 the estimate

$$\|\int_{\sigma}^{\sigma+\eta} D[F(y(\tau),t)-F(x(\tau),t)]\|\leq$$

$$\leq \int_{\sigma}^{\sigma+\eta}\omega(\|y(\tau)-x(\tau)\|)\,dh(\tau)=$$

$$=\lim_{\alpha\to 0+}[\int_{\sigma}^{\sigma+\alpha}\omega(\|y(\tau)-x(\tau)\|)\,dh(\tau)+$$

$$+\int_{\sigma+\alpha}^{\sigma+\eta}\omega(\|y(\tau)-x(\tau)\|)\,dh(\tau)]=$$

$$=\omega(\|y(\sigma)-x(\sigma)\|)(h(\sigma+)-h(\sigma))+$$

$$+\lim_{\alpha\to 0+}\int_{\sigma+\alpha}^{\sigma+\eta}\omega(\|y(\tau)-x(\tau)\|)\,dh(\tau)=$$

$$=\lim_{\alpha\to 0+}\int_{\sigma+\alpha}^{\sigma+\eta}\omega(\|y(\tau)-x(\tau)\|)\,dh(\tau)\leq$$

$$\leq \sup_{\rho\in(\sigma,\sigma+\eta]}\omega(\|y(\rho)-x(\rho)\|)\lim_{\alpha\to 0+}(h(\sigma+\eta)-h(\sigma+\alpha))=$$

$$= \sup_{\rho\in(\sigma,\sigma+\eta]} \omega(\|y(\rho)-x(\rho)\|)(h(\sigma+\eta)-h(\sigma+)), \tag{10.9}$$

because $y(\sigma)=x(\sigma)$ and $\omega(\|y(\sigma)-x(\sigma)\|)=0$.

For $\rho\in[\sigma,\sigma+\eta_2(\sigma)]$ we have

$$y(\rho)-x(\rho)=y(\rho)-y(\sigma)-\int_\sigma^\rho DF(x(\tau),t)$$

and therefore

$$\lim_{\rho\to\sigma+}(y(\rho)-x(\rho)) =$$

$$= y(\sigma+)-y(\sigma)-\lim_{\rho\to\sigma+}(F(x(\sigma),\rho)-F(x(\sigma),\sigma)) =$$

$$= y(\sigma+)-y(\sigma)-(F(x(\sigma),\sigma+)-F(x(\sigma),\sigma)) = P(\sigma+)-P(\sigma)$$

and also

$$\lim_{\rho\to\sigma+}\|y(\rho)-x(\rho)\| = \|P(\sigma+)-P(\sigma)\|. \tag{10.10}$$

For every $\varepsilon>0$ we define

$$\alpha=\frac{\varepsilon}{K(h(t_1)-h(t_0)+1)}>0 \tag{10.11}$$

and assume that $r=r(\alpha)>0$ is such that $\omega(r)<\alpha$. Further, we choose $\gamma\in(0,\frac{r}{2})$.

Since (10.10) holds, there is an $\eta_3(\sigma)\in(0,\eta_2(\sigma))$ such that

$$\|y(\rho)-x(\rho)\|\le\|P(\sigma+)-P(\sigma)\|+\gamma \tag{10.12}$$

for $\rho\in(\sigma,\sigma+\eta_3(\sigma))$ and also

$$\omega(\|y(\rho)-x(\rho)\|)\le\omega(\|P(\sigma+)-P(\sigma)\|+\gamma) \tag{10.13}$$

for $\rho\in(\sigma,\sigma+\eta_3(\sigma))$.

Let us denote

$$N(\alpha) = \{\sigma \in [t_0, t_1];\ \|P(\sigma+) - P(\sigma)\| \geq \frac{r}{2}\}.$$

Since P is of bounded variation on $[t_0, t_1]$ the set $N(\alpha)$ is finite and we denote by $l(\alpha)$ the number of elements of $N(\alpha)$.

If $\sigma \in [t_0, t_1] \setminus N(\alpha)$ and $\rho \in (\sigma, \sigma + \eta_3(\sigma))$ then by (10.13) we have

$$\omega(\|y(\rho) - x(\rho)\|) \leq \omega(\frac{r}{2} + \gamma) < \omega(\frac{r}{2} + \frac{r}{2}) = \omega(r) < \alpha$$

and by (10.9) also

$$\|\int_{\sigma}^{\sigma+\eta} D[F(y(\tau), t) - F(x(\tau), t)]\| \leq \alpha(h(\sigma + \eta) - h(\sigma+)) \tag{10.14}$$

whenever $\eta \in (0, \eta_3(\sigma))$.

If $\sigma \in [t_0, t_1] \cap N(\alpha)$ then there exists $\eta_4(\sigma) \in (0, \eta_3(\sigma))$ such that for $\eta \in (0, \eta_4(\sigma))$ we have

$$h(\sigma + \eta) - h(\sigma+) = |h(\sigma + \eta) - h(\sigma+)| <$$

$$< \frac{\alpha}{(l(\alpha) + 1)\omega(\|P(\sigma+) - P(\sigma)\| + \gamma)}$$

and $(\sigma, \sigma + \eta_4(\sigma)) \cap N(\alpha) = \emptyset$. Hence (10.9) and (10.13) yield

$$\|\int_{\sigma}^{\sigma+\eta} D[F(y(\tau), t) - F(x(\tau), t)]\| \leq$$

$$\leq \omega(\|P(\sigma+) - P(\sigma)\| + \gamma)\frac{\alpha}{(l(\alpha) + 1)\omega(\|P(\sigma+) - P(\sigma)\| + \gamma)} =$$

$$\frac{\alpha}{l(\alpha) + 1} \tag{10.15}$$

for every $\eta \in (\sigma, \sigma + \eta_4(\sigma))$.

Define now

$$\widehat{h}_\alpha(t) = \frac{\alpha}{l(\alpha)+1} \sum_{\sigma \in N(\alpha)} H_\sigma(t)$$

for $t \in [t_0, t_1]$ where $H_\sigma(t) = 0$ for $t \le \sigma$ and $H_\sigma(t) = 1$ for $t > \sigma$. The function $\widehat{h}_\alpha : [t_0, t_1] \to \mathbb{R}$ is nondecreasing, continuous from the left and

$$\operatorname{var}_{t_0}^{t_1} \widehat{h}_\alpha = \widehat{h}_\alpha(t_1) - \widehat{h}_\alpha(t_0) = \frac{\alpha}{l(\alpha)+1} l(\alpha) < \alpha. \tag{10.16}$$

The points of discontinuity of the function $\widehat{h}_\alpha$ are clearly only the points belonging to $N(\alpha)$, and for $t \in N(\alpha)$ we have

$$\widehat{h}_\alpha(t+) - \widehat{h}_\alpha(t) = \frac{\alpha}{l(\alpha)+1}.$$

Using the function $\widehat{h}_\alpha$ we can set

$$h_\alpha(t) = \alpha h_c(t) + \widehat{h}_\alpha(t)$$

for $t \in [t_0, t_1]$ where by h_c the continuous part of the function h is denoted. By definition the function h_α is nondecreasing and continuous from the left on $[t_0, t_1]$ and by (10.16) and (10.11) we obtain

$$h_\alpha(t_1) - h_\alpha(t_0) = \alpha[h_c(t_1) - h_c(t_0)] + \widehat{h}_\alpha(t_1) - \widehat{h}_\alpha(t_0) <$$

$$< \alpha[h(t_1) - h(t_0) + 1] = \frac{\varepsilon}{K}. \tag{10.17}$$

If $\sigma \in [t_0, t_1] \setminus N(\alpha)$ then set $\delta(\sigma) = \eta_3(\sigma) > 0$ and if $\sigma \in [t_0, t_1] \cap N(\alpha)$ then set $\delta(\sigma) = \eta_4(\sigma) > 0$. By this choice of $\delta(\sigma) >$

0 for $\sigma \in [t_0, t_1]$, by (10.14), (10.15) and by the definition of h_α we obtain the inequality

$$\|\int_{\sigma}^{\sigma+\eta} D[F(y(\tau),t) - F(x(\tau),t)]\| \le h_\alpha(\sigma+\eta) - h_\alpha(\sigma)$$

for $\eta \in [0, \delta(\sigma)]$, and (10.8) gives

$$V(\sigma+\eta, y(\sigma+\eta)) - V(\sigma, x(\sigma)) \le$$

$$\le K(\operatorname{var}_{t_0}^{\sigma+\eta} P - \operatorname{var}_{t_0}^{\sigma} P) + \eta M +$$

$$+\eta\varepsilon + K(h_\alpha(\sigma+\eta) - h_\alpha(\sigma)) = g(\sigma+\eta) - g(\sigma) \qquad (10.18)$$

for all $\sigma \in [t_0, t_1]$ and $\eta \in [0, \delta(\sigma)]$ where

$$g(t) = K \operatorname{var}_{t_0}^{t} P + Mt + \varepsilon t + K h_\alpha(t).$$

The function g is of bounded variation on $[t_0, t_1]$ and continuous from the left on $(t_0, t_1]$. From Proposition 10.11 and from (10.17) we obtain by (10.18) the inequality

$$V(t_1, y(t_1)) - V(t_0, y(t_0)) \le g(t_1) - g(t_0) =$$

$$= K \operatorname{var}_{t_0}^{t_1} P + M(t_1 - t_0) + \varepsilon(t_1 - t_0) + K(h_\alpha(t_1) - h_\alpha(t_0)) <$$

$$< K \operatorname{var}_{t_0}^{t_1} P + M(t_1 - t_0) + \varepsilon(t_1 - t_0) + \varepsilon.$$

Since $\varepsilon > 0$ can be arbitrary small, we obtain from this inequality the result given in (10.7). $\square$

Ljapunov type theorems

10.13 Theorem. *Assume that* $V : [0,+\infty) \times \overline{B_a} \to \mathbb{R}$, $0 < a < c$ *is such that for every* $x \in \overline{B_a} = \{y \in \mathbb{R}^n; \|y\| \leq a\}$ *the function* $V(\cdot, x)$ *is continuous from the left.*

Assume that the function $V(t,x)$ *is positive definite, i.e. there exists a continuous increasing real function* $b : [0,+\infty) \to \mathbb{R}$ *such that* $b(\rho) = 0$ *if and only if* $\rho = 0$ *and*

$$V(t,x) \geq b(\|x\| \tag{10.19}$$

for all $(t,x) \in [0,+\infty) \times \overline{B_a}$,

$$V(t,0) = 0 \tag{10.20}$$

and

$$\|V(t,x) - V(t,y)\| \leq K\|x - y\| \tag{10.21}$$

for $x, y \in \overline{B_a}$, $K > 0$ *being a constant.*

If the function $V(t, x(t))$ *is nonincreasing along every solution* $x(t)$ *of the equation (10.2) then the trivial solution* $x \equiv 0$ *of (10.2) is variationally stable.*

Proof. Since we assume that the function $V(t, x(t))$ is nonincreasing whenever $x : [\alpha, \beta] \to \mathbb{R}^n$ is a solution of (10.2) we have

$$\limsup_{\eta \to 0+} \frac{V(t+\eta, x(t+\eta)) - V(t, x(t))}{\eta} \leq 0 \tag{10.22}$$

for $t \in [\alpha, \beta]$.

Let us check that under these assumptions the properties required in Definition 10.1 are satisfied.

Let $\varepsilon > 0$ be given and let $y : [t_0, t_1] \to \mathbb{R}^n$ be of bounded variation on $[t_0, t_1]$ and continuous from the left in $(t_0, t_1]$. Since the function V satisfies the assumptions of Lemma 10.12 with $\Phi \equiv 0$ in the relation (10.6) (see (10.22)) we obtain by (10.7),(10.20) and (10.21) the inequality

$$V(r, y(r)) \le V(t_0, y(t_0)) + K \operatorname{var}_{t_0}^r (y(s) - \int_{t_0}^s DF(y(\tau), t)) \le$$

$$\le K\|y(t_0)\| + K \operatorname{var}_{t_0}^r (y(s) - \int_{t_0}^s DF(y(\tau), t)) \qquad (10.23)$$

which holds for every $r \in [t_0, t_1]$.

Let us define $\alpha(\varepsilon) = \inf_{r \le \varepsilon} b(r)$. Then $\alpha(\varepsilon) > 0$ for $\varepsilon > 0$ and $\lim_{\varepsilon \to 0+} \alpha(\varepsilon) = 0$. Further, choose $\delta(\varepsilon) > 0$ such that $2K\delta(\varepsilon) < \alpha(\varepsilon)$.

If in this situation the function y is such that $\|y(t_0)\| < \delta(\varepsilon)$ and

$$\operatorname{var}_{t_0}^{t_1} (y(s) - \int_{t_0}^s DF(y(\tau), t)) < \delta(\varepsilon)$$

then by (10.23) we obtain the inequality

$$V(r, y(r)) \le 2K\delta(\varepsilon) \qquad (10.24)$$

provided $r \in [t_0, t_1]$.

If there exists a $\hat{t} \in [t_0, t_1]$ such that $\|y(\hat{t})\| \ge \varepsilon$ then by (10.19) we get the inequality

$$V(\hat{t}, y(\hat{t})) \ge b(\|y(\hat{t})\|) \ge \inf_{r \le \varepsilon} b(r) = \alpha(\varepsilon)$$

which contradicts (10.24). Hence $\|y(t)\| < \varepsilon$ for all $t \in [t_0, t_1]$ and by Definition 10.1 the solution $x \equiv 0$ is variationally stable. □

10.14 Theorem. *Let $V : [0, +\infty) \times \overline{B_a} \to \mathbb{R}$, $0 < a < c$ be a function with the properties given in Theorem 10.13.*

If for every solution $x : [t_0, t_1] \to B_a$ of the equation (10.2) the inequality

$$\limsup_{\eta \to 0+} \frac{V(t+\eta, x(t+\eta)) - V(t, x(t))}{\eta} \leq -\Phi(x(t)) \tag{10.25}$$

holds for every $t \in [t_0, t_1]$, where $\Phi : \mathbb{R}^n \to \mathbb{R}$ is continuous, $\Phi(0) = 0$, $\Phi(x) > 0$ for $x \neq 0$, then the solution $x \equiv 0$ of (10.2) is variationally–asymptotically stable.

Proof. From (10.25) it is clear that the function $V(t, x(t))$ is nonincreasing along every solution $x(t)$ of (10.2) and therefore by Theorem 10.13 the trivial solution $x \equiv 0$ of (10.2) is variationally stable. By Definition 10.3 it remains to show that the solution $x \equiv 0$ of (10.2) is variationally attracting in the sense of Definition 10.2.

From the variational stability of the solution $x \equiv 0$ of (10.2) there is a $\delta_0 \in (0, a)$ such that if $y : [t_0, t_1] \to \mathbb{R}^n$ is of bounded variation on $[t_0, t_1]$ where $0 \leq t_0 < t_1 < +\infty$, y is continuous from the left on $(t_0, t_1]$ and such that $\|y(t_0)\| < \delta_0$,

$$\operatorname{var}_{t_0}^{t_1}(y(s) - \int_{t_0}^{s} DF(y(\tau), t)) < \delta_0,$$

then $\|y(t)\| < a$ for $t \in [t_0, t_1]$, i.e. $y(t) \in B_a$ for every $t \in [t_0, t_1]$.

Let $\varepsilon > 0$ be arbitrary. From the variational stability of the trivial solution we obtain that there is a $\delta(\varepsilon) > 0$ such that for every $y : [t_2, t_3] \to \mathbb{R}^n$ of bounded variation on $[t_2, t_3]$ where $0 \leq t_2 < t_3 < +\infty$, y continuous from the left on $(t_2, t_3]$ and such that

$$\|y(t_0)\| < \delta(\varepsilon) \tag{10.26}$$

and

$$\operatorname{var}_{t_0}^{t_1}(y(s) - \int_{t_0}^{s} DF(y(\tau), t)) < \delta(\varepsilon), \tag{10.27}$$

we have

$$\|y(t)\| < \varepsilon \tag{10.28}$$

for $t \in [t_2, t_3]$. Define $\gamma(\varepsilon) = \min(\delta_0, \delta(\varepsilon))$ and
$T(\varepsilon) = -K\dfrac{\delta_0 + \gamma(\varepsilon)}{M} > 0$ where

$$M = \sup\{-\Phi(x); \gamma(\varepsilon) \leq \|x\| < \varepsilon\} =$$

$$= -\inf\{\Phi(x); \gamma(\varepsilon) \leq \|x\| < \varepsilon\} < 0$$

and assume that $y : [t_0, t_1] \to \mathbb{R}^n$ is of bounded variation on $[t_0, t_1]$ where $0 \leq t_0 < t_1 < +\infty$, y is continuous from the left on $(t_0, t_1]$ and such that $\|y(t_0)\| < \delta_0$,

$$\operatorname{var}_{t_0}^{t_1}(y(s) - \int_{t_0}^{s} DF(y(\tau), t)) < \gamma(\varepsilon). \tag{10.29}$$

Assume that $T(\varepsilon) < t_1 - t_0$, i.e. $t_0 + T(\varepsilon) < t_1$. We show that there exists a $t^* \in [t_0, t_0 + T(\varepsilon)]$ such that $\|y(t^*)\| < \gamma(\varepsilon)$. Assume the contrary, i.e. $\|y(s)\| \geq \gamma(\varepsilon)$ for every $s \in [t_0, t_0 + T(\varepsilon)]$. Lemma 10.12 yields

$$V(t_0 + T(\varepsilon), y(t_0 + T(\varepsilon))) - V(t_0, y(t_0)) \leq$$

$$\leq \operatorname{var}_{t_0}^{t_0+T(\varepsilon)}(y(s) - \int_{t_0}^{s} DF(y(\tau), t)) + MT(\varepsilon) <$$

$$< K\gamma(\varepsilon) + M\frac{-K(\delta_0 + \gamma(\varepsilon))}{M} = -K\delta_0.$$

Hence

$$V(t_0 + T(\varepsilon), y(t_0 + T(\varepsilon))) \leq V(t_0, y(t_0)) - K\delta_0 \leq$$

$$\leq K\|y(t_0)\| - K\delta_0 < K\delta_0 - K\delta_0 = 0$$

and this contradicts the inequality

$$V(t_0 + T(\varepsilon), y(t_0 + T(\varepsilon))) \geq b(\|y(t_0 + T(\varepsilon))\|) \geq b(\gamma(\varepsilon)) > 0.$$

Hence necessarily there is a $t^* \in [t_0, t_0 + T(\varepsilon)]$ such that

$$\|y(t^*)\| < \gamma(\varepsilon)$$

and by (10.29) we have $\|y(t)\| < \varepsilon$ for $t \in [t^*, t_1]$, because (10.26), (10.27) hold in view of the choice of $\gamma(\varepsilon)$ and (10.28) is satisfied for the case $t_1 = t^*$, $t_3 = t_1$. Consequently, also $\|y(t)\| < \varepsilon$ for $t > t_0 + T(\varepsilon)$, because $t^* \in [t_0, t_0 + T(\varepsilon)]$ and therefore the trivial solution $x \equiv 0$ is a variationally attracting solution of (10.2). □

Converse Ljapunov theorems

This part is devoted to the conversion of the Ljapunov type stability results, namely Theorems 10.13 and 10.14. Our goal is to show that the variational stability and the asymptotic variational stability imply the existence of Ljapunov functions with the properties described in Theorems 10.13 and 10.14.

First we show some auxiliary results.

Let us introduce a modified notion of the variation of a function.

10.15 Definition. Assume that $-\infty < a < b < +\infty$ and that $G : [a, b] \to \mathbb{R}^n$ is given.

For a given decomposition

$$D : a = \alpha_0 < \alpha_1 < \cdots < \alpha_k = b$$

of the interval $[a, b]$ and for every $\lambda \geq 0$ define

$$\sum_{j=1}^{k} e^{-\lambda(b-\alpha_{j-1})} |G(\alpha_j) - G(\alpha_{j-1})| = v_\lambda(G, D)$$

and set

$$e_\lambda \operatorname{var}_a^b G = \sup_D v_\lambda(G, D)$$

where the supremum is taken over all finite decompositions D of the interval $[a, b]$. The number $e_\lambda \operatorname{var}_a^b G$ is called the e_λ–variation of the function G over the interval $[a, b]$.

10.16 Lemma. *If* $-\infty < a < b < +\infty$ *and* $G : [a, b] \to \mathbb{R}^n$ *then for every* $\lambda \geq 0$ *we have*

$$e^{-\lambda(b-a)} \operatorname{var}_a^b G \leq e_\lambda \operatorname{var}_a^b G \leq \operatorname{var}_a^b G. \tag{10.30}$$

If $a \leq c \leq b$ *then for* $\lambda \geq 0$ *the identity*

$$e_\lambda \operatorname{var}_a^b G = e^{-\lambda(b-c)} e_\lambda \operatorname{var}_a^c G + e_\lambda \operatorname{var}_c^b G \tag{10.31}$$

holds.

Proof. For every $\lambda \geq 0$ and every decomposition D of $[a, b]$ we have

$$e^{-\lambda(b-a)} \leq e^{-\lambda(b-\alpha_{j-1})} \leq e^0 = 1$$

for $j = 1, 2, \dots k$.

Therefore

$$e^{-\lambda(b-a)} v_0(G, D) \leq v_\lambda(G, D) \leq$$

$$\leq v_0(G, D) = \sum_{j=1}^{k} |G(\alpha_j) - G(\alpha_{j-1})|$$

and passing to the supremum over all finite decompositions D of $[a, b]$ we obtain the inequality (10.30).

It is easy to see that for proving the second statement we can restrict ourselves to the case of decompositions D which contain the point c as a node, i.e.

$$D : a = \alpha_0 < \alpha_1 < \cdots < \alpha_{l-1} < \alpha_l = c < \alpha_{l+1} < \cdots < \alpha_k = b.$$

Then

$$v_\lambda(G,D) = \sum_{j=1}^{k} e^{-\lambda(b-\alpha_{j-1})}|G(\alpha_j) - G(\alpha_{j-1})| =$$

$$= \sum_{j=1}^{l} e^{-\lambda(b-\alpha_{j-1})}|G(\alpha_j) - G(\alpha_{j-1})| +$$

$$+ \sum_{j=l+1}^{k} e^{\lambda(b-\alpha_{j-1})}|G(\alpha_j) - G(\alpha_{j-1})| =$$

$$= e^{-\lambda(b-c)} \sum_{j=1}^{l} e^{-\lambda(c-\alpha_{j-1}}|G(\alpha_j) - G(\alpha_{j-1})| +$$

$$+ \sum_{j=l+1}^{k} e^{\lambda(b-\alpha_{j-1})}|G(\alpha_j) - G(\alpha_{j-1})| =$$

$$= e^{-\lambda(b-c)} v_\lambda(G, D_1) + v_\lambda(G, D_2) \tag{10.32}$$

where

$$D_1 : a = \alpha_0 < \alpha_1 < \cdots < \alpha_{l-1} < \alpha_l = c$$

and

$$D_2 : c = \alpha_l < \alpha_{l+1} < \cdots < \alpha_k = b$$

are decompositions of $[a, c]$ and $[c, b]$, respectively. On the other hand, any two such decompositions D_1 and D_2 form a decomposition D of the interval $[a, b]$. The equality (10.31) now easily follows from (10.32) when we pass to the corresponding suprema.
□

10.17 Corollary. *If $a \le c \le b$ and $\lambda \ge 0$ then*

$$e_\lambda \operatorname{var}_a^c G \le e_\lambda \operatorname{var}_a^b G. \tag{10.33}$$

For $a > 0,\ t > 0,\ x \in B_a$ denote by $A_a(t,x)$ the set of all functions $\varphi : [0,+\infty) \to \mathbb{R}^n$ which are locally of bounded variation on $[0,+\infty)$, $\varphi(0) = 0$, $\varphi(t) = x$, φ is continuous from the left and $\sup_{s\in[0,t]} \|\varphi(s)\| < a$.

Moreover, for $\lambda \ge 0$. $s \ge 0$ and $x \in B_a$ define

$$V_\alpha(s,x) = \begin{cases} \inf\limits_{\varphi\in A_a(s,x)} \{e_\lambda \operatorname{var}_0^s(\varphi(\sigma) - \int_0^\sigma DF(\varphi(\tau),t))\} \\ \qquad \text{if } s > 0 \\ \|x\| \quad \text{if } s = 0. \end{cases} \tag{10.34}$$

Note that this definition of $V_\alpha(s,x)$ makes sense because for $\varphi \in A_a(s,x)$ the integral $\int_0^\sigma DF(\varphi(\tau),t)$ is a function of bounded variation in the variable σ and therefore the function $\varphi(\sigma) - \int_0^\sigma DF(\varphi(\tau),t)$ is of bounded variation on $[0,s]$ as well and the e_λ–variation of this function is bounded.

The function $\varphi \equiv 0$ evidently belongs to $A_a(s,0)$ and therefore we have

$$V_\lambda(s,0) = 0 \tag{10.35}$$

for every $s \ge 0$ and $\lambda \ge 0$ because $\varphi(\sigma) - \int_0^\sigma DF(\varphi(\tau),t) = 0$ for $\sigma > 0$.

Since $e_\lambda \operatorname{var}_0^s(\varphi(\sigma) - \int_0^\sigma DF(\varphi(\tau),t)) \ge 0$ for every $\varphi \in A_a(s,x)$ we have by the definition (10.34) also the inequality

$$V_\lambda(s,x) \ge 0 \tag{10.36}$$

for every $s \ge 0$ and $x \in \mathbb{R}^n$.

10.18 Lemma. *For $x, y \in B_a = \{x \in \mathbb{R}^n; \|x\| < a\}$, $s \in [0, +\infty)$ and $\lambda \geq 0$ the inequality*

$$|V_\lambda(s,x) - V_\lambda(s,y)| \leq \|x - y\| \tag{10.37}$$

holds.

Proof. Assume that $s > 0$ and $0 < \eta < s$. Let $\varphi \in A_a(s,x)$ be arbitrary. Define

$$\varphi_\eta(\sigma) = \varphi(\sigma) \text{ for } \sigma \in [0, s - \eta],$$

and

$$\varphi_\eta(\sigma) = \varphi(\sigma - \eta) + \frac{1}{\eta}(y - \varphi(\sigma - \eta))(\sigma - s + \eta) \text{ for } \sigma \in [s - \eta, s].$$

The function φ_η coincides with φ on $[0, s - \eta]$ and is linear with $\varphi_\eta(s) = y$ on $[s - \eta, s]$. By definition clearly $\varphi_\eta \in A_a(s,y)$ and by (10.31) from Lemma 10.15 we obtain

$$V_\lambda(s,y) \leq e_\lambda \operatorname{var}_0^s(\varphi_\eta(\sigma) - \int_0^\sigma DF(\varphi_\eta(\tau),t)) =$$

$$= e^{-\lambda\eta} e_\lambda \operatorname{var}_0^{s-\eta}(\varphi(\sigma) - \int_0^\sigma DF(\varphi(\tau),t)) +$$

$$+ e_\lambda \operatorname{var}_{s-\eta}^s(\varphi_\eta(\sigma) - \int_0^\sigma DF(\varphi_\eta(\tau),t)) \leq$$

$$\leq e^{-\lambda\eta} e_\lambda \operatorname{var}_0^{s-\eta}(\varphi(\sigma) - \int_0^\sigma DF(\varphi(\tau),t)) +$$

$$+ \operatorname{var}_{s-\eta}^s \varphi_\eta + \operatorname{var}_{s-\eta}^s(\int_0^\sigma DF(\varphi(\tau),t)) \leq$$

$$\leq e^{-\lambda\eta} e_\lambda \operatorname{var}_0^{s-\eta}(\varphi(\sigma) - \int_0^\sigma DF(\varphi(\tau),t))+$$

$$+|y-\varphi(s-\eta)| + h(s) - h(s-\eta).$$

Since for every $\eta > 0$ we have

$$e^{-\lambda\eta} e_\lambda \operatorname{var}_0^{s-\eta}(\varphi(\sigma) - \int_0^\sigma DF(\varphi(\tau),t)) =$$

$$= e_\lambda \operatorname{var}_0^{s}(\varphi(\sigma) - \int_0^\sigma DF(\varphi(\tau),t))-$$

$$-e_\lambda \operatorname{var}_{s-\eta}^{s}(\varphi(\sigma) - \int_0^\sigma DF(\varphi(\tau),t)) \leq$$

$$\leq e_\lambda \operatorname{var}_0^{s}(\varphi(\sigma) - \int_0^\sigma DF(\varphi(\tau),t))$$

by (10.35), we obtain for every $\eta > 0$ the inequality

$$V_\lambda(s,y) \leq e_\lambda \operatorname{var}_0^{s}(\varphi(\sigma) - \int_0^\sigma DF(\varphi(\tau),t))+$$

$$+|y-\varphi(s-\eta)| + h(s) - h(s-\eta).$$

The functions φ and h are assumed to be continuous from the left and therefore $\lim_{\tau\to s-}\varphi(\tau) = \varphi(s) = x$; moreover the last inequality is valid for every $\eta > 0$ and consequently we can pass to the limit $\eta \to 0+$ in order to obtain

$$V_\lambda(s,y) \leq e_\lambda \operatorname{var}_0^{s}(\varphi(\sigma) - \int_0^\sigma DF(\varphi(\tau),t)) + |y-x|$$

for every $\varphi \in A_a(s,x)$. Taking the infimum for all $\varphi \in A_a(s,x)$ on the right hand side of the last inequality we arrive at

$$V_\lambda(s,y) \leq V_\lambda(s,x) + |x-y|. \tag{10.38}$$

Since this reasoning is fully symmetric with respect to x and y we similarly obtain also

$$V_\lambda(s,x) \le V_\lambda(s,y) + |x-y|$$

and this together with (10.38) yields (10.37) for $s>0$.

If $s=0$, then we have by definition

$$|V_\lambda(0,y) - V_\lambda(0,x)| = ||y|-|x|| \le |x-y|.$$

This proves the statement. □

10.19 Corollary. *Since $V_\lambda(s,0)=0$ for every $s \ge 0$, we have by (10.35) and (10.37)*

$$0 \le V_\lambda(s,x) \le \|x\|. \tag{10.39}$$

10.20 Lemma. *For $y \in B_a$, $s,r \in [0,+\infty)$ and $\lambda \ge 0$ the inequality*

$$|V_\lambda(r,y) - V_\lambda(s,y)| \le (1-e^{-\lambda|r-s|})a + |h(r)-h(s)| \tag{10.40}$$

holds.

Proof. Suppose that $0 < s \le r$ and let $\varphi \in A_a(r,y)$ be given. Then by Lemma 10.16 we have

$$e_\lambda \operatorname{var}_0^r\Big(\varphi(\sigma) - \int_0^\sigma DF(\varphi(\tau),t)\Big) =$$

$$= e^{-\lambda(r-s)} e_\lambda \operatorname{var}_0^s\Big(\varphi(\sigma) - \int_0^\sigma DF(\varphi(\tau),t)\Big) +$$

$$+ e_\lambda \operatorname{var}_s^r\Big(\varphi(\sigma) - \int_0^\sigma DF(\varphi(\tau),t)\Big) \ge$$

$$\geq e^{-\lambda(r-s)}V_\lambda(s,\varphi(s)) + e_\lambda \operatorname{var}_s^r(\varphi(\sigma) - \int_0^\sigma DF(\varphi(\tau),t)) \geq$$

$$\geq e^{-\lambda(r-s)}[V_\lambda(s,\varphi(s)) + \operatorname{var}_s^r \varphi - \operatorname{var}_s^r \int_0^\sigma DF(\varphi(\tau),t)] \geq$$

$$\geq e^{-\lambda(r-s)}[V_\lambda(s,\varphi(s)) + |y-\varphi(s)| + (h(r)-h(s))] \geq$$

$$\geq e^{-\lambda(r-s)}[V_\lambda(s,y) + (h(r)-h(s))]. \tag{10.41}$$

The inequality (10.36) from Lemma 10.18 leads to

$$V_\lambda(s,\varphi(s)) + |y-\varphi(s)| \geq V_\lambda(s,y).$$

Taking the infimum over $\varphi \in A_a(r,y)$ on the left hand side of (10.41) we obtain

$$V_\lambda(r,y) \geq e^{-\lambda(r-s)}[V_\lambda(s,y) + (h(r)-h(s))] \geq$$

$$\geq e^{-\lambda(r-s)}V_\lambda(s,y) + (h(r)-h(s)). \tag{10.42}$$

Now let $\varphi \in A_a(s,y)$ be arbitrary. Let us define

$$\varphi^*(\sigma) = \begin{cases} \varphi(\sigma) & \text{for } \sigma \in [0,s], \\ y & \text{for } \sigma \in [s,r]. \end{cases}$$

We have evidently $\varphi^*(s) = \varphi(s) = y$, $\varphi^* \in A_a(r,y)$ and by (10.30), (10.35) we obtain

$$V_\lambda(r,y) \leq e_\lambda \operatorname{var}_0^r(\varphi^*(\sigma) - \int_0^\sigma DF(\varphi^*(\tau),t)) =$$

$$= e^{-\lambda(r-s)}e_\lambda \operatorname{var}_0^s(\varphi(\sigma) - \int_0^\sigma DF(\varphi(\tau),t)) +$$

$$+e_\lambda \operatorname{var}_s^r(\varphi^*(\sigma) - \int_0^\sigma DF(\varphi^*(\tau),t)) \leq$$

$$\leq e^{-\lambda(r-s)} e_\lambda \operatorname{var}_0^s(\varphi(\sigma) - \int_0^\sigma DF(\varphi(\tau),t)) +$$

$$+ \operatorname{var}_s^r \varphi^* + \operatorname{var}_s^r \int_0^\sigma DF(\varphi^*(\tau),t) \leq$$

$$\leq e^{-\lambda(r-s)} e_\lambda \operatorname{var}_0^s(\varphi(\sigma) - \int_0^\sigma DF(\varphi(\tau),t)) + h(r) - h(s).$$

Taking the infimum over all $\varphi \in A_a(s,y)$ on the right hand side of this inequality we obtain

$$V_\lambda(r,y) \leq e^{-\lambda(r-s)} V_\lambda(s,y) + (h(r) - h(s)).$$

Together with (10.42) we have

$$|V_\lambda(r,y) - e^{-\lambda(r-s)} V_\lambda(s,y)| \leq h(r) - h(s).$$

Hence by (10.39) we get the inequality

$$|V_\lambda(r,y) - V_\lambda(s,y)| \leq$$

$$\leq |V_\lambda(r,y) - e^{-\lambda(r-s)} V_\lambda(s,y)| + |1 - e^{-\lambda(r-s)}||V_\lambda(s,y)| \leq$$

$$\leq h(r) - h(s) + (1 - e^{-\lambda(r-s)})|y| \leq$$

$$\leq h(r) - h(s) + (1 - e^{-\lambda(r-s)})a$$

because $|y| \leq a$. In this way we have obtained (10.40).

Assume that $s = 0$ and $r > 0$. Then by (10.39) and by the definition given in (10.34) we get

$$V_\lambda(r,y) - V_\lambda(s,y) = V_\lambda(r,y) - V_\lambda(0,y) =$$

$$= V_\lambda(r, y) - |y| \leq 0. \tag{10.43}$$

We derive an estimate from below. Assume that $\varphi \in A_a(r, y)$. We have

$$e_\lambda \operatorname{var}_0^r(\varphi(\sigma) - \int_0^\sigma DF(\varphi(\tau), t)) \geq$$

$$\geq e_\lambda \operatorname{var}_0^r \varphi - e_\lambda \operatorname{var}_0^r(\int_0^\sigma DF(\varphi(\tau), t)) \geq$$

$$\geq e^{-\lambda r} \operatorname{var}_0^r \varphi - \operatorname{var}_0^r(\int_0^\sigma DF(\varphi(\tau), t)) \geq$$

$$\geq e^{-\lambda r}|\varphi(r) - \varphi(0)| - (h(r) - h(0)) =$$

$$= e^{-\lambda r}|y| - (h(r) - h(0))$$

by (10.30), Lemma 10.16 and Lemma 3.9. Passing again to the infimum for $\varphi \in A_a(r, y)$ on the left hand side of this inequality we get

$$V_\lambda(r, y) \geq e^{-\lambda r}|y| - (h(r) - h(0))$$

and

$$V_\lambda(r, y) - V_\lambda(0, y) = V_\lambda(r, y) - |y| \geq$$

$$\geq (e^{-\lambda r} - 1)|y| - (h(r) - h(0)) =$$

$$= -(1 - e^{-\lambda r})|y| - (h(r) - h(0)).$$

This together with (10.43) yields

$$|V_\lambda(r, y) - V_\lambda(0, y)| \leq (1 - e^{-\lambda r})a - (h(r) - h(0)),$$

and this means that the inequality (10.40) holds in this case, too. The remaining case of $r = s = 0$ is evident. Finally, let us mention that the case when $r < s$ can be dealt with in the same way because the situation is symmetric in s and r. □

By the previous Lemmas 10.18 and 10.20 we immediately conclude that the following holds.

10.21 Corollary. *For $x, y \in B_a = \{x \in \mathbb{R}^n; \|x\| < a\}$, $r, s \in [0, +\infty)$ and $\lambda \geq 0$ the inequality*

$$|V_\lambda(s,x) - V_\lambda(r,y)| \leq \|x - y\| + (1 - e^{-\lambda|r-s|})a + |h(r) - h(s)| \tag{10.44}$$

holds.

Now we will discuss the behaviour of the function $V_\lambda(t,x)$ defined by (10.34) along the solutions of the generalized ordinary differential equation

$$\frac{dx}{d\tau} = DF(x,t). \tag{10.2}$$

We still assume that the assumptions given at the beginning of this chapter are satisfied for the right hand side $F(x,t)$.

The next statement will be of interest for the forthcoming considerations.

10.22 Lemma. *If $\psi : [s, s + \eta(s)] \to \mathbb{R}^n$ is a solution of the generalized ordinary differential equation (10.2), $s \geq 0$, $\eta(s) > 0$, then for every λ the inequality*

$$\limsup_{\eta \to 0+} \frac{V_\lambda(s + \eta, \psi(s + \eta)) - V_\lambda(s, \psi(s))}{\eta} \leq -\lambda V_\lambda(s, \psi(s)) \tag{10.45}$$

holds.

Proof. Let $s \in [0, +\infty)$ and $x \in \mathbb{R}^n$ be given. Let us choose $a > 0$ such that $a > |x| + h(s+1) - h(s)$. Assume that $\varphi \in A_a(s,x)$ is given and let $\psi : [s, s + \eta(s)] \to \mathbb{R}^n$ be a solution of (10.2) on $[s, s + \eta(s)]$ with $\psi(s) = x$ where $0 < \eta(s) < 1$. The existence of such a solution is guaranteed by the local existence theorem 4.2. For $0 < \eta < \eta(s)$ define

$$\varphi_\eta(\sigma) = \begin{cases} \varphi(\sigma) & \text{for } \sigma \in [0, s], \\ \psi(\sigma) & \text{for } \sigma \in [s, s + \eta]; \end{cases}$$

we have $\varphi(s) = \psi(s) = \varphi_\eta(s) = x$. Evidently $\varphi_\eta \in A_a(s+\eta, \psi(s+\eta))$ since ψ is continuous from the left and by the definition of a solution we have

$$\|\psi(\sigma)\| = \|x + \int_s^\sigma DF(\psi(\tau), t)\| \le$$

$$\le \|x\| + h(\sigma) - h(s) \le \|x\| + h(s+1) - h(s) < a$$

for $\sigma \in [s, s+\eta]$ and

$$V_\lambda(s+\eta, \psi(s+\eta)) \le$$

$$\le e_\lambda \operatorname{var}_0^{s+\eta}(\varphi_\eta(\sigma) - \int_0^\sigma DF(\varphi_\eta(\tau), t)) =$$

$$= e^{-\lambda\eta} e_\lambda \operatorname{var}_0^s(\varphi(\sigma) - \int_0^\sigma DF(\varphi(\tau), t)) +$$

$$+ e_\lambda \operatorname{var}_s^{s+\eta}(\psi(\sigma) - \int_0^s DF(\varphi(\tau), t) - \int_s^\sigma DF(\psi(\tau), t)) =$$

$$= e^{-\lambda\eta} e_\lambda \operatorname{var}_0^s(\varphi(\sigma) - \int_0^\sigma DF(\varphi(\tau), t)) +$$

$$+ e_\lambda \operatorname{var}_s^{s+\eta}(x - \int_0^s DF(\varphi(\tau), t)) =$$

$$= e^{-\lambda\eta} e_\lambda \operatorname{var}_0^s(\varphi(\sigma) - \int_0^\sigma DF(\varphi(\tau), t)).$$

Taking the infimum for all $\varphi \in A_a(s, x)$ on the right hand side of this inequality we obtain

$$V_\lambda(s+\eta, \psi(s+\eta)) \le e^{-\lambda\eta} V_\lambda(s, x) = e^{-\lambda\eta} V_\lambda(s, \psi(s)).$$

This inequality yields

$$V_\lambda(s+\eta,\psi(s+\eta)) - V_\lambda(s,\psi(s)) \le (e^{-\lambda\eta}-1)V_\lambda(s,\psi(s))$$

and also

$$\frac{V_\lambda(s+\eta,\psi(s+\eta)) - V_\lambda(s,\psi(s))}{\eta} \le \frac{e^{-\lambda\eta}-1}{\eta}V_\lambda(s,\psi(s))$$

for every $0<\eta<\eta(s)$.

Since $\lim_{\eta\to 0}\dfrac{e^{-\lambda\eta}-1}{\eta} = -\lambda$ we immediately obtain (10.45).

□

Now we are in position when the converse theorems to Theorems 10.13 and 10.14 can be proved.

10.23 Theorem. *If the trivial solution $x \equiv 0$ of the generalized ordinary differential equation (10.2) is variationally stable then for every $0 < a < c$ there exists a function $V : [0,+\infty)\times B_a \to \mathbb{R}$ satisfying the following conditions:*

1) *for every $x \in B_a$ the function $V(\cdot,x)$ is continuous from the left and $V(\cdot,x)$ is locally of bounded variation on $[0,+\infty)$,*
2) *$V(t,0)=0$ and*

$$|V(t,x)-V(t,y)| \le \|x-y\|$$

for $x,y\in B_a$, $t\in[0,+\infty)$,

3) *the function V is nonincreasing along the solutions of the equation (10.2),*
4) *the function $V(t,x)$ is positive definite, i.e. there is a continuous nondecreasing real–valued function $b:[0,+\infty)\to\mathbb{R}$ such that $b(\rho)=0$ if and only if $\rho=0$ and*

$$b(\|x\|)\le V(t,x)$$

for every $x\in B_a$, $t\in[0,+\infty)$.

Proof. The candidate for the function V is the function $V_0(s,x)$ defined by (10.34) for $\lambda = 0$, i.e. we take $V(s,x) = V_0(s,x)$. The properties stated in 1) are easy consequences of Corollary 10.21. The properties given in 2) follow from (10.35) and from Lemma 10.18. By Lemma 10.22 for every solution $\psi : [s, s+\delta] \to \mathbb{R}^n$ of the generalized ordinary differential equation (10.2) we have

$$\limsup_{\eta \to 0+} \frac{V_\lambda(s+\eta, \psi(s+\eta)) - V_\lambda(s, \psi(s))}{\eta} \le 0$$

and therefore 3) is also satisfied. It remains to show that the function $V(t,x)$ given in this way is positive definite. This is the only point where the variational stability of the solution $x \equiv 0$ of the equation (10.2) is used. Assume that there is an ε, $0 < \varepsilon < a$, and a sequence (t_k, x_k), $k = 1, 2, \dots$, $\varepsilon \le \|x_k\| < a$, $t_k \to \infty$ for $k \to \infty$ such that $V(t_k, x_k) \to 0$ for $k \to \infty$.

Let $\delta(\varepsilon) > 0$ correspond to ε by Definition 10.4 of stability with respect to perturbations (the variational stability of $x \equiv 0$ is equivalent to the stability with respect to perturbations of this solution by Theorem 10.8).

Assume that $k_0 \in \mathbb{N}$ is such that for $k > k_0$ we have $V(t_k, x_k) < \delta(\varepsilon)$. Then there exists $\varphi_k \in A_a(t_k, x_k)$ such that

$$\operatorname{var}_0^{t_k}\Big(\varphi_k(\sigma) - \int_0^\sigma DF(\varphi_k(\tau), t)\Big) < \delta(\varepsilon).$$

Let us set

$$P(\sigma) = \varphi_k(\sigma) - \int_0^\sigma DF(\varphi_k(\tau), t) \text{ for } \sigma \in [0, t_k],$$

$$P(\sigma) = x_k - \int_0^{t_k} DF(\varphi_k(\tau), t) \text{ for } \sigma \in [t_k, +\infty).$$

We evidently have

$$\operatorname{var}_0^\infty P = \operatorname{var}_0^{t_k}\Big(\varphi_k(\sigma) - \int_0^\sigma DF(\varphi_k(\tau),t)\Big) < \delta(\varepsilon).$$

and the function P is continuous from the left. For $\sigma \in [0,t]$ we have

$$\varphi_k(\sigma) = \int_0^\sigma DF(\varphi_k(\tau),t) + \varphi_k(\sigma) - \int_0^\sigma DF(\varphi_k(\tau),t) =$$

$$= \int_0^\sigma DF(\varphi_k(\tau),t) + P(\sigma) - P(0) =$$

$$= \varphi_k(0) + \int_0^\sigma D[F(\varphi_k(\tau),t) + P(t)]$$

because $\varphi_k(0) = 0$. Hence φ_k is a solution of the equation

$$\frac{dy}{d\tau} = D[F(y,t) + P(t)]$$

and therefore, by the variational stability we have $\|\varphi_k(s)\| < \varepsilon$ for every $s \in [0,t_k]$. Hence we also have $\|\varphi_k(t_k)\| = \|x_k\| < \varepsilon$ and this contradicts our assumption. In this way we obtain that the function $V(t,x)$ is positive definite and 4) is also satisfied. □

The next statement is the converse for Theorem 10.14 on variational–asymptotic stability.

10.24 Theorem. *If the trivial solution $x \equiv 0$ of the generalized ordinary differential equation (10.2) is variationally–asymptotically stable then for every $0 < a < c$ there exists a function $U : [0,+\infty) \times B_a \to \mathbb{R}$ satisfying the following conditions:*

1) *for every $x \in B_a$ the function $U(\cdot,x)$ is continuous from the left and $U(\cdot,x)$ is locally of bounded variation on $[0,+\infty)$,*

2) $U(t,0) = 0$ and

$$|U(t,x) - U(t,y)| \le \|x - y\|$$

for $x, y \in B_a,\ t \in [0, +\infty)$,

3) *for every solution* $\psi(\sigma)$ *of the equation (10.2) defined for* $\sigma \ge t$, *where* $\psi(t) = x \in B_a$ *the relation*

$$\limsup_{\eta \to 0+} \frac{U(t+\eta, \psi(t+\eta)) - U(t,x)}{\eta} \le -U(t,x)$$

holds,

4) *the function* $U(t,x)$ *is positive definite.*

Proof. For $x \in B_a,\ s \ge 0$ let us set

$$U(s,x) = V_1(s,x)$$

where $V_0(s,x)$ is the function defined by (10.34) for $\lambda = 1$. In the same way as in the proof of Theorem 10.23 we can see that the function U satisfies 1), 2) and 3). (The item 3) is exactly the statement given in Lemma 10.22.)

It remains to show that 4) is satisfied for this choice of the function U.

Since the solution $x \equiv 0$ of (10.2) is assumed to be variationally attracting, it is by Theorem 10.8 also attracting with respect to perturbations and therefore there exists $\delta_0 > 0$ and for every $\varepsilon > 0$ there is a $T = T(\varepsilon) \ge 0$ and $\gamma = \gamma(\varepsilon) > 0$ such that if $\|y_0\| < \delta_0$, $y_0 \in \mathbb{R}^n$ and $P \in BV([t_0, t_1])$ is a function of bounded variation on $[t_0, t_1]$, continuous from the left on $(t_0, t_1]$ and such that $\mathrm{var}_{t_0}^{t_1} P < \gamma(\varepsilon)$ then

$$\|y(t, t_0, y_0)\| < \varepsilon$$

for all $t \in [t_0, t_1] \cap [t_0 + T(\varepsilon), +\infty)$ and $t_0 \geq 0$ where $y(t, t_0, y_0)$ is a solution of

$$\frac{dx}{d\tau} = D[F(x,t) + P(t)] \tag{10.3}$$

with $y(t_0, t_0, y_0) = y_0$.

Assume that U is not positive definite. Then there exists ε, $0 < \varepsilon < a = \delta_0$ and a sequence (t_k, x_k), $k = 1, 2, \ldots$, $\varepsilon \leq \|x_k\| < a$, $t_k \to \infty$ for $k \to \infty$ such that $U(t_k, x_k) \to 0$ for $k \to \infty$.

Let us choose $k_0 \in \mathbb{N}$ such that for $k \in \mathbb{N}$, $k > k_0$ we have $t_k > T(\varepsilon) + 1$ and

$$U(t_k, x_k) < \gamma(\varepsilon) e^{-(T(\varepsilon)+1)}.$$

According to the definition of U let us choose $\varphi \subset A_a(t_k, x_k)$ such that

$$e_1 \operatorname{var}_0^{t_k}(\varphi(\sigma) - \int_0^\sigma DF(\varphi(\tau), t)) < \gamma(\varepsilon) e^{-(T(\varepsilon)+1)}.$$

Define $t_0 = t_k - (T(\varepsilon) + 1)$. Then $t_0 > 0$ because $t_k > T(\varepsilon) + 1$ and also $t_k = t_0 + T(\varepsilon) + 1 > t_0 + T(\varepsilon)$. Further, evidently

$$e_1 \operatorname{var}_{t_0}^{t_k}(\varphi(\sigma) - \int_0^\sigma DF(\varphi(\tau), t)) < \gamma(\varepsilon) e^{-(T(\varepsilon)+1)},$$

and by (10.30) (see Lemma 10.16) also

$$e^{-(T(\varepsilon)+1)} \operatorname{var}_{t_0}^{t_k}(\varphi(\sigma) - \int_0^\sigma DF(\varphi(\tau), t)) =$$

$$= e^{-(t_k - t_0)} \operatorname{var}_{t_0}^{t_k}(\varphi(\sigma) - \int_0^\sigma DF(\varphi(\tau), t)) < \gamma(\varepsilon) e^{-(T(\varepsilon)+1)}$$

and therefore

$$\operatorname{var}_{t_0}^{t_k}(\varphi(\sigma) - \int_0^\sigma DF(\varphi(\tau), t)) < \gamma(\varepsilon). \tag{10.46}$$

For $\sigma \in [t_0, t_k]$ we define

$$P(\sigma) = \varphi(\sigma) - \int_0^\sigma DF(\varphi(\tau), t)).$$

The function $P : [t_0, t_k] \to \mathbb{R}^n$ is evidently continuous from the left and by the inequality (10.46) we have

$$\operatorname{var}_{t_0}^{t_k} P < \gamma(\varepsilon).$$

Moreover,

$$\varphi(\sigma) = \int_0^\sigma DF(\varphi(\tau), t) + \varphi(\sigma) - \int_0^\sigma DF(\varphi(\tau), t) =$$

$$= \int_0^\sigma DF(\varphi(\tau), t) + P(\sigma)$$

and also

$$\varphi(s) - \varphi(t_0) = \int_{t_0}^s DF(\varphi(\tau), t) + P(s) - P(t_0) =$$

$$= \int_{t_0}^s D[F(\varphi(\tau), t) + P(t)],$$

and this means that the function $\varphi : [t_0, t_k] \to \mathbb{R}^n$ is a solution of the equation (10.3) with $\|\varphi(t_0)\| \leq a = \delta_0$ because $\varphi \in A_a(t, k, x_k)$. By the definition of the attractivity the inequality $\|\varphi(t_0)\| < \varepsilon$ holds for every $t > t_0 + T(\varepsilon)$. This is of course valid also for the value $t = t_k > t_0 + T(\varepsilon)$, i.e. $\|\varphi(t_k)\| = \|x_k\| < \varepsilon$ and this contradicts the assumption $\|x_k\| \geq \varepsilon$. This yields the positive definiteness of U. □

References

[1] **Z. Artstein**, *Continuous dependence on parameters: On the best possible results*, J. Differential Equations *19* (1975), 214 – 225.

[2] **Z. Artstein**, *Topological dynamics of ordinary differential equations and Kurzweil equations*, J. Differential Equations *23* (1977), 224 – 243.

[3] **Z. Artstein**, *Topological dynamics of an ordinary differential equation*, J. Differential Equations *23* (1977), 216 – 223.

[4] **Z. Artstein**, *The limiting equations of nonautonomous ordinary differential equations*, J. Differential Equations *25* (1977), 184 – 202.

[5] **F.V. Atkinson**, *Discrete and Continuous Boundary Value Problems*, Academic Press, New York, London, 1964.

[6] **G. Aumann**, *Reelle Funktionen*, Springer Verlag, Berlin, Heidelberg, New York, 1969.

[7] **P. Binding**, *Bounded variation evolution equations*, J. of Math. Anal. Appl. *48* (1974), 70 – 94.

[8] **R.C. Brown, M. Tvrdý**, *Generalized boundary value problems with abstract side conditions and their adjoints I*, Czech. Math. Journal *30 (105)* (1980), 7 – 27.

[9] **R.C. Brown, M. Tvrdý**, *Generalized boundary value problems with abstract side conditions and their adjoints II*, Czech. Math. Journal *31 (106)* (1981), 501 – 509.

[10] **R.C. Brown, M. Tvrdý, O. Vejvoda**, *Duality theory for linear n-th order integro-differential operators with domain in L determined by interface side conditions*, Czech. Math. Journal *32 (107)* (1982), 183 – 196.

[11] **P. S. Bullen**, *A survey on integration by parts for Perron integrals*, J. Austral. Math. Soc. (Series A) *40* (1986), 343 – 363.

[12] **P.S. Bullen**, *Some applications of a theorem of Marcinkie-*

wicz, in New Integrals, Proceedings Coleraine 1988, Springer Verlag, Berlin, Heidelberg, 1990, pp. 10 – 18.

[13] **P. S. Bullen , R. Výborný**, *Some applications of a theorem of Marcinkiewicz*, Canad. Math. Bull. *32* no. 2 (1991), 165 – 174.

[14] **C. Carathéodory**, *Vorlesungen über reelle Funktionen*, B. G. Teubner, Berlin, Leipzig, 1918.

[15] **T.S. Chew**, *On Kurzweil generalized ordinary differential equations*, J. Differential Equations *76* no. 2 (1988), 286 – 293.

[16] **T. S. Chew, F. Flordeliza**, *On $x' = f(t,x)$ and Henstock – Kurzweil integrals*, Differential and Integral Equations *4* (1991), 861 – 868.

[17] **S. – N. Chow, J.A. Yorke**, *Ljapunov theory and perturbation of stable and asymptotically stable systems*, J. Differential Equations *15* (1977), 308 – 321.

[18] **E.A. Coddington, N. Levinson**, *Theory of Ordinary Differential Equations*, McGraw–Hill, New York, Toronto, London, 1955.

[19] **P.C. Das, R.R. Sharma**, *Existence and stability of measure differential equations*, Czech. Math. Journal *22 (97)* (1972), 145 – 158.

[20] **J. D. Dollard, C.N. Friedman**, *Product Integration with Applications to Differential Equations*, Addison–Wesley Publ. Company, Reading, Massachusetts, 1979.

[21] **N. Dunford, J.T. Schwartz**, *Linear operators I., II.*, Interscience Publishers, New York, London, 1958, 1963.

[22] **V. Doležal**, *On non-unicity of solutions of a system of differential equations*, Časopis pěst. mat. *85* (1960), 311 – 337.

[23] **N.P. Erugin, I.Z. Shtokalo et al.**, *Lectures on Ordinary Differental Equations*, Golovnoe izd., Kiev, 1974. (Russian)

[24] **A.F. Filippov**, *Differential Equations with Discontinuous Right Hand Side*, Moskva Nauka, 1985. (Russian)

[25] **J. Foran**, *Fundamentals of Real Analysis*, Marcel Dekker, Inc., New York Basel Hong Kong, 1991.

[26] **D. Fraňková**, *Continuous dependence on a parameter of solutions of generalized differential equations*, Časopis pěst. mat. *114* (1989), 230 - 261.

[27] **D. Fraňková, Š. Schwabik**, *Generalized Sturm - Liouville equations II*, Czech. Math. Journal *38 (113)* (1988), 531 - 553.

[28] **D. Fraňková**, *Regulated functions*, Mathematica Bohemica *116* (1991), 20 - 59.

[29] **D. Fraňková**, *Nonabsolutely convergent series*, Mathematica Bohemica *116* (1991), 248 - 267.

[30] **D. Fraňková**, *Substitution method for generalized linear differential equations*, Mathematica Bohemica *116* (1991), 337 - 359.

[31] **I.I. Gichman**, *On the origins of a theorem of N.N. Bogoljubov*, Ukr. mat. žurnal *IV* (1952), 215 - 219. (Russian)

[32] **R.D. Gill, S. Johansen**, *A survey of product-integration with a view toward application in survival analysis*, Annals of Statistics *18* no. 4 (1990), 1501 - 1555.

[33] **J. Groh**, *A nonlinear Volterra-Stieltjes equation and a Gronwall inequality in one dimension*, Illinois J. Math. *24* (1980), 244 - 263.

[34] **I. Győri**, *A generalization of Bellman's inequality for Stieltjes integrals and a uniqueness theorem*, Studia Sci. Math. Hungarica *6* (1971), 137 - 145.

[35] **A. Halanay, D. Wexler**, *Qualitative Theory of Systems with Impulses*, Editura Acad. Rep. Soc. România, Bucureşti, 1968. (Romanian)

[36] **Ph. Hartman**, *Ordinary Differential Equations*, J. Wiley & Sons, New York, London, Sydney, 1964.

[37] **R. Henstock**, *Theory of Integration*, Butterworths, London, 1963.

[38] **R. Henstock**, *A Riemann type integral of Lebesgue power*, Canad. J. Math. *20* (1967), 79 – 87.

[39] **R. Henstock**, *Integration by parts*, Aequationes Math. *9* (1973), 1 – 18.

[40] **R. Henstock**, *Lectures on the Theory of Integration*, World Scientific, Singapore, 1988.

[41] **R. Henstock**, *The General Theory of Integration*, Clarendon Press, Oxford, 1991.

[42] **T.H. Hildebrandt**, *On systems of linear differentio–Stieltjes integral equations*, Illinois J. Math. *3* (1959), 352 – 373.

[43] **Ch. S. Hönig**, *Volterra Stieltjes – Integral Equations*, North – Holland Publ. Comp., Amsterdam, 1975.

[44] **Ch. S. Hönig**, *Equations integrales generalisees et applications*, Publ. Math. d'Orsay. no. Exp. no. 5 (1981 – 82).

[45] **Ch. S. Hönig**, *On linear Kurzweil – Henstock – integral equations*, Seminario Brasileiro de Análise *1990*, 283 – 299.

[46] **J. Jarník**, *On some assumptions of the theorem on the continuous dependence on a parameter*, Časopis pěst. mat. *86* (1961), 404 – 414.

[47] **J. Jarník**, *On a certain modification of the theorem on the continuous dependence on a parameter*, Časopis pěst. mat. *86* (1961), 415 – 423.

[48] **J. Jarník**, *Dependence of solutions of a class of differential equations of the second order on a parameter*, Czech. Math. J. *15 (90)* (1965), 124 – 160.

[49] **J. Jarník, J. Kurzweil**, *Continuous dependence on a parameter*, Contr. to the theory of nonlinear oscillations *5* (1960), 25 – 35.

[50] **J. Jarník, J. Kurzweil**, *A nonabsolutely convergent integral which admits C1 – transformations*, Časopis pěst. mat. *109* (1984), 157 – 167.

[51] **J. Jarník, J. Kurzweil**, *Integral of multivalued mapping and its connection with differential relations*, Časopis pěst.

mat. *108* (1983), 8 – 28.

[52] **J. Jarník, J. Kurzweil**, *A non absolutely convergent integral which admits transformation and can be used for integration on manifolds*, Czechoslovak Mathematical Journal *35(110)* (1985), 116 – 139.

[53] **J. Jarník, J. Kurzweil**, *Perron integral, Perron product integral and ordinary linear differential equations*, EQUADIFF 6, Lecture Notes in Mathematics 1192, Springer, Berlin (1986), 149 – 154.

[54] **J. Jarník, J. Kurzweil**, *On some extensions of the Perron integral on one – dimensional intervals. An approach by integral sums fulfilling a symmetry condition*, Functiones et Approximatio, Commentarii Mathematici *17* (1987), 49 – 55.

[55] **J. Jarník, J. Kurzweil**, *A new and more powerful concept of the PU – integral*, Czechoslovak Mathematical Journal *38 (113)* (1988), 8 – 48.

[56] **J. Jarník, J. Kurzweil**, *The PU – integral and its properties*, Real Analysis Exchange *14* (1988 – 89), 34 – 43.

[57] **J. Jarník, J. Kurzweil**, *The PU – integral: its definition and some basic properties. In: New Integrals, Proceedings Coleraine 1988. Lecture Notes in Mathematics 1419, Springer, Berlin 1990, 66 – 81.*.

[58] **J. Jarník, J. Kurzweil**, *On a generalization of the Perron integral on one – dimensional intervals*, Annales Polonici Mathematici *51* (1990), 17 – 30.

[59] **J. Jarník, J. Kurzweil**, *Equiintegrability and controlled convergence of Perron – type integrable functions*, Real Analysis Exchange *17* no. 1 (1991 – 92), 110-139.

[60] **J. Jarník, J. Kurzweil**, *Differentiability and integrability in n dimensions with respect to α – regular intervals*, Results of Mathematics *21* (1992), 138 – 151.

[61] **J. Jarník, J. Kurzweil**, *Equivalent definitions of regular generalized Perron integral*, Czech. Math.J. *42(117)* (1992),

365 – 378.

[62] **J. Jarník, J. Kurzweil**, *Generalized multidimensional Perron integral involving a new regularity condition*, Results of Mathematics (to appear).

[63] **J. Jarník, J. Kurzweil**, *Pfeffer integrability does not imply M1 – integrability*, Czechoslovak Math. J. (to appear).

[64] **J. Jarník, J. Kurzweil, Š. Schwabik**, *On Mawhin's approach to multiple nonabsolutely convergent integral*, Časopis pěst. mat. *108* (1983), 356 – 380.

[65] **J. Jarník, J. Kurzweil**, *A general form of the product integral and linear ordinary differential equations*, Czech. Math. Journal *37* (1987), 642 – 659.

[66] **M.A. Krasnoselskij, S.G. Krein**, *On the averaging principle in nonlinear mechanics*, Uspehi mat. nauk *10* no. 3 (1955), 147 – 152. (Russian)

[67] **K. Kreith**, *Conjugate points of fourth order differential equations with impulses*, Diff. an Int. Equations *1* (1988), 391 – 398.

[68] **J. Kurzweil**, *Generalized ordinary differential equations and continuous dependence on a parameter*, Czech. Math. Journal *7 (82)* (1957), 418 – 449.

[69] **J. Kurzweil**, *Generalized ordinary differential equations*, Czech. Math. Journal *8 (83)* (1958), 360 – 388.

[70] **J. Kurzweil**, *On generalized ordinary differential equations with discontinuous solutions*, Prikl. Mat. Meh. *XXII* (1958), 27 –45. (Russian)

[71] **J. Kurzweil**, *Unicity of solutions of generalized differential equations*, Czechoslovak Mathematical Journal *8(83)* (1958), 502 – 509.

[72] **J. Kurzweil**, *Addition to my paper "Generalized ordinary differential equations and continuous dependence on a parameter"*, Czech. Math. Journal *9(84)* (1959), 564 – 573.

[73] **J. Kurzweil**, *Linear differential equations with distributions*

as coefficients, Bulletin de l'Academie Polonaise des Sciences, Serie des Sciences Math., Astr. et Phys. *7* (1959), 557 – 560.

[74] **J. Kurzweil**, *On integration by parts*, Czech. Math. Journal *8 (83)* (1958), 356 – 359.

[75] **J. Kurzweil**, *Problems which lead to a generalization of the concept of an ordinary differential equation*, in Differential Equations and their Applications, Proc. of the conference held in Prague in September 1962, Publ. House of the Czech. Acad. of Sciences, Prague, 1963.

[76] **J. Kurzweil**, *On Fubini theorem for general Perron integral*, Czech. Math. Journal *23 (98)* (1973), 286 – 297.

[77] **J. Kurzweil**, *The averaging principle and its applications in partial differential equations*, In: Nonlinear Vibration Problems (1963), 142 – 148.

[78] **J. Kurzweil**, *Exponentially stable integral manifolds, averaging principle and continuous dependence on a parameter*, Czech. Math. Journal *16(91)* (1966), 380 – 423, 463 – 492.

[79] **J. Kurzweil**, *Invariant manifolds for differential systems*, In: Atti VIII. Congr. U. M. I., Trieste 1967, 291 – 292.

[80] **J. Kurzweil**, *Invariant manifolds for flows, differential equations and dynamical systems*, Proceedings Internat. Symp. Mayaguez, Academic Press (1967), 431 – 468.

[81] **J. Kurzweil**, *Invariant sets off differential systems sistem*, Differencialnye uravnenia *4* (1968), 785 – 797. (Russian)

[82] **J. Kurzweil**, *On multiplication of Perron – integrable functions*, Czech. Math. Journal *23 (98)* (1973), 542 – 566.

[83] **J. Kurzweil**, *Appendix A: The Perron – Ward integral and related concepts*, K. Jacobs: Measure and Integral, Academic Press, 1978, pp. 515 – 533.

[84] **J. Kurzweil**, *Nichtabsolut konvergente Integrale*, BSB B. G. Teubner Verlagsgesellschaft, Leipzig, 1980.

[85] **J. Kurzweil**, *The integral as the limit of integral sums*, Jahrbuch Überblicke Mathematik, Bibl. Institut Mannheim 1984,

105 – 136.

[86] **J. Kurzweil**, *Ordinary Differential Equations*, Elsevier, Amsterdam, 1986.

[87] **J. Kurzweil, Š. Schwabik**, *Ordinary differential equations the solution of which are ACG^* - functions*, Archivum Mathematicum (Brno) *26* no. 2–3 (1990), 129–136.

[88] **J. Kurzweil, Z. Vorel**, *Continuous dependence of solutions of differential equations on a parameter*, Czech. Math. J. *23 (82)* (1957), 568 – 583.

[89] **J. Ligęza**, *Cauchy's problem for systems of linear differential equations with distributional coefficients*, Colloquium Math. *33* (1975), 295 – 303.

[90] **J. Ligęza**, *The existence and uniqueness of distributional solutions of some systems of non–linear differential equations*, Časopis pěst. mat. *102* (1977), 30 – 36.

[91] **J. Ligęza**, *On distributional solutions of some systems of linear differential equations*, Časopis pěst. mat. *102* (1977), 37 – 41.

[92] **Lee Peng-Yee**, *Lanzhou Lectures on Henstock Integration*, World Scientific, Singapore, 1989.

[93] **J. Mawhin**, *Introduction a l'Analyse*, CABAY, Libraire – éditeur, Louvain – La – Neuve, 1984.

[94] **J. Mawhin**, *Generalized multiple Perron integrals and the Green – Goursat theorem for differentiable vector fields*, Czech. Math. J. *31 (106)* (1981), 614 – 632.

[95] **J. Mawhin**, *Classical problems in analysis and new integrals*, Real Analysis Exchange *12* (1986), 69 – 84.

[96] **R. M. McLeod**, *The Generalized Riemann Integral*, Math. Assoc. of America, 1980.

[97] **E. J. McShane**, *A Riemann – type integral that includes Lebesgue – Stieltjes, Bochner and stochastic integrals*, Memoirs of AMS *88* (1969).

[98] **E. J. McShane**, *Unified Integration*, Academic Press, New York, London, 1983.

[99] **V.D. Mil'man, A.D. Myškis**, *On the stability of motion in the presence of impulses*, Sibirskij mat. žur. *1* (1960), 233 – 237. (Russian)

[100] **A.B. Mingarelli**, *Volterra – Stieltjes Integral Equations and Generalized Differential Expressions*, Lecture Notes in Math., Springer Verlag, Berlin, 1983.

[101] **Ju. A. Mitropolskij, A.M. Samojlenko, N.A. Perestjuk**, *To the problem of the averaging of second order systems with impulsive action*, Ukr. mat. ž. *29* (1977), 750 – 762. (Russian)

[102] **Ju. A. Mitropolskij, A.M. Samojlenko, N.A. Perestjuk**, *The method of averaging in systems with impulsive action*, Ukr. mat. ž. *37* (1985), 56 – 64. (Russian)

[103] **A.D. Myškis , A.M. Samojlenko**, *Systems with impulses at given time moments*, Mat. sbornik *74* (1967), 202 – 208. (Russian)

[104] **I.P. Natanson**, *Theory of Functions of a Real Variable*, Frederick Ungar, New York, 1955, 1960.

[105] **A.I. Panasjuk**, *Integration of generalized ordinary differential equations of approximate type in R^n*, Differencialnye uravnenija no. 11 (1991), 1901 – 1914. (Russian)

[106] **S.G. Pandit**, *Differential systems with impulsive perturbations*, Pacific J. Math. *86* (1980), 553 – 560.

[107] **S.G. Pandit, S.G. Deo**, *Differential Systems Involving Impulses*, Lecture Notes in Math., 954, Springer Verlag, Berlin, 1982.

[108] **N.A. Perestjuk, O.S. Černikova**, *On the problem of stability of solutions of systems of differential equations with impulsive action*, Ukr. mat. ž. *36* (1984), 190 – 195. (Russian)

[109] **N.A. Perestjuk, V.N. Šovkopljas**, *Periodic solutions of nonlinear differential equations with impulsive action*, Ukr.

mat. žurnal *31* (1979), 517 - 524. (Russian)

[110] **J. Persson**, *Fundamental theorems for linear measure differential equations*, Math. Scand. *62* (1988), 19 - 43.

[111] **R. Pfaff**, *Gewöhnliche lineare Differentialgleichungen zweiter Ordnung mit Distributionskoeffizienten*, Arch. Math. (Basel) *32* (1979), 469 - 478.

[112] **R. Pfaff**, *Gewöhnliche lineare Differentialgleichungen n–ter Ordnung mit Distributionskoeffizienten*, Proc. Roy. Soc. Edinburgh Sect. A *85 - A* (1980), 291 - 298.

[113] **R. Pfaff**, *Generalized systems of linear differential equations*, Proc. Roy. Soc. Edinburgh Sect. A *89 - A* (1981), 1 - 14.

[114] **M. Rama Mohana Rao, V. Sree Hari Rao**, *Stability of impulsively perturbed systems*, Bull. Austral. Math. Soc. *16* (1977), 99 - 110.

[115] **S. Saks**, *Theory of the Integral*, Monografie Matematyczne, Warszawa, Lwow, 1937.

[116] **A.M. Samojlenko, N.A. Perestjuk**, *Stability of solutions of differential equations with impulsive action*, Diff. uravnenija *13* (1977), 1981 - 1992. (Russian)

[117] **A.M. Samojlenko, N.A. Perestjuk**, *Differential equations with impulsive action*, Kiev State University, 1980. (Russian)

[118] **A.M. Samojlenko, N.A. Perestjuk**, *Periodic and almost periodic solutions of differential equations with impulsive action*, Ukr. mat. ž. *34* (1982), 66 - 73. (Russian)

[119] **A.M. Samojlenko, N.A. Perestjuk**, *Stability of solutions of differential equations with impulsive action*, Diff. uravnenija *17* (1981), 1995 - 2001. (Russian)

[120] **A.M. Samojlenko, N.A. Perestjuk**, *Differential Equations with Impulsive Action*, Golovnoe izd., Kiev, 1987. (Russian)

[121] **M. Schechter**, *Principles of Functional Analysis*, Academic Press, New York, London, 1973.

[122] **L. Schlesinger**, *Vorlesungen über lineare Differentialgleichungen*, Teubner, Leipzig, 1908.

[123] **L. Schlesinger**, *Neue Grundlagen für einen Infinitesimalkalkul der Matrizen*, Math. Zeitschrift *33* (1931), 33 - 61.

[124] **W.W. Schmaedeke**, *Optimal control theory for nonlinear vector differential equations containing measures*, SIAM J., Ser. A, Control *3* (1965), 231 - 280.

[125] **G. Schmidt**, *On multiplicative Lebesgue integration and families of evolution operators*, Math. Scand. *29* (1971), 113 - 133.

[126] **Š. Schwabik**, *Über ein Differentialgleichungssystem mit unstetigen Lösungen endlicher Variation*, ZAMM, Sonderheft GAMM *48* (1968), T31 - T32.

[127] **Š. Schwabik**, *Stetige Abhängigkeit von einem Parameter und invariante Mannigfaltigkeiten für verallgemeinerte Differentialgleichungen*, Czech. Math. Journal *19 (94)* (1969), 398 - 427.

[128] **Š. Schwabik**, *Verallgemeinerte gewöhnliche Differentialgleichungen; Systeme mit Impulsen auf Flächen I*, Czech. Math. Journal *20 (95)* (1970), 468 - 490.

[129] **Š. Schwabik**, *Verallgemeinerte gewöhnliche Differentialgleichungen; Systeme mit Impulsen auf Flächen II*, Czech Math. Journal *21 (96)* (1971), 172 - 197.

[130] **Š. Schwabik**, *Bemerkungen zu Stabilitätsfragen für verallgemeinerte Differentialgleichungen*, Časopis pěst. mat. *96* (1971), 57 - 66.

[131] **Š. Schwabik**, *Verallgemeinerte lineare Differentialgleichungssysteme*, Časopis pěst. mat. *96* (1971), 183 - 211.

[132] **Š. Schwabik**, *Stetige Abhängigkeit von einem Parameter für ein Differentialgleichungssystem mit Impulsen*, Czech. Math. Journal *21 (96)* (1971), 198 - 212.

[133] **Š. Schwabik**, *On an integral operator in the space of functions with bounded variation*, Časopis pěst. mat. *97* (1972), 297 - 330.

[134] **Š. Schwabik**, *On Fredholm - Stieltjes integral equations*, Comm. Math. Univ. Carolinae *12,4* (1971), 845 - 849.

[135] **Š. Schwabik**, *On the relation between Young's and Kurzweil's concept of Stieltjes integral*, Časopis pěst. mat. *98* (1973), 237 - 251.

[136] **Š. Schwabik**, *On a modified sum integral of Stieltjes type*, Časopis pěst. mat. *98* (1973), 274 - 277.

[137] **Š. Schwabik**, *Remark on linear equations in Banach space*, Časopis pěst. mat. *99* (1974), 115 - 122.

[138] **Š. Schwabik**, *On Volterra - Stieltjes integral equations*, Časopis pěst. mat. *99* (1974), 255 - 278.

[139] **Š. Schwabik**, *On an integral operator in the space of functions with bounded variation II*, Časopis pěst. mat. *102* (1977), 189 - 202.

[140] **Š. Schwabik**, *Note on Volterra - Stieltjes integral equations*, Časopis pěst. mat. *102* (1977), 275 - 279.

[141] **Š. Schwabik**, *Differential equations with interface conditions*, Časopis pěst. mat. *105* (1980), 391 - 408.

[142] **Š. Schwabik**, *Generalized Volterra integral equations*, Czech. Math. Journal *32 (107)* (1982), 245 - 270.

[143] **Š. Schwabik**, *Uniqueness for generalized differential equations*, Teubner Texte zur Math. *50*, 263 - 269..

[144] **Š. Schwabik**, *Variational stability for generalized differential equations*, Časopis pěst. mat. *109* (1984), 389 - 420.

[145] **Š. Schwabik**, *Generalized Sturm - Liouville equations*, Archivum Mathematicum *23* (1987), 95 - 107.

[146] **Š. Schwabik**, *A survey of some new results for regulated functions*, 28 Seminario Brasileiro de Analise, Sao Paulo 1988, 201 - 209.

[147] **Š. Schwabik**, *Generalized differential equations – a survey*, Teubner Texte zur Math. Leipzig *118* (1990), 59 – 70.

[148] **Š. Schwabik**, *The Perron product integral and generalized linear differential equations*, Časopis pěst. mat. *115* (1990), 368 – 404.

[149] **Š. Schwabik**, *Linear operators in the space of regulated functions*, Mathematica Bohemica *117* (1992), 79 – 92.

[150] **Š. Schwabik**, *Generalized Differential Equations. Fundamental Results*, Rozpravy ČSAV *95* no. 6 (1985).

[151] **Š. Schwabik**, *Generalized Differential Equations. Special Results*, Rozpravy ČSAV *99* no. 3 (1989).

[152] **Š. Schwabik, M. Tvrdý**, *On the generalized linear ordinary differential equation*, Časopis pěst. mat. *98* (1973), 206 – 211.

[153] **Š. Schwabik, M. Tvrdý**, *On linear problems in the space BV*, Lecture Notes in Math. 703, Springer Verlag Berlin, 1979, pp. 366 – 373.

[154] **Š. Schwabik, M. Tvrdý**, *Boundary value problems for systems of generalized linear differential equations*, Czech. Math. Journal *29 (104)* (1979), 451 – 477.

[155] **Š. Schwabik, M. Tvrdý, O. Vejvoda**, *Differential and Integral Equations. Boundary Value Problems and Adjoints*, Academia Praha & Reidel Dordrecht, 1979.

[156] **B. S. Thomson**, *Derivation bases on the real line (I)*, Real Analysis Exchange *8* (1982 – 83), 67 – 207.

[157] **B. S. Thomson**, *Derivation bases on the real line (II)*, Real Analysis Exchange *8* (1982 – 83), 278 – 442.

[158] **M. Tvrdý**, *Boundary value problems for linear generalized differential equations and their adjoints*, Czech. Math. Journal *23 (98)* (1973), 183 – 217.

[159] **M. Tvrdý**, *Boundary value problems for generalized linear integrodifferential equations with left – continuous solutions*, Časopis pěst. mat. *99* (1974), 147–157.

[160] **M. Tvrdý**, *Note on functional – differential equations with initial functions of bounded variation*, Czech. Math. Journal *25 (100)* (1975), 67 – 70.

[161] **M. Tvrdý**, *Linear boundary value type problems for functional – diferential equations and their adjoints*, Czech. Math. Journal *25 (100)* (1975), 37 – 66.

[162] **M. Tvrdý**, *Linear functional – differential operators: normal solvability and adjoints*, In: Colloquia Mathematica Societatis János Bolyai *15* (1975), 379 – 389.

[163] **M. Tvrdý**, *Fredholm – Stieltjes integral equations with linear constraints: duality theory and Green's function*, Časopis pěst. mat. *104* (1979), 357 – 369.

[164] **M. Tvrdý**, *Regulated functions and the Perron – Stieltjes integral*, Časopis pěst. mat. *114* (1990), 187 – 209.

[165] **M. Tvrdý**, *Generalized differential equations in the space of regulated functions (Boundary value problems and controllability)*, Mathematica Bohemica *116* (1991), 225 – 244.

[166] **M. Tvrdý, O. Vejvoda**, *General boundary value problem for an integrodifferential system and its adjoint I*, Časopis pěst. mat. *97* (1972), 399 – 419.

[167] **M. Tvrdý, O. Vejvoda**, *General boundary value problem for an integrodifferential system and its adjoint II*, Časopis pěst. mat. *98* (1973), 26 – 42.

[168] **O. Vejvoda, M. Tvrdý**, *Existence of solutions to a linear integro – boundary – differential equation with additional conditions*, Annali di Mat.Pura ed Appl. *89* (1971), 169 – 216.

[169] **V. Volterra**, *Sulle equazioni differenziali lineari*, Rend. Accademia dei Lincei *3* (1887), 393 – 396.

[170] **V. Volterra, B. Hostinský**, *Operations infinitesimales linéaires*, Gauthier–Villars, Paris, 1938.

[171] **I. Vrkoč**, *A note to the unicity of generalized differential equations*, Czech. Math. J. *8 (83)* (1958), 510 – 511.

[172] **I. Vrkoč**, *Integral stability*, Czech. Math. J. *9 (84)* (1959), 71 – 128. (Russian)

[173] **S.T. Zavališčin, A.N. Seseskin A.N.**, *Impulsive processes, models and applications*, Nauka Moskva, 1991. (Russian)

Index